Lines and Fields in ELECTRONIC TECHNOLOGY

Lines and Fields in ELECTRONIC TECHNOLOGY

WILLIAM D. STANLEY
Old Dominion University

RICHARD F. HARRINGTON
Old Dominion University

PRENTICE HALL

Upper Saddle River, New Jersey
Columbus, Ohio

LIBRARY OF CONGRESS CATALOGING-IN-PUBLICATION DATA
Stanley, William D.
Lines and fields in electronic technology/William D. Stanley, Richard F. Harrington.
p. cm.
Includes index.
ISBN 0-02-415654-X
1. Electric lines. 2. Wave guides. 3. Antennas (Electronics) 4. Telecommunication lines.
I. Harrington, Richard F.
II. Title.
TK3201.S74 1994 93-48719
621.319—dc20 CIP

Cover Photo: Doug Johnson/Science Photo Library
Editor: Dave Garza
Production Editor: Stephen C. Robb
Art Coordinator: Peter A. Robison
Text Designer: Russ Maselli
Production Manager: Pamela D. Bennett
Illustrations: Diphrent Strokes, Inc.

This book was set in Century Schoolbook and Helvetica Bold Condensed by Bi-Comp, Incorporated, and was printed and bound by Book Press, Inc., a Quebecor America Book Group Company. The cover was printed by Phoenix Color Corp.

Printed in the United States of America

10 9 8 7 6 5 4

ISBN 0-02-415654-X

PRENTICE-HALL INTERNATIONAL (UK) LIMITED, *LONDON*
PRENTICE-HALL OF AUSTRALIA PTY. LIMITED, *SYDNEY*
PRENTICE-HALL CANADA INC., *TORONTO*
PRENTICE-HALL HISPANOAMERICANA, S.A., *MEXICO*
PRENTICE-HALL OF INDIA PRIVATE LIMITED, *NEW DELHI*
PRENTICE-HALL OF JAPAN, INC., *TOKYO*
PEARSON EDUCATION ASIA PTE. LTD., *SINGAPORE*
EDITORA PRENTICE-HALL DO BRASIL, LTDA., *RIO DE JANEIRO*

DEDICATION

To our families
Karen Louise
Pat, Susan, Janey, Courtney, David, and Shannon

Preface

The transmission of electrical energy between two points on the earth or in space was one of the earliest problems to be confronted in the evolution of the electronics industry, and it remains a challenging process today. Recent developments in fiber optic technology and satellite repeater systems have opened up new vistas of transmission capabilities. Nevertheless, the needs for transmission reliability and speed have grown at such a rapid rate that the emerging technology is still struggling to meet the more sophisticated demands of the present and the future. To fully utilize the newer technology, students must examine all aspects of transmission to understand its relation to the underlying principles of this technology.

The major objectives of this text are to provide a broad treatment of electronic transmission technology at an *engineering technology* level. This includes traditional transmission line, waveguide, and antenna concepts as well as more recent developments in fiber optics and satellite communications. An additional subject area contained here, but rarely found in textbooks at this level, is a descriptive discussion of Maxwell's equations and their importance in the electronics field. Although the full implications of Maxwell's equations remain elusive to most applied engineers and technologists, some appreciation of their impact is necessary as part of a broad general education in electricity.

The full scope of the book may be employed at about the junior or senior level in a four-year electrical or electronic technology program. However, most portions of the book may be satisfactorily used as the text for a course in transmission lines in the second year of a two-year electronic technology program.

Primary prerequisites for full understanding of the book are basic dc and ac circuit theory. Some background in differential and integral calculus is necessary to understand the derivations fully; however, the applications and results may be applied in the solutions of practical problems using only algebra and trigonometry.

Chapter 1 deals with the basic properties of transmission lines. This includes the classifications of lines, the inductance and capacitance of lines, and a thorough discussion of characteristic impedance and its properties.

Chapter 2 provides a thorough development of the transient behavior of lossless lines with a dc or pulse input. This treatment, which precedes steady-state ac analysis, is designed to impart the most intuitive concepts of wave propagation and is essential in developing a comprehensive understanding of traveling wave phenomena. Important applications to time domain reflectometry are also included.

Chapter 3 provides a full development of the steady-state ac analysis of transmission lines, the area of application most widely utilized in both communications and power system transmission. Both lossy and lossless lines are considered, and both matched and mismatched conditions are explored in depth.

Chapter 4 is devoted to a comprehensive analysis of lossless lines under steady-state ac conditions. The assumption of the lossless line is widely employed in the design of RF communications systems with moderately short transmissions lengths, and it serves as the basis for designing many high frequency and microwave components and systems.

The widely employed Smith chart is introduced in Chapter 5, and many of its practical properties are studied. This includes such applications as impedance matching and stub matching.

Chapter 6 provides the important transition from distributed voltage and current concepts to electromagnetic field concepts. Electric and magnetic fields are introduced, and a descriptive discussion of Maxwell's equations and the associated boundary conditions follows. The phenomenon of skin depth is presented, and plane wave propagation is discussed in detail.

Chapter 7 introduces high frequency transmission lines. It begins with a presentation of boundary conditions for electromagnetic waves and a description of modes that exist on transmission lines. Practical transmission lines, including parallel lines, coaxial cables, stripline, and microstrip, are presented in detail.

Chapter 8 covers the subject of waveguide transmission lines. Rectangular waveguides are presented along with the means for analyzing and designing waveguides. The emphasis is on the applied analysis using an algebraic basis for predicting waveguide performance.

Chapter 9 is devoted to a general development of antenna technology. The emphasis is on *system* properties (e.g., gain, directivity, radiation patterns, beamwidth, etc.) rather than on the field theory point of view. Descriptions of dipoles, monopoles, waveguide horn antennas, and parabolic reflector antennas are presented.

Chapter 10 deals with some of the classical properties of wave propagation. After a discussion of reflection and refraction, the Friis transmission formula is developed. Line-of-sight, surface wave, and ionospheric propagation are discussed in general terms.

Chapter 11 provides an introduction to the contemporary subject of fiber optic technology. Optical transmitters, receivers, and fiber optic transmission lines are presented.

Chapter 12 presents an overview of satellite communications that includes a discussion of spacecraft systems, ground terminals, system noise effects, and communication link design.

The authors would like to express their deep appreciation to Ms. Michele Boulden for her assistance in typing many portions of the text.

Contents

1

Basic Transmission Line Properties

OBJECTIVES

After completing this chapter, the reader should be able to:

1. Discuss the phenomenon of electrical traveling waves.
2. State the relationship governing **velocity of propagation, frequency,** and **wavelength.**
3. State the relationship governing **propagation time, distance,** and **velocity of propagation.**
4. Perform rough estimates to determine when propagation time effects are significant for both **pulse signals** and **sinusoidal signals.**
5. Discuss the physical characteristics of **parallel-line** and **coaxial cable** transmission lines.
6. Discuss the difference between **balanced** and **unbalanced** lines and give an example of each type.
7. Discuss the effects of **inductance** and **capacitance** for a lossless line.
8. Define the **characteristic impedance** of a transmission line.
9. Determine the **characteristic impedance** of a lossless line in terms of the inductance and capacitance per unit length.
10. Discuss qualitatively the terms **permittivity** and **permeability.**
11. Determine the velocity of propagation of a lossless line in terms of inductance and capacitance per unit length.
12. Show quantitatively how permittivity and permeability affect the velocity of propagation.

1–1 ■ ELECTRICAL WAVES

A major topic considered throughout this book is the concept of **electrical waves.** An electrical traveling wave is a quantity of electrical energy that moves at a specific velocity through either a guided or an unguided medium. Examples of guided media are transmission lines, waveguides, and optical fibers, all of which will be considered in this book. Examples of unguided media are free space, the atmosphere, and a body of water. The speed at which the wave travels is called the **velocity of propagation.** The velocity of propagation may be the same as the speed of light, or it may be somewhat slower.

A fundamental point of view in wave theory is that time is always required to relay an effect from one point to another. Thus, things do not happen instantaneously at all points as usually assumed in lumped circuit theory.

1–2 ■ WAVELENGTH AND FREQUENCY

Some fundamental definitions concerning the wavelength and frequency of traveling waves must be established. For this purpose, a sinusoidal propagating wave such as shown in Fig. 1–1 will be assumed. Let f represent the frequency of this sinusoidal propagating wave, and let T represent its period. The period and frequency are related by

$$f = \frac{1}{T} \tag{1–1}$$

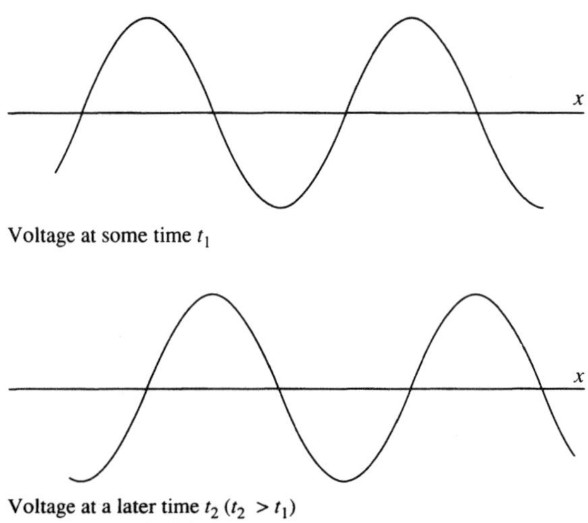

FIGURE 1–1 Illustration of traveling sinusoidal wave.

Let ν represent the velocity of propagation of the wave. *The wavelength* λ *is defined as the distance the wave travels in one cycle.* Since the duration of one cycle is the period T, the wavelength is

$$\lambda = \text{velocity} \times \text{time} = \nu T = \frac{\nu}{f} \tag{1–2}$$

Solving for ν, we have

$$\nu = f\lambda \tag{1–3}$$

Equation (1–3) is a fundamental equation that will appear throughout the book, so it should be memorized at this point. The units of the quantities in (1–3) can be adjusted but should be consistent on both sides. The quantity f should always be expressed in cycles per second or **hertz (Hz).** The units for ν are length units/second, and the corresponding units for λ are the same length units. For example, if ν is expressed in meters/second (m/s), the units for λ will be meters (m).

Equation (1–3) applies for any arbitrary velocity of propagation ν. However, in many situations, the velocity of propagation is assumed to be the same as that of light in free space. By common convention, the symbol c is used to represent the speed of light. When the speed of light is assumed, (1–3) is written as

$$c = f\lambda \tag{1–4}$$

Both forms (1–3) and (1–4) will be used extensively throughout this book.

The value of c to the accuracy of three significant figures in both meters/second (m/s) and miles/second (mi/s) can be expressed as

$$c = 3 \times 10^8 \text{ m/s} = 186{,}000 \text{ mi/s} \tag{1–5}$$

Light waves in the atmosphere travel at a slightly lower velocity than in free space. However, for most practical purposes, the free-space velocity may be assumed for air, and that practice will be followed throughout the book.

EXAMPLE 1–1

Calculate the free-space wavelength in meters for each of the following frequencies: (a) 10 kHz, (b) 1 MHz, (c) 100 MHz, and (d) 10 GHz.

Solution

From (1–4), the wavelength is

$$\lambda = \frac{c}{f} = \frac{3 \times 10^8}{f} \tag{1–6}$$

The various frequencies in hertz may be substituted in (1–6) to yield the following results:

(a)
$$\lambda = \frac{3 \times 10^8}{10^4} = 30 \text{ km} \tag{1–7}$$

(b)
$$\lambda = \frac{3 \times 10^8}{10^6} = 300 \text{ m} \tag{1–8}$$

(c)
$$\lambda = \frac{3 \times 10^8}{10^8} = 3 \text{ m} \tag{1–9}$$

(d)
$$\lambda = \frac{3 \times 10^8}{10^{10}} = 0.03 \text{ m} = 3 \text{ cm} \tag{1–10}$$

Note the extreme range in wavelengths for the frequencies considered. All of these frequencies appear in various types of electrical signals.

EXAMPLE 1–2

Calculate the free-space wavelength in miles for 60-Hz power transmission.

Solution

From (1–4) with c expressed in miles/seconds, we have

$$\lambda = \frac{c}{f} = \frac{186{,}000}{60} = 3100 \text{ mi} \tag{1–11}$$

At the standard power frequency of 60 Hz, one wavelength extends almost across the length of the United States!

1–3 ■ PROPAGATION AND TRANSMISSION LINE EFFECTS

Consider the simple circuit of Fig. 1–2 in which a 100-V dc source is to be connected through a pair of wires to a 50-Ω resistance. The resistance of the wires is assumed to be negligible in comparison to the 50-Ω lumped resistance. In this situation, as will be assumed in many other cases throughout the book, the time $t = 0$ will be used as the reference time at which the switch is closed. The length of the pair of wires is d, and any connections at the ends are assumed to be negligible in comparison to d

We were taught in basic circuit theory that the current I that would flow through the resistance should be $I = 100 \text{ V}/50\ \Omega = 2$ A. However, that idealized case is based on the assumption that propagation time effects are negligible.

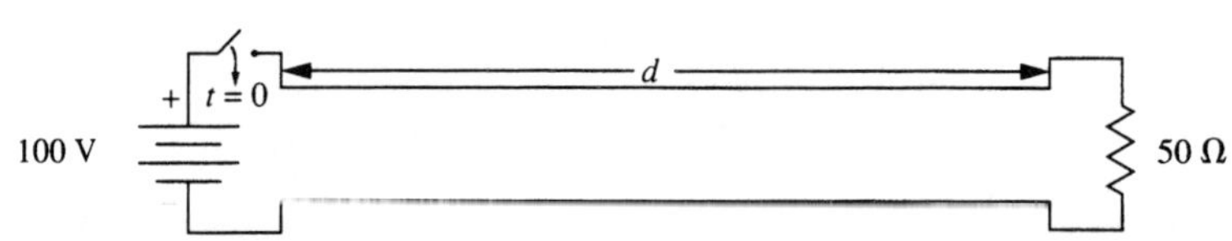

FIGURE 1–2 Simple circuit used in discussing propagation effects.

The fact is that the 2-A current is not immediately established in the resistor. When the switch is closed, a wave consisting of both voltage and current will propagate from the left-hand end of the line toward the load. The value of the current in the wave may or may not be 2 A. (It does not have to be 2 A because the dc source does not yet "know" that the load resistance is 50 Ω.) Upon reaching the load, if the current is 2 A, the required final condition is met, and the circuit is fully settled. However, if the current is not 2 A, a new wave will be generated, and it will travel back toward the source. This process will continue until the final required value is established at the load. This process is referred to as the **transient response** of the line.

In case the reader's faith in basic circuit theory has been shaken, the current will eventually settle to the dc value of 2 A as required in basic circuit theory. The time required to settle is based on the propagation time of the line and certain conditions regarding the match between the line and the load. These conditions will be explored in great detail later, but the important point to emphasize now is that voltages and currents do not establish themselves instantaneously whenever there are connecting wires.

The quantity t_1 will be defined as the **one-way propagation time,** which is the time required for a wave to propagate from one end of a line to the other end. In terms of the length d and the velocity of propagation, this time is

$$t_1 = \frac{d}{v} \tag{1–12}$$

The velocity of propagation may or may not be equal to the speed of light c.

When are propagation time effects important, and when must they be considered in analyzing a circuit? Based on the reader's previous experience with circuit theory, it is clear that many situations do not require this type of microscopic detail. In general, propagation and/or wave effects must be considered in either of the following situations:

1. Pulse waveforms are employed, and the propagation times are of the same order of magnitude as the widths or durations of the pulses.
2. Sinusoidal waveforms are employed, and the lengths of the connections are of the same order of magnitude as the wavelength.

The first effect is referred to as a **time-domain effect,** and the second as a **frequency-domain** effect. Both effects are results of the same phenomenon, but they are different ways of viewing the problem.

Time-Domain Effects

Time-domain effects are significant in circuitry involving dc and pulse waveforms, e.g., digital circuits. Although there are no exact rules, as a means of conveying a quantitative measure to the reader, we will assume the somewhat arbitrary rule that propagation time effects are potentially important when a given time delay exceeds one-tenth of a pulse waveform or one-tenth the propagation time of an active device such as a logic gate. When time delays exceed lengths of this range, the possibility exists that a given pulse will not reach a gate in sufficient time to convey the proper required level for the gate to function properly.

Frequency-Domain Effects

Frequency-domain effects are significant in circuitry involving sinusoidal and nearly sinusoidal waveforms such as in radio and television communications systems. Again, as a somewhat arbitrary rule of thumb, we will assume that frequency-domain or steady-state ac effects are potentially important when the length of a given line exceeds one-tenth the wavelength at the operating frequency of the system.*

EXAMPLE 1–3

A digital system utilizes pulses whose minimum widths are about 20 ns. If the speed of light is assumed, determine the longest lengths of wire-pairs that can be allowed based on the "10% rule."

Solution

Based on the 10% rule, the longest propagation time allowed is $t_1 = 0.1 \times 20$ ns $= 2$ ns. At the speed of light, this corresponds to a length d of

$$d = ct_1 = (3 \times 10^8) \times (2 \times 10^{-9}) = 0.6 \text{ m} \tag{1–13}$$

Understand that this result is simply an estimate of the range of length that could produce difficulties. If there are many gates that have to be sequentially activated, this length could be excessive, and much shorter lengths would be necessary.

EXAMPLE 1–4

A radio transmitter operates at a frequency of 150 MHz. If the speed of light is assumed, determine the longest length of connecting line that could be used without considering frequency-domain effects based on the 10% rule.

Solution

The wavelength λ at a frequency of 150 MHz is

$$\lambda = \frac{c}{f} = \frac{3 \times 10^8}{150 \times 10^6} = 2 \text{ m} \tag{1–14}$$

*In some microwave systems, a length of $\lambda/16$ is assumed. Since all rules of this nature are arbitrary, we will choose the more easily remembered rule of 0.1λ.

Based on the 10% rule, the maximum length d is $d = 0.1 \times 2 = 0.2$ m. Typically, transmission lines used to connect between a transmitter and an antenna will be much longer than this value. However, this result means that such a connecting line must be selected to satisfy certain rigid requirements based on transmission line effects, and this will be considered later in the book.

1-4 ■ TRANSMISSION LINES

Up to this point, we have considered a transmission line in terms of any pair of connecting wires. However, the usual interpretation of the term is with regard to those lines that are constructed to possess certain specified characteristics. The two most common types of lines will be discussed qualitatively in this section and in significant detail in Chapter 7. The two types to be considered are (1) parallel lines and (2) coaxial cable.

Parallel Lines

Parallel lines represent the simplest type of geometry in that the two conductors are of equal size and are spaced apart by a constant separation as shown in Fig. 1–3. The medium between the conductors may be air ("open-wire" lines), or it may be a material such as a polyethylene ("twin-lead"). Parallel lines have been used extensively in the telephone industry and for 60-Hz power transmission.

A parallel line is referred to as a **balanced line.** This description means that when the parallel line is operated properly, all electric and magnetic fields are symmetrical with respect to ground and the impedance of each wire to

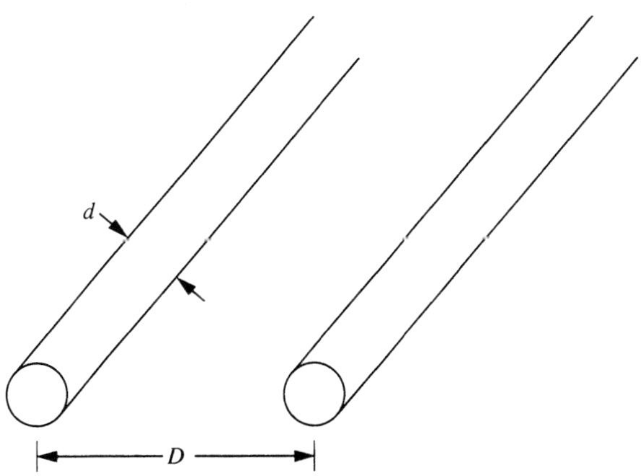

FIGURE 1–3 Geometry of parallel-line transmission line.

ground is the same. The balance minimizes the amount of radiation from the line and other undesirable coupling effects. However, a balanced line should not be driven by an unbalanced source (i.e., a source with one side grounded), unless there is a circuit between the source and the line to convert between the forms. (A transformer can be used in many cases.)

The fields of a parallel line extend some distance from the wires. As the frequency increases, the losses due to unwanted radiation on the line increase. Although parallel lines are often used domestically at TV frequencies for coupling between antenna and receiver, where some losses are acceptable, their use is generally restricted to operation below about 100 MHz.

Coaxial Cable

The most common type of coaxial cable has the geometry shown in Fig. 1–4. The two required conductors are realized by a center conductor and an enclosing conducting shield. The shield and conductor are separated by an insulating material, which is referred to as a **dielectric.** The complete cable is enclosed with an insulating **jacket** such as vinyl.

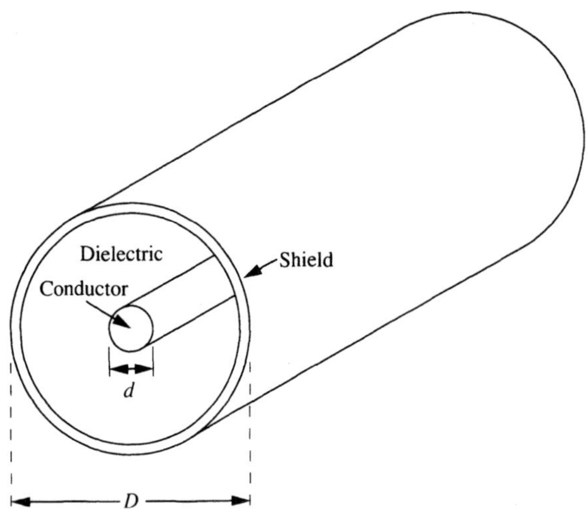

FIGURE 1–4 Geometry of coaxial cable transmission line.

The common form of coaxial cable that has been described is an example of an **unbalanced line;** i.e., the two conductors do not have symmetrical electric and magnetic properties with respect to ground. Instead the conducting shield is connected to the system ground. Ideally, this should result in an effective shield around the inner conductor and prevent any radiation losses. However, some losses in the dielectric material may still appear.

In general, coaxial cables are used in a wide number of applications at frequencies extending into the lower microwave region. Because they are unbalanced, coaxial cables may be readily connected to unbalanced sources. (If the load is balanced, however, there may still be a need for a conversion circuit.) The most significant disadvantages of coaxial cable are the cost and, in some applications, the physical size.

1–5 ■ TRANSMISSION LINE INDUCTANCE AND CAPACITANCE

For the purpose of this development, a **lossless** transmission line will be assumed. That is, for now any series resistance and shunt conductance will be neglected (but their effects will be considered later).

A transmission line is an example of a **distributed-parameter** circuit, in which the effects of various parameters such as capacitance and inductance are spread out as a function of length. This circuit is in contrast to a **lumped-parameter** circuit, which is the circuit model employed in basic circuit theory texts.

Because distributed-parameter effects are more difficult to analyze, early researchers focused on **incremental** models, in which lumped components are assumed for an infinitesimally short distance.

Throughout the body of transmission line theory, the symbols L and C are used to denote, respectively, the **inductance and capacitance per unit length.** Thus, common units for L are henries/meter (H/m), and the corresponding units for C are farads/meter (F/m). This is in sharp contrast to lumped circuit theory where L and C represent simply total inductance in henries and total capacitance in farads, respectively. Since most of this book is devoted to transmission effects, the reader may assume L and C to be expressed in parameter values per unit length unless otherwise stated.

The basic form of the incremental model of a lossless line of length Δx is shown in Fig. 1–5. The model shown is an unbalanced model, which is easier to describe, but it could be readily adapted to a balanced form by placing half of the inductance in each of the two horizontal conductors. An incremental section of the line is assumed to possess a series inductance and a shunt

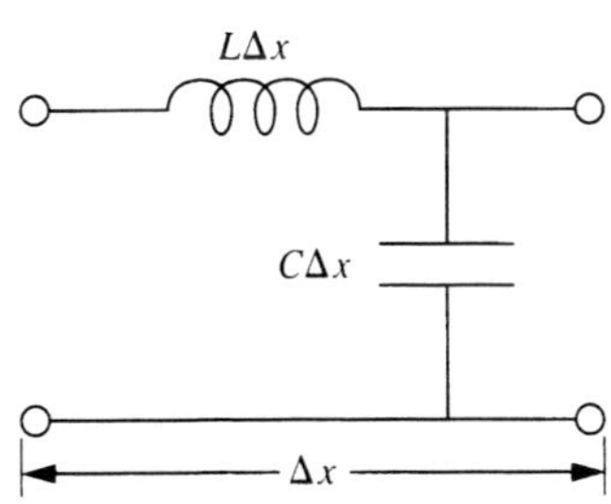

FIGURE 1–5 Incremental model of lossless transmission line.

capacitance. Since L and C are each expressed in parameter units/length, the net inductance of the section is $L\Delta x$, and the net capacitance is $C\Delta x$.

Let us briefly return to basic circuit theory to reinforce the concepts of inductance and capacitance.

Inductance

Inductance is a property of any circuit that encloses magnetic flux. In general, magnetic flux is measured in webers (Wb) and is a direct result of current flow. The definition of inductance is

$$\text{inductance (H)} = \frac{\text{flux (Wb)}^{\dagger}}{\text{current (A)}} \tag{1–15}$$

Capacitance

Capacitance is a property of any circuit in which an electric field exists between two conductors as a result of the movement of charge. In general, charge is measured in coulombs (C) and is a direct result of the potential difference established between the conductors. The definition of capacitance is

$$\text{capacitance (F)} = \frac{\text{charge (C)}}{\text{voltage (V)}} \tag{1–16}$$

Referring again to Fig. 1–5, note that current flow along the two conductors results in the creation of magnetic flux all around the region, and this effect is represented as series inductance. Voltage across the two conductors results in the displacement of charge and the creation of an electric field, and this is represented as shunt capacitance.

Formulas for the inductance and capacitance per unit length for various transmission line configurations will be developed in detail in Chapter 7. For the next several chapters, however, the reader should be aware of the existence of these effects for all transmission lines considered.

1–6 ■ CHARACTERISTIC IMPEDANCE

One of the most important parameters used in describing a transmission line is the **characteristic impedance.** The symbol Z_0 is the most widely used symbol for characteristic impedance, and it will be employed as appropriate throughout the book. The basic definition of characteristic impedance is

$$Z_0 = \frac{\text{voltage wave value}}{\text{current wave value}} \tag{1–17}$$

†In certain configurations where the flux is linked several times, such as a coil, the flux concept is modified to become **flux linkages** in **weber-turns.**

The significance of these waves will be established in Chapter 2, so do not be disturbed if they seem vague now. Instead concentrate here on the fact that any uniform transmission line will have a particular value of characteristic impedance associated with it.

In general, Z_0 may be a complex number of the form encountered in steady-state ac circuits. However, in a variety of situations Z_0 may be a real number. These authors have chosen to use the symbol R_0 to represent the characteristic impedance in all cases where it is assumed to be real. Thus,

$$Z_0 = R_0 \qquad \text{(when } Z_0 \text{ is real)} \tag{1–18}$$

This convention differs from that used in many books on the subject, but we believe that it will be clearer to the reader when a real value of Z_0 is assumed.

Rather interestingly, the *characteristic impedance of a transmission line is always a real number whenever the line is lossless.* This may seem puzzling since series resistance and shunt conductance are both assumed to be zero in the lossless case. Remember, however, that characteristic impedance is not a property of energy dissipation. Rather, it is a voltage-to-current ratio, and it is a real number when no dissipation occurs.

The values of characteristic impedance range from about 50 Ω to 75 Ω for coaxial cable and from about 300 Ω to 600 Ω for parallel lines.

The value of the characteristic impedance of a line is a function of all the incremental parameters. For a **lossless line,** it will be shown later in the book that the value is given by

$$R_0 = \sqrt{\frac{L}{C}} \tag{1–19}$$

where L and C are the inductance and capacitance per unit length as established earlier.

In general, the inductance increases as the spacing between the lines increases, and the capacitance decreases for the same condition. Thus, the characteristic impedance increases as the spacing between the wires increases, and vice versa.

EXAMPLE 1-5

An assumed-lossless transmission line has L = 400 nH/m and C = 100 pF/m. Determine the characteristic impedance.

Solution

Since the line is assumed to be lossless, the characteristic impedance will be real, and the symbol R_0 will be used. From (1–19), we have

$$R_0 = \sqrt{\frac{L}{C}} = \sqrt{\frac{400 \times 10^{-9}\ \text{H/m}}{100 \times 10^{-12}\ \text{F/m}}} = 63.25\ \Omega \tag{1–20}$$

Note that the basic inductance and capacitance units of henries and farads were used in the evaluation of (1–20). The per-unit length units, however, can be in any distance units provided that *both numerator and denominator have the same distance units*. For example, if L is expressed in henries/foot (H/ft) and C is expressed in farads/foot (F/ft), the values may be directly substituted in the equation.

1–7 ■ PERMITTIVITY AND PERMEABILITY

Certain properties of materials need to be introduced at this point to enhance the reader's vocabulary. These quantities will be explained in greater detail much later in the book when the background is sufficient. However, some qualitative measure of their significance will be suggested at this point.

Dielectric

The term **dielectric** was introduced earlier in conjunction with the physical description of a transmission line. Basically, a dielectric material is an insulating material, but the term is usually applied to those low-loss materials used to separate the conductors in a transmission line or a capacitor. Examples of dielectric materials are polyethylene, mylar, and, of course, free space.

Permittivity

The term **permittivity** refers to a basic property of a dielectric medium which will be given quantitatively in Chapter 6. For the moment, accept the fact that as the permittivity of a material increases, the volume of electrical charge produced by a fixed configuration and fixed voltage increases.

The basic unit of permittivity is farads/meter (F/m). In general, three different quantities are used in describing permittivity:

ϵ = permittivity of arbitrary material in farads/meter (F/m)

ϵ_0 = permittivity of free space

$= \dfrac{1}{36\pi} \times 10^{-9}$ F/m $= 8.842 \times 10^{-12}$ F/m

ϵ_r = dielectric constant or relative permittivity

The three quantities are related by

$$\epsilon = \epsilon_r \epsilon_0 \tag{1–21}$$

or

$$\epsilon_r = \frac{\epsilon}{\epsilon_0} \tag{1–22}$$

Note that the dielectric constant ϵ_r has no units since it is simply the ratio of two actual permittivity values. Since $\epsilon = \epsilon_0$ in free space, the dielectric constant of free space is simply $\epsilon_r = 1$. This value is also assumed as the dielectric constant for air in most applications.

Values of dielectric constant for most materials used in transmission lines range from 1 (for air lines) to about 10 or so.

Permeability

The term **permeability** refers to a magnetic property of any arbitrary material and is a measure of the flux density produced by a magnetizing current. As in the case of permittivity, the full significance of permeability will be developed in Chapter 6. However, for the moment, accept the fact that as the permeability of a material increases, the value of magnetic flux produced by a given current increases.

The basic unit of permeability is henries/meter (H/m). In general, three different quantities are used in describing permeability:

μ = permeability of arbitrary material, henries/meter (H/m)

μ_0 = permeability of free space = $4\pi \times 10^{-7}$ H/m

μ_r = relative permeability

The three quantities are related by

$$\mu = \mu_r \mu_0 \tag{1–23}$$

or

$$\mu_r = \frac{\mu}{\mu_0} \tag{1–24}$$

Note that the relative permeability has no dimensions since it is the ratio of two permeability values. Since $\mu = \mu_0$ in free space, the relative permeability of free space is $\mu_r = 1$. Unlike permittivity, where many common materials have values different from that of free space, the vast majority of common materials have permeability values equal to that of free space. The primary exceptions are the ferromagnetic materials such as iron, nickel, and cobalt. Such materials are never used in the dielectric material of a transmission line. Therefore, *to simplify notation throughout the book, the assumption will be made in a large number of formulas that $\mu = \mu_0$ unless otherwise indicated.*

EXAMPLE 1–6

The dielectric constant of polyethylene is listed as 2.25. Determine the permittivity.

Solution

From (1–21) and the value of ϵ_0 given in the text, we have

$$\epsilon = \epsilon_r \epsilon_0 = 2.25 \times \frac{1}{36\pi} \times 10^{-9} = 19.89 \times 10^{-12} \text{ F/m} \tag{1–25}$$

1–8 ■ VELOCITY OF PROPAGATION

The velocity of propagation of a wave traveling on a transmission line is a function of the inductance and capacitance per unit length. It will be shown in Chapter 3 that the velocity ν of propagation for a **lossless** transmission line is given by

$$\nu = \frac{1}{\sqrt{LC}} \tag{1–26}$$

where L is the inductance per unit length and C is the capacitance per unit length. The units for ν depend on the length units for L and C, which must be the same. For example, if L and C are given in henries/meter and farads/meter, respectively, the units for ν will be in meters/second.

It can also be shown that the velocity of propagation is related to the permittivity and permeability of the dielectric medium inside the line. For a lossless line, the velocity is given by

$$\nu = \frac{1}{\sqrt{\mu\epsilon}} \tag{1–27}$$

where the previous comments made about the distance units in (1–26) apply as well.

Let us establish some plausibility for (1–27) by assuming that $\mu = \mu_0 = 4\pi \times 10^{-7}$ H/m and $\epsilon = \epsilon_0 = (\frac{1}{36}\pi) \times 10^{-9}$ F/m and substituting these values in (1–27). The result is

$$\nu = \frac{1}{\sqrt{4\pi \times 10^{-7} \times \dfrac{1}{36\pi} \times 10^{-9}}} = \frac{1}{\sqrt{\dfrac{1}{9} \times 10^{-16}}} = 3 \times 10^{8} \text{ m/s} = c \tag{1–28}$$

Thus, the free-space velocity of light is obtained as expected.

In general, when $\mu = \mu_r\mu_0$ and $\epsilon = \epsilon_r\epsilon_0$ are substituted, the result may be expressed as

$$\nu = \frac{3 \times 10^8}{\sqrt{\mu_r\epsilon_r}} = \frac{c}{\sqrt{\mu_r\epsilon_r}} \tag{1–29}$$

From (1–29), we deduce that the velocity of propagation decreases as either the permeability or the dielectric constant increases, with the decrease being an inverse square root function. For example, if the dielectric constant is increased by a factor of four, the velocity of propagation is halved.

As previously discussed, rarely is any value of μ other than that of μ_0 encountered in transmission line work. Consequently, throughout this book, unless otherwise noted, the velocity of propagation will be assumed as

$$\nu = \frac{3 \times 10^8}{\sqrt{\epsilon_r}} = \frac{c}{\sqrt{\epsilon_r}} \tag{1–30}$$

A final interesting point is that L and C may be determined from a knowledge of the characteristic impedance and the velocity of propagation for a lossless line. Consider Eqs. (1–19) and (1–26). It will be left as an exercise for the reader to show that L and C may be expressed in terms of R_0 and ν by the following equations:

$$L = \frac{R_0}{\nu} \tag{1–31}$$

$$C = \frac{1}{R_0 \nu} \tag{1–32}$$

The per-unit distance units for L and C are the same as the distance units for ν. For example, if ν is expressed in meters/second, L and C will be expressed in henries/meter and farads/meter, respectively (assuming, of course, that R_0 is expressed in ohms).

EXAMPLE 1–7

Consider the lossless transmission line of Example 1–5 in which $L = 400$ nH/m and $C = 100$ pF/m. Determine the velocity of propagation.

Solution

From (1–26), we have

$$\nu = \frac{1}{\sqrt{(400 \times 10^{-9}) \times (100 \times 10^{-12})}} = 1.581 \times 10^8 \text{ m/s} \tag{1–33}$$

This value is slightly greater than half of the speed of light.

EXAMPLE 1–8

For the transmission line of Example 1–7, determine the dielectric constant. (Assume that $\mu = \mu_0$, as will always be the case, unless otherwise indicated.)

Solution

The velocity is related to the dielectric constant ϵ_r by (1–30). Using the result of Example 1–7, we have the following relationship:

$$1.581 \times 10^8 = \frac{3 \times 10^8}{\sqrt{\epsilon_r}} \tag{1–34}$$

Solving for ϵ_r yields

$$\epsilon_r = 3.6 \tag{1–35}$$

EXAMPLE 1–9

One of the most widely used types of coaxial cables is designated as RG–8/U. One variation of this cable type has the following specifications:

nominal impedance = 50 Ω

nominal velocity of propagation = 78%

Although no line is completely lossless, the losses will be ignored for this problem. Determine (a) L, (b) C, and (c) ϵ_r for the dielectric.

Solution

The velocity of propagation as given in many handbooks is actually the velocity relative to the speed of light, which is often called the **velocity factor.** Thus, the actual velocity is

$$\nu = 0.78c = 0.78(3 \times 10^8) = 2.34 \times 10^8 \text{ m/s} \tag{1–36}$$

(a) The value of L is determined from (1–31) as

$$L = \frac{R_0}{\nu} = \frac{50}{2.34 \times 10^8} = 213.7 \text{ nH/m} \tag{1–37}$$

(b) The value of C is determined from (1–32) as

$$C = \frac{1}{R_0\nu} = \frac{1}{50 \times (2.34 \times 10^8)} = 85.47 \text{ pF/m} \tag{1–38}$$

Incidentally, the table from which the given problem data were obtained lists the capacitance as 26 pF/ft. When that value is converted to pF/m (1 ft = 0.3048 m), the value obtained is 85.30 pF/m, which differs from the calculated value by less than 0.2%. The insignificant difference could be a result of roundoff or a slight difference due to the actual presence of some losses.

(c) Determination of the dielectric constant is made in the same manner as in Example 1–8. However, it is even simpler in this case since

$$\frac{\nu}{c} = \frac{1}{\sqrt{\epsilon_r}} = 0.78 \tag{1–39}$$

The ratio ν/c is the velocity factor given, which is 0.78 on a decimal basis. Solving for ϵ_r, we have

$$\epsilon_r = 1.64 \tag{1–40}$$

■ PROBLEMS

1–1. Calculate the free-space wavelength in meters for the following frequencies:

(a) 2 kHz
(b) 200 kHz
(c) 20 MHz
(d) 2 GHz

1–2. Calculate the free-space wavelength in meters for the following frequencies:
(a) 80 kHz
(b) 8 MHz
(c) 800 MHz
(d) 8 GHz

1–3. Calculate the free-space wavelength in miles for a frequency of 400 Hz.

1–4. Calculate the free-space wavelength in miles for a frequency of 1.5 kHz.

1–5. A sinusoidal signal has a free-space wavelength of 80 m. Calculate the frequency.

1–6. A sinusoidal signal has a free-space wavelength of 6 m. Calculate the frequency.

1–7. A digital signal utilizes pulses whose minimum widths are about 3 ns. Assuming the speed of light, determine the longest lengths of wire-pairs that can be allowed based on the 10% rule.

1–8. The longest connecting wires in a digital system are about 20 cm. Assuming the speed of light, determine the shortest acceptable pulse width based on the 10% rule.

1–9. A communication system operates at a frequency of 800 MHz. Assuming the speed of light, determine the length of connecting line that could be used without considering frequency-domain effects based on the 10% rule.

1–10. The length of a connecting cable between two points in a radio-frequency system is 50 cm. Assuming the speed of light, determine the highest operating frequency that should be used without considering frequency-domain effects based on the 10% rule.

1–11. In a coil, a current of 100 mA results in a magnetic flux of 50 μWb. Determine the inductance.

1–12. A current of 4 mA is flowing in a 20-μH coil. Determine the magnetic flux.

1–13. In a capacitor, a voltage of 20 V results in charge storage of 5 μC. Determine the capacitance.

1–14. A capacitance of 40 μF is charged to a voltage of 12 V. Determine the electric charge.

1–15. A lossless transmission line has an inductance of 320 nH/m and a capacitance of 57 pF/m. Determine the characteristic impedance.

1–16. A lossless transmission line has an inductance of 1.2 μH/m and a capacitance of 15 pF/m. Determine the characteristic impedance.

1–17. The dielectric constant of mica is 6. Determine the permittivity.

1–18. The permittivity of a material is 14×10^{-12} F/m. Determine the dielectric constant.

1–19. The relative permeability of nickel is 800. Determine the actual permeability.

1–20. The permeability of a ferromagnetic material is 10^{-4} H/m. Determine the relative permeability.

1–21. Determine the velocity of propagation of the transmission line of Problem 1–15.

1–22. Determine the velocity of propagation of the transmission line of Problem 1–16.

1–23. The dielectric constant in a transmission line is 4.7, and $\mu = \mu_0$. Determine the velocity of propagation.

1–24. The dielectric constant in a certain transmission line is 3, and $\mu = \mu_0$. Determine the velocity of propagation.

1–25. A coaxial cable has the following specifications: $R_0 = 73\ \Omega$, and velocity of propagation $= 2.1 \times 10^8$ m/s. Determine L and C.

1–26. A transmission line has the following specifications: $R_0 = 150\ \Omega$, and velocity factor $= 0.8$. Determine L and C.

1–27. The lower end of the commercial amplitude-modulation (AM) band is about 550 kHz. AM stations use "quarter-wave" vertical antennas whose lengths are 0.25λ. Determine the length in meters of a vertical antenna operating at the lower end.

1–28. The upper end of the commercial AM band referred to in Problem 1–27 is about 1610 kHz. Determine the length in meters of a vertical antenna operating at the upper end.

1–29. One popular simple antenna is the "half-wave" horizontal antenna whose theoretical length is 0.5λ at the operating frequency. In practice, however, the antenna is usually shortened by about 5% due to end effects. Determine the practical length in meters for a half-wave antenna to provide optimum reception at the lower end of the commercial FM band, which is about 88 MHz.

1–30. Based on the discussion of Problem 1–29, determine the practical length in meters for a half-wave antenna to provide optimum reception at the upper end of the FM band, which is about 108 MHz.

1–31. Show that the free-space velocity of light in feet/second is very close to $c = 984 \times 10^6$ ft/s.

1–32. Show that the free-space wavelength in feet can be expressed as

$$\lambda(\text{ft}) = \frac{982}{f(\text{MHz})}$$

You may use the result of Problem 1–31.

2

Transients on Lossless Transmission Lines

OBJECTIVES

After completing this chapter, the reader should be able to:

1. Discuss the process of charging a transmission line from a combination of forward and reverse waves.
2. Define **voltage** and **current reflection coefficients** and calculate their values.
3. Construct a **bounce diagram** for a lossless transmission line and use it to plot the voltage or current response anywhere on the line.
4. Define the conditions for a **matched line** and discuss the significance.
5. Draw the dc steady-state model for a transmission line and determine the final values of voltage and current for the pulse.

2–1 ■ TRANSIENT PROCESS ON TRANSMISSION LINE

This section is devoted to a qualitative discussion of the process that occurs when an uncharged transmission line is excited by a dc voltage at one end. The exact mathematical details will be developed later, but it is important that the physical process be fully understood first.

Consider the uncharged lossless transmission line shown in Fig. 2–1(a). A dc voltage of value E is to be connected to the left-hand side of the line through a resistance R_1, and by popular convention, this will be referred to as the **generator.** The right-hand end of the line is connected to a resistance R_2, and this end will be referred to as the **load.**

At $t = 0$, the switch is closed as shown in Fig. 2–1(b). A dc voltage appears instantaneously across the generator end of the line as depicted by the + and − signs. (As will be seen later, this voltage is, in general, *not* equal to E.) Current begins to flow outward along the top line, and the same value starts flowing back toward the source on the bottom as indicated by the small arrows. What is happening here is that a traveling wave is being created, and this wave will travel from source to load with velocity ν. This wave will be referred to as the **first forward wave.**

It is important to understand that although the wave is traveling in only one direction, the current itself is flowing toward the load along the top and flowing toward the source along the bottom. This continuity of current flow is required by Kirchhoff's current law.

After some time has passed, the first forward wave reaches the load as shown in Fig. 2–1(c). Depending on the conditions at the load (to be discussed later), a new wave may be generated as shown in Fig. 2–1(d). This new wave will travel from the load toward the generator and will be referred to as the **first reverse wave.** The voltage and current associated with the first reverse wave add algebraically to the first forward wave, which remains on the line.

At a still later time, the first reverse wave reaches the generator as shown in Fig. 2–1(e). Depending on the conditions at the source (to be discussed later), a new wave may be generated as shown in Fig. 2–1(f). This new wave will travel from the source toward the load and will be referred to as the **second forward wave.** The voltage and current associated with the second forward wave add algebraically to the previous two waves, which remain on the line.

The process keeps repeating itself in the same pattern as described in the first round trip. Thus, the second forward wave travels to the load, and a second reverse wave is created. The second reverse wave travels to the source, and a third forward wave is created, and so on. In theory, this process continues indefinitely.

In every practical case, the magnitudes of voltage and current for the waves diminish, and after a reasonable number of round trips, the voltage and current reach practical limiting values. In this case, the line is said to be **charged.** This process is similar in some ways to that of charging a capacitor through a resistor. In theory, the exponential function never reaches its final value, but in a practical sense, it becomes infinitesimally close.

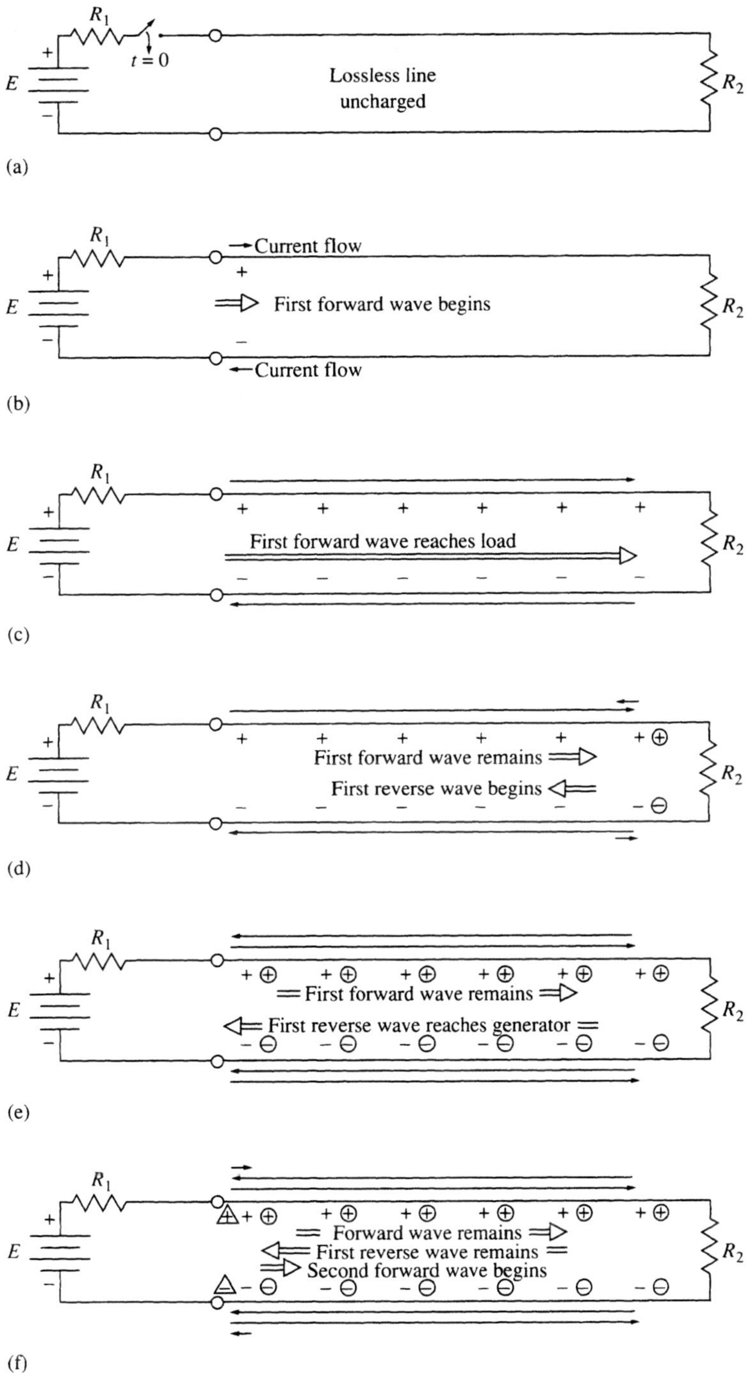

FIGURE 2–1 Illustration of transients on transmission lines.

2–2 ■ REFLECTION COEFFICIENT

As discussed in the last section, a new wave will, in general, be created when a wave reaches either end of the line. For either the source or the load end, the wave that arrives is referred to as the **incident wave,** and the new wave that is created is referred to as the **reflected wave.** The value of the reflected wave may be predicted from a knowledge of the incident wave and the conditions at the end of the line.

A significant parameter to establish these relationships is the **reflection coefficient.** It is possible to define separate reflection coefficients for voltage and current, but they are closely related, as will be seen shortly. For either end of the line, the voltage reflection coefficient can be calculated as follows:

$$\text{voltage reflection coefficient} = \frac{\text{terminating resistance} - R_0}{\text{terminating resistance} + R_0} \tag{2–1}$$

Referring to Fig. 2–2, the terminating resistance for the generator end is R_1, and the terminating resistance for the load end is R_2.

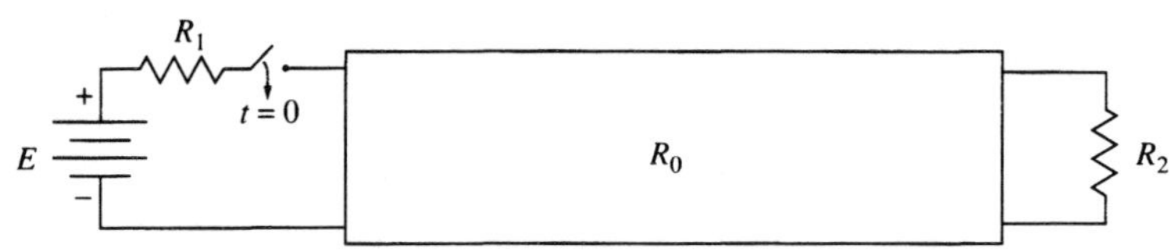

FIGURE 2–2 Transmission line with terminating resistances.

The current reflection coefficient is related to the voltage reflection coefficient simply as

$$(\text{current reflection coefficient}) = -(\text{voltage reflection coefficient}) \tag{2–2}$$

The minus sign is a result of the fact that the reflected current has a direction sense opposite that of the reflected voltage.

Rather than define separate symbols for both voltage and current reflection symbols, it is customary to use one symbol to represent the voltage reflection coefficient and simply to refer to its negative as the current reflection coefficient.

Let Γ_1 represent the voltage reflection coefficient at the source end. It is given by

$$\Gamma_1 = \frac{R_1 - R_0}{R_1 + R_0} \tag{2–3}$$

Let Γ_2 represent the voltage reflection coefficient at the load end. It is given by

$$\Gamma_2 = \frac{R_2 - R_0}{R_2 + R_0} \tag{2–4}$$

The current reflection coefficients at the ends of the line are simply $-\Gamma_1$ and $-\Gamma_2$, respectively. When there is no adjective in front of the term *reflection coefficient,* it will be understood to mean **voltage reflection coefficient.**

Formulas have been given for calculating the reflection coefficient, but its exact definition should be clearly understood. At either end of the line, the definitions are

$$\Gamma = \frac{\text{reflected voltage wave}}{\text{incident voltage wave}} \tag{2–5}$$

$$-\Gamma = \frac{\text{reflected current wave}}{\text{incident current wave}} \tag{2–6}$$

Thus, the incident voltage wave is multiplied by Γ to determine the reflected voltage wave, and the incident current wave is multiplied by $-\Gamma$ to determine the reflected current wave.

For the situation of dc input voltage, lossless line, and resistive terminations, the reflection coefficient is a real number, which may be either positive or negative. Later in the text, however, we will see that the reflection coefficient may be complex when ac circuits, lossy lines, or complex loads are considered.

It can be shown that the magnitude of the reflection coefficient is bounded by unity; i.e.,

$$0 \le |\Gamma| \le 1 \tag{2–7}$$

Thus, any reflected wave can never be greater than the corresponding incident wave. In most real situations, the reflected wave will be smaller.

Before completing this section, let us investigate the possibility of the reflection coefficient having a value of zero. From (2–3), we note that

$$\Gamma_1 = 0 \qquad \text{when } R_1 = R_0 \tag{2–8}$$

This condition is referred to as a **matched generator.** From (2–4), we note that

$$\Gamma_2 = 0 \qquad \text{when } R_2 = R_0 \tag{2–9}$$

This condition is referred to as a **matched load.**

When both $\Gamma_1 = 0$ and $\Gamma_2 = 0$, the complete line is referred to as a **matched line.** This is a desirable (but not always attainable) situation in a large number of practical applications. Thus, *for a matched line,* $R_1 = R_2 = R_0$.

What are the implications of a matched generator and/or a matched load? First, if the load is matched, $\Gamma_2 = 0$, and there will be no reflected wave when the first forward wave arrives. Thus, the line will be fully charged at the end of a single one-way trip. This is true irrespective of whether the generator is matched because the presence of a matched load prevents any reflection from arising at the load.

If the generator is matched but the load is mismatched, the first reverse wave will be created due to the load, but when it arrives back at the generator, no further forward wave will be generated. Thus, the line is fully charged at the end of one round trip in this case.

EXAMPLE 2–1

The 50-Ω lossless line shown in Fig. 2–3 is connected through a 30-Ω resistance to a battery. The load is a 200-Ω resistance. Determine the voltage and current reflection coefficients at the two ends.

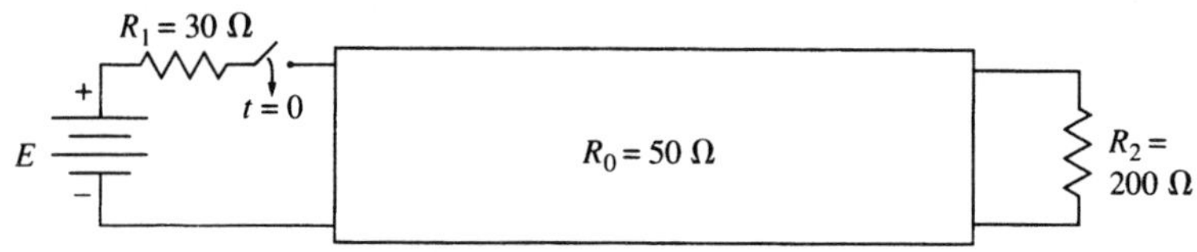

FIGURE 2–3 Circuit of Example 2–1.

Solution

At the generator end, the voltage reflection coefficient is

$$\Gamma_1 = \frac{R_1 - R_0}{R_1 + R_0} = \frac{30 - 50}{30 + 50} = \frac{-20}{80} = -0.25 \tag{2–10}$$

At the load end, the voltage reflection coefficient is

$$\Gamma_2 = \frac{R_2 - R_0}{R_2 + R_0} = \frac{200 - 50}{200 + 50} = \frac{150}{250} = 0.6 \tag{2–11}$$

Note that Γ_1 is negative and Γ_2 is positive in this particular case, but all combinations are possible.

The current reflection coefficients at the two ends are simply the negatives of the voltage reflection coefficients; i.e.,

$$\text{current reflection coefficient at source} = -\Gamma_1 = 0.25 \tag{2–12}$$

$$\text{current reflection coefficient at load} = -\Gamma_2 = -0.6 \tag{2–13}$$

EXAMPLE 2–2

Assume in the system of Example 2–1 that the first forward wave of voltage has a value of 40 V. Determine (a) the value of the first forward wave of current, (b) the value of the first reverse wave of voltage, and (c) the value of the first reverse wave of current.

Solution

(a) Since the characteristic impedance of the line is 50 Ω, the first forward current wave has the value

$$\text{first forward current} = \frac{40\ \text{V}}{50\ \Omega} = 0.8\ \text{A} \tag{2–14}$$

(b) The first reverse wave of voltage has the value

$$\Gamma_2 \times \text{first forward voltage} = 0.6 \times 40\ \text{V} = 24\ \text{V} \tag{2–15}$$

(c) The first reverse wave of current has the value

$$-\Gamma_2 \times \text{first forward current} = -0.6 \times 0.8\text{ A} = -0.48\text{ A} \qquad \textbf{(2–16)}$$

Note that the signs of both incident and reflected voltage waves are the same (both positive), so the reverse voltage wave will act to increase the magnitude of the voltage along the line. However, the signs of the incident and reflected current waves are different, so the reverse current wave will act to decrease the magnitude of the current along the line. If the voltage reflection coefficient had been negative (meaning that the current reflection coefficient would be positive), the opposite pattern would be true.

2-3 ■ INITIAL SOURCE/LINE MODEL

From the work of the preceding section, we have seen how to determine the reflected wave of either voltage or current from a knowledge of the incident voltage or current and the reflection coefficient. The only remaining question that must be answered is how to predict the first waves of voltage and current that arise when the switch is first closed.

The key to this analysis is the basic property that the ratio of voltage to current for a single forward wave on an *uncharged line* is given by the characteristic impedance R_0. This is equivalent to the assumption that the uncharged line acts as if it were a pure resistance of value R_0 at the instant the switch is closed. Thus, an equivalent circuit of the form shown in Fig. 2–4 is valid *immediately after the switch is closed.* (The circuit will no longer be valid after the first reverse wave returns to the source.)

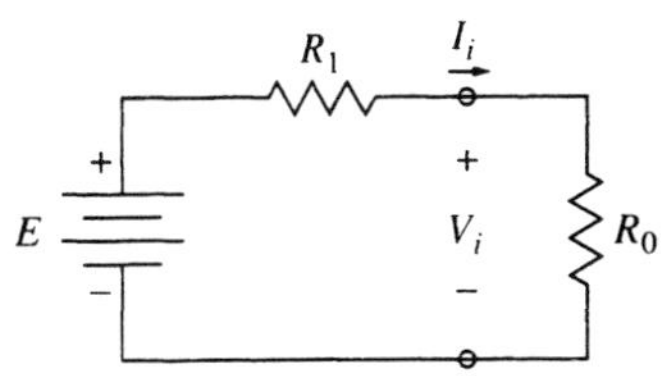

FIGURE 2–4 Equivalent circuit of uncharged transmission line at generator end immediately after excitation.

For the purpose of this development, let V_i represent the initial value of the voltage wave, and let I_i represent the initial value of the current wave. From a straightforward analysis of this simple circuit,

$$V_i = \frac{R_0}{R_0 + R_1} E \qquad \textbf{(2–17)}$$

and

$$I_i = \frac{E}{R_0 + R_1} \qquad \textbf{(2–18)}$$

Of course, it is immediately evident that

$$\frac{V_i}{I_i} = R_0 \tag{2–19}$$

EXAMPLE 2–3

Assume in the circuit of Examples 2–1 and 2–2 that the battery has a value E = 64 V. Determine the values of the first forward waves of voltage and current.

Solution

The equivalent circuit applicable at $t = 0$ is shown in Fig. 2–5. A simple analysis yields

$$V_i = \frac{50}{50 + 30} \times 64 = 0.625 \times 64 = 40 \text{ V} \tag{2–20}$$

and

$$I_i = \frac{64}{50 + 30} = \frac{64}{80} = 0.8 \text{ A} \tag{2–21}$$

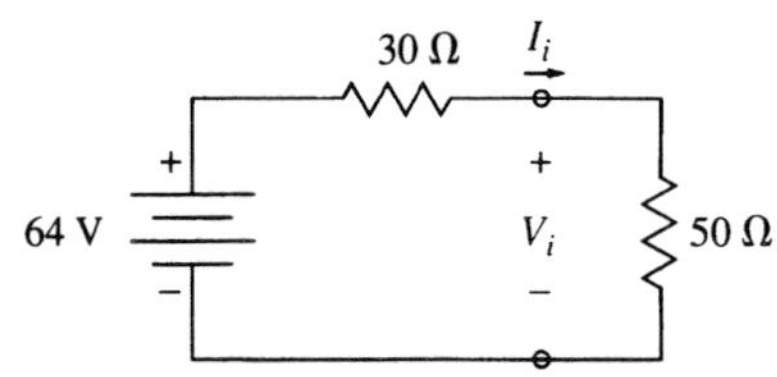

FIGURE 2–5 Circuit of Example 2–3.

2–4 ■ STEADY-STATE dc MODEL

As we have seen, the initial voltage and current waves are determined from a model in which the line acts as a resistance of value R_0. In some sense, the generator does not "know" what the value of the load resistance is since it takes the first reverse wave one round trip to return to the generator and "provide information" about its value. Thus, the initial reaction is transient in nature and is determined by the properties of the line and generator rather than the load.

If the line is lossless, and if the input generator is a dc voltage, the transient conditions on the line will eventually disappear in most practical situations. This limiting case is referred to as the **steady-state dc circuit,** and the final values of voltage and current may be readily predicted. The model depicting this condition is shown in Fig. 2–6.

In the dc circuit model, the line simply acts like a pair of wires, and since the line is assumed to be lossless, no resistance appears in the wires. Let V_f

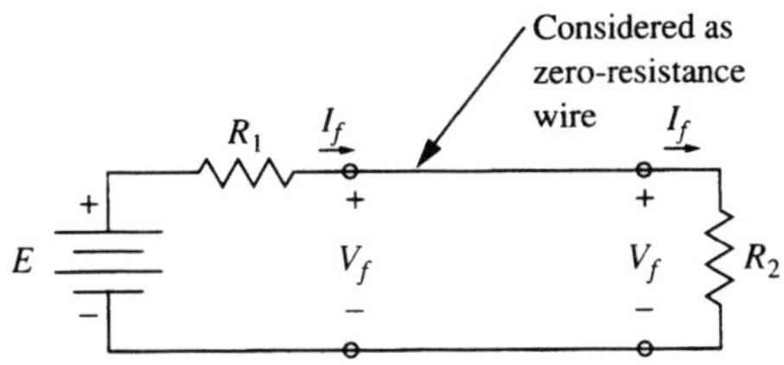

FIGURE 2–6 Equivalent circuit of lossless transmission line with dc source after steady-state conditions are reached.

represent the final value of voltage, and let I_f represent the final value of current. The values of V_f and I_f are readily determined as

$$V_f = \frac{R_2}{R_2 + R_1} E \tag{2–22}$$

$$I_f = \frac{E}{R_2 + R_1} \tag{2–23}$$

Note that the characteristic impedance does not even enter into the calculations for the dc steady-state case. Characteristic impedance applies to a single wave of voltage and current, and when dc steady-state conditions are reached (corresponding to a charged line), it is no longer a property of primary interest.

EXAMPLE 2–4

For the circuit of Examples 2–1, 2–2, and 2–3, determine the dc steady-state values of the voltage and current on the line.

Solution

The steady-state dc equivalent circuit is shown in Fig. 2–7. The final values of voltage and current are determined as

$$V_f = \frac{200}{200 + 30} \times 64 = \frac{200}{230} \times 64 = 55.65 \text{ V} \tag{2–24}$$

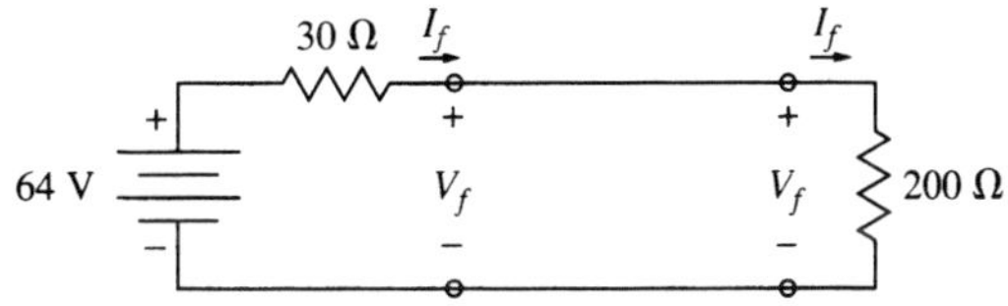

FIGURE 2–7 Circuit of Example 2–4.

and

$$I_f = \frac{64}{200 + 30} = 0.2783 \text{ A} \tag{2–25}$$

It should be stressed that in the dc steady state, these are the values of voltage and current *at all points on the line.*

2–5 ■ COMPLETE TRANSIENT SOLUTION WITH dc INPUT

The stage has now been set for obtaining the complete solution for the transient voltage and current anywhere along a lossless line with a dc excitation. Thus far, very simple notation has been adequate, but for a complete solution applicable for any point on the line and at any time, more complex notation is required.

Consider the system depicted in Fig. 2–8. The quantity x is a distance variable defined in a positive sense from left to right. The value $x = 0$ defines the input to the line, and for line length d, the value $x = d$ defines the load.

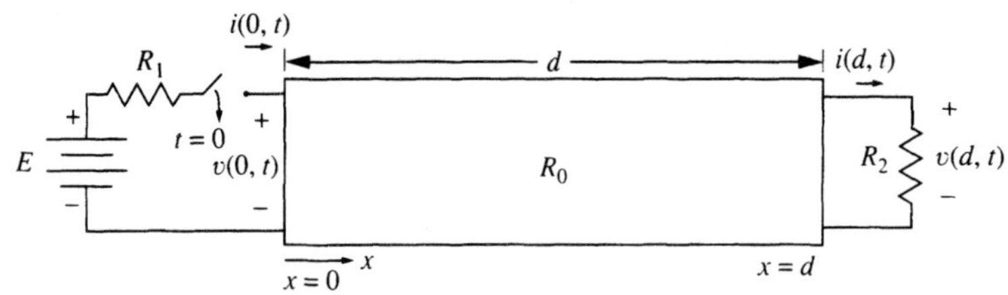

FIGURE 2–8 Form of transmission line with notation shown.

The quantity t defines the time variable. The switch will be assumed to close at $t = 0$. The one-way delay time represents a specific value of t that will be denoted as t_1; i.e.,

$$t_1 = \frac{d}{\nu} \tag{2–26}$$

where ν is the velocity of propagation. When sufficient time has passed that dc steady-state conditions exist on the line, the time will be referred to as $t = \infty$. Understand, however, that the practical implication is that a sufficiently long time has passed during which steady-state conditions exist.

From the preceding discussion, the voltage and current are each functions of *two* independent variables: x and t. This is indicated mathematically by defining the voltage v and current i by the following forms:

$$v \equiv v(x, t) \tag{2–27}$$

$$i \equiv i(x, t) \tag{2–28}$$

The quantities on the right-hand sides of the preceding equations can be read as "v as a function of x and t" and "i as a function of x and t." The quantity (x, t) is referred to as the **argument.** It will not always be necessary to carry along the argument, but it will be shown when it is desirable to emphasize the dependency on both x and t. It should be understood in all situations where it is not explicitly shown.

The generator end of the line corresponds to $x = 0$, so the voltage and current there are denoted as $v(0, t)$ and $i(0, t)$, as shown in Fig. 2–8. The load corresponds to $x = d$, so the voltage and current there are denoted as $v(d, t)$ and $i(d, t)$, respectively, as shown in Fig. 2–8. The *initial* values of the voltage and current at the source end are denoted as $v(0, 0)$ and $i(0, 0)$, respectively. The steady-state values of the voltage and current anywhere on the line are denoted as $v(x, \infty)$ and $i(x, \infty)$.

An important graphical aid in predicting the voltage and/or current anywhere along the line is the **bounce diagram.** A bounce diagram may be constructed for either voltage or current by using the values of voltage or current and the appropriate reflection coefficients. The basic form for either voltage or current is shown in Fig. 2–9.

The horizontal line at the top corresponds to the displacement x along the line with $x = 0$ at the left and $x = d$ at the right. The vertical line along the left corresponds to time $t = 0$ at the top. Increasing time corresponds to downward movement, and this may be increased to any arbitrary value.

Based on the procedure discussed in Section 2–3, the first wave of either voltage or current is calculated, and the value is labeled on the first tilted line. This line is drawn at an angle such that it reaches the load end of the line $(x = d)$ at a time corresponding to a one-way time t_1.

The first reflected wave is determined by multiplying the first forward wave by the load reflection coefficient as discussed in Section 2–2. This value is labeled on the second tilted line, which has a negative slope. This line is drawn at an angle such that it reaches the generator end of the line $(x = 0)$ at a time corresponding to a two-way delay time $2t_1$.

The second forward wave is determined by multiplying the first reverse wave by the source reflection coefficient as discussed in Section 2–2. This value is labeled on the next tilted line. The procedure of the first round trip is repeated for as many round trips as desirable. Except for certain assumed-unrealistic cases, the values of the waves diminish as successive round trips continue. The actual voltage or current at any point on the line as a function of time may be graphed by the procedure that follows. For the purpose of discussion, assume that the bounce diagram is for voltage, although the procedure for current is the same.

First, consider the process of plotting $v(0, t)$, i.e., the voltage at the source end. Starting at $x = 0$ and $t = 0$, move downward along the t-axis. Immediately, the first forward wave is passed, so the value of $v(0, t)$ for $0 \le t \le 2t_1$ is that value. The first change occurs at $t = 2t_1$, at which time the first reverse wave arrives and the second forward wave is generated. These two waves add algebraically to the earlier wave at the same instant of time, and the process

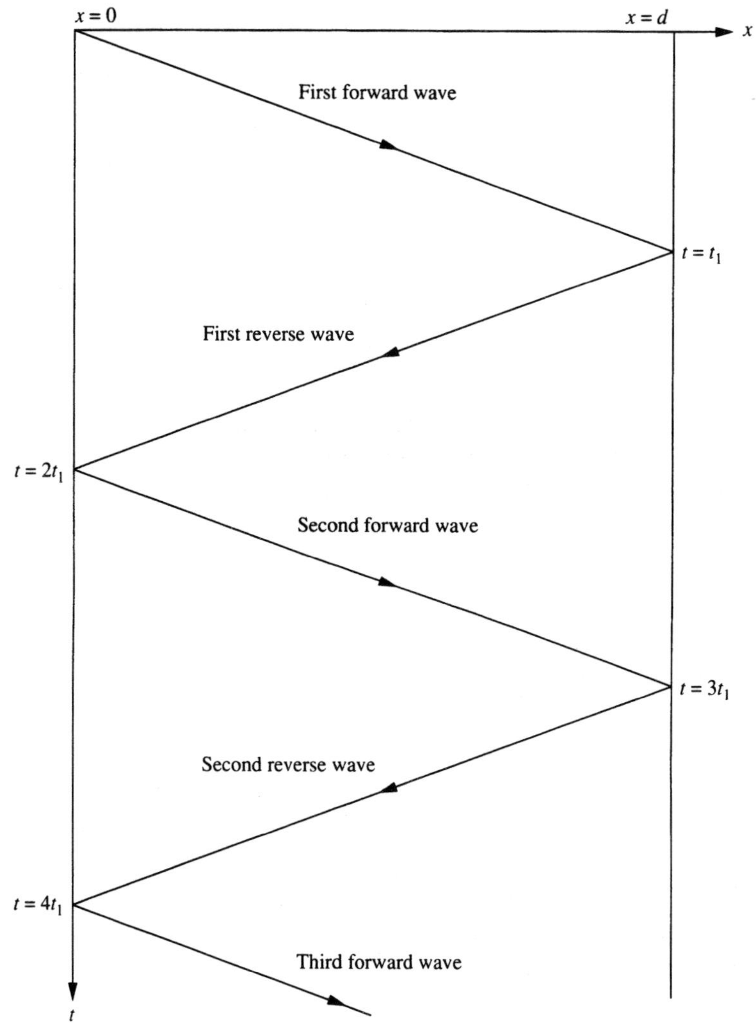

FIGURE 2–9 Form of bounce diagram.

is accumulative. Thus, immediately after $2t_1$, the net voltage is the sum of three different waves, and this value remains until $4t_1$, at which time two more waves are added. This process can be continued as long as desired.

Next, consider the process of plotting $v(d, t)$, i.e., the voltage at the load. Starting at $x = d$, move downward along the line parallel to the t-axis at $x = d$. There is no voltage at the load until $t = t_1$ since it takes the first wave that long to arrive. However, at $t = t_1$, the first forward wave arrives and the first reverse wave is created. These two waves add at the same instant of time, so the voltage immediately after $t = t_1$ is the sum of the first two waves. The load voltage remains at this value until $3t_1$, at which time the second forward wave and the second reverse wave are simultaneously added to the earlier voltage. This process may be continued as long as desired.

We have seen that at the two ends of the line, the waves add in pairs after the first wave is created. However, at any point other than the ends, the waves add one by one as they pass the point of reference. For example, assume that the voltage halfway down the line is desired, i.e., $v(0.5d,\ t)$. This may be determined by drawing a vertical line halfway down the x-axis as shown on the bounce diagram of Fig. 2–10. The first forward wave passes at $t = 0.5t_1$, and the first reflected wave passes at $t = 1.5t_1$. The second forward wave passes at $t = 2.5t_1$; the second reflected wave passes at $t = 3.5t_1$; and so on. The voltages are added as they occur. Thus, changes take place at $t = 0.5t_1$, $1.5t_1$, $2.5t_1$, $3.5t_1$, and so on. For the lossless line, the final value must equal the value at the two ends.

We will now study the phenomena through the analysis of several problems. We will begin with the simple case of a matched line and proceed gradually to consider more complex cases.

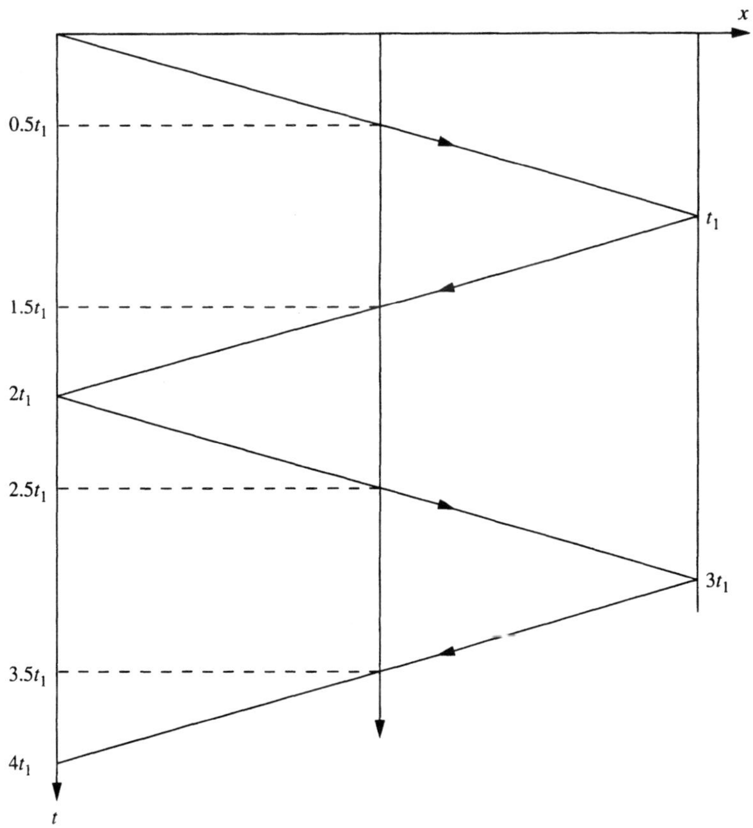

FIGURE 2–10 Use of the bounce diagram to determine voltage or current at the midpoint of the line.

EXAMPLE 2–5

The initially uncharged line in Fig. 2–11(a) is excited at $t = 0$ by a 60-V voltage as shown. Draw the voltage bounce diagram, and use it to plot the voltages at the two ends of the line, i.e., $v(0, t)$ and $v(200, t)$.

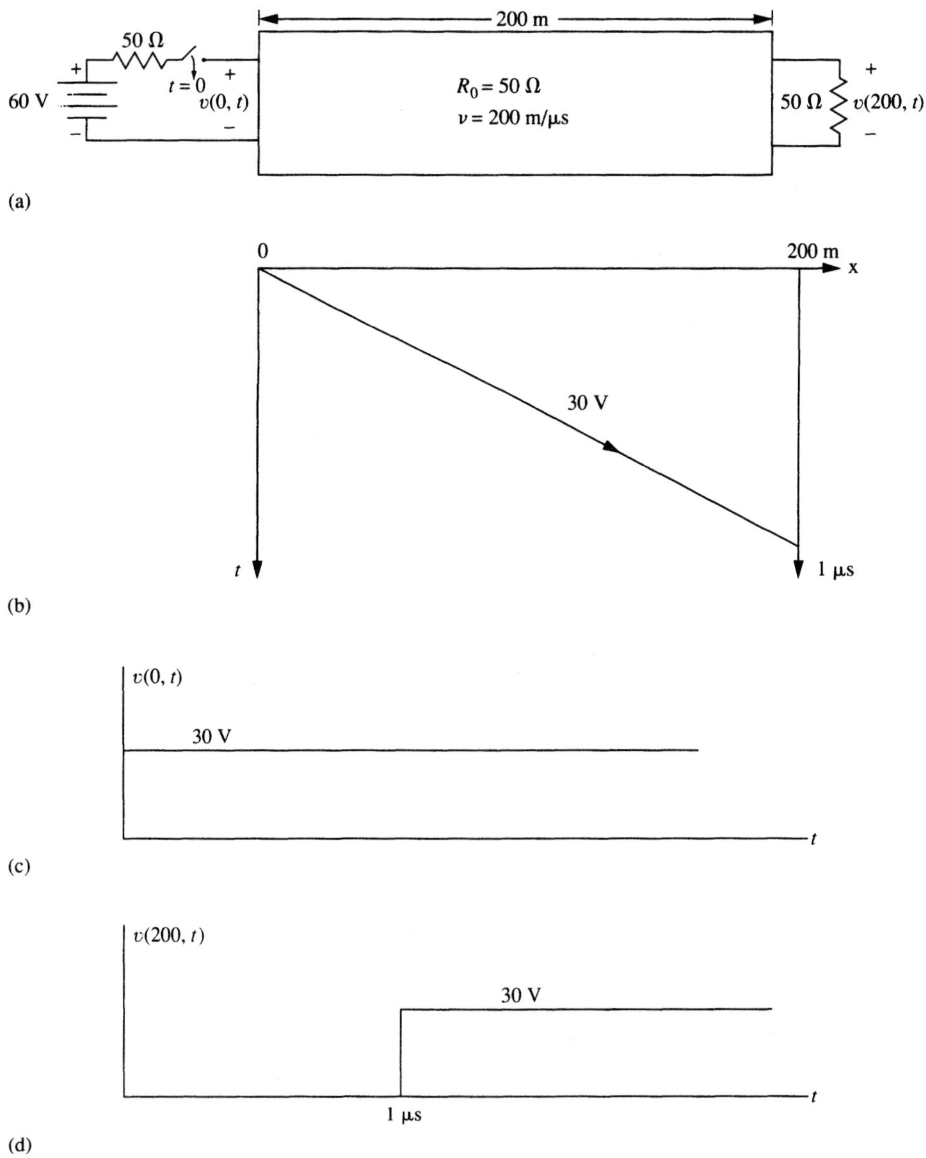

FIGURE 2–11 Circuit, bounce diagram, and waveforms of Example 2–5.

Solution

The one-way delay time is

$$t_1 = \frac{d}{\nu} = \frac{200\ \text{m}}{200\ \text{m}/\mu\text{s}} = 1\ \mu\text{s} \tag{2–29}$$

The reflection coefficients at the two ends of the line are readily noted to be

$$\Gamma_1 = \frac{R_1 - R_0}{R_1 + R_0} = \frac{50 - 50}{50 + 50} = 0 \tag{2–30}$$

$$\Gamma_2 = \frac{R_2 - R_0}{R_2 + R_0} = \frac{50 - 50}{50 + 50} = 0 \tag{2–31}$$

In subsequent problems, we will simply state that $\Gamma = 0$ without calculation where a line is terminated in its characteristic impedance.

Since the line initially acts as a resistance of 50 Ω, the 60-V source value is divided into two equal parts, and the initial generator-end line voltage is

$$v(0, 0) = \frac{50}{50 + 50} \times 60\ \text{V} = 30\ \text{V} \tag{2–32}$$

This voltage represents the value of the first wave as indicated on the bounce diagram of Fig. 2–11(b). Since $\Gamma_2 = 0$, there is no reflected voltage at the load, and the bounce diagram reduces to a single forward wave.

The voltage at the generator end of the line is obtained by moving down in time along the left-hand end of the bounce diagram, and it consists of a voltage instantaneously reaching a level of 30 V at $t = 0$ and remaining there indefinitely as shown in Fig. 2–11(c). The voltage at the load end of the line is obtained by moving down in time along the right-hand end of the bounce diagram. The value is zero until $t = 1\ \mu\text{s}$, at which time it jumps to a value of 30 V and remains there indefinitely as shown in Fig. 2–11(d).

This case is one of the simplest situations resulting from a matched line and could be easily predicted without the use of the bounce diagram. However, the bounce diagram will be promoted as the basis for analysis, and its utility will be apparent in later, more complex examples.

EXAMPLE 2–6

Assume that the circuit of Example 2–5 is modified by changing the generator resistance to 25 Ω as shown in Fig. 2–12(a). Repeat the analysis of Example 2–5.

Solution

The reflection coefficient at the load is still $\Gamma_2 = 0$. However, the source reflection coefficient is now

$$\Gamma_1 = \frac{R_1 - R_0}{R_1 + R_0} = \frac{25 - 50}{25 + 50} = -\frac{1}{3} \tag{2–33}$$

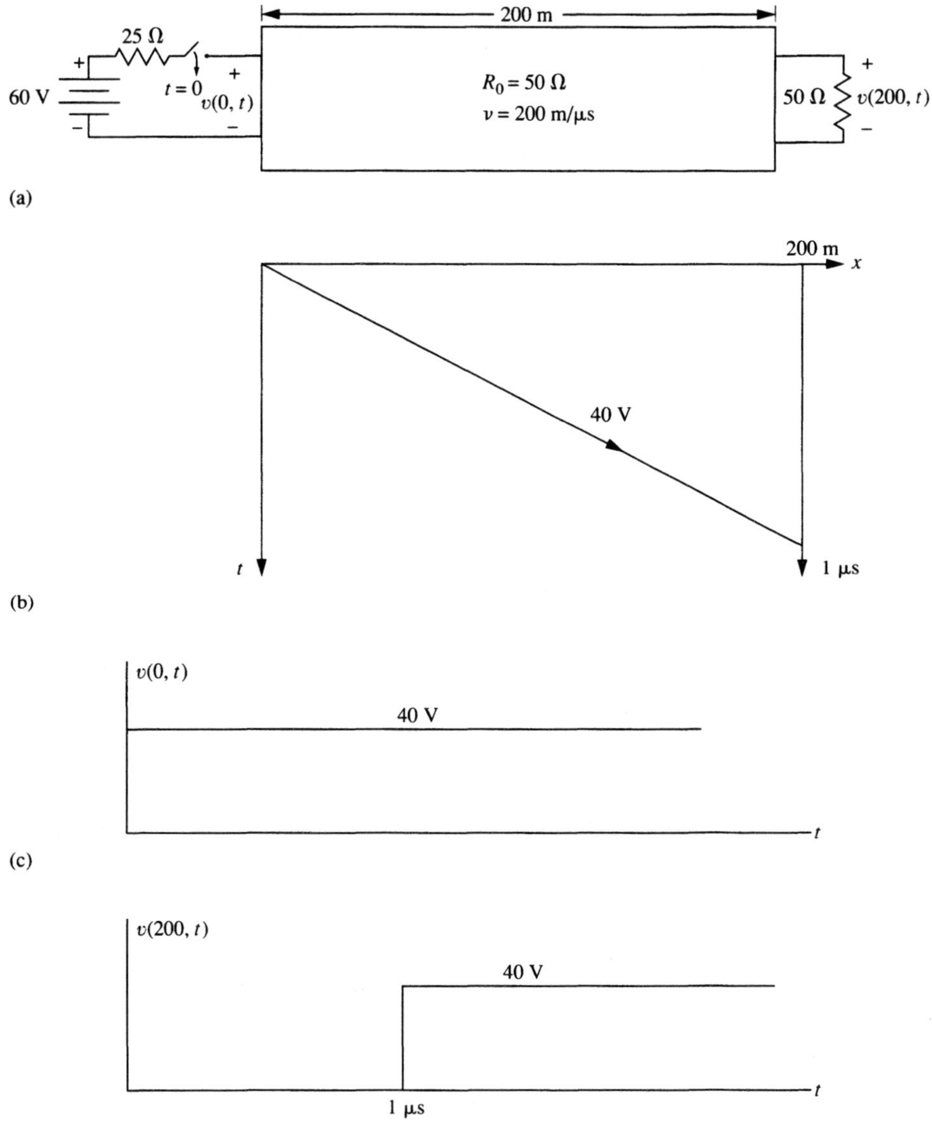

FIGURE 2–12 Circuit, bounce diagram, and waveforms of Example 2–6.

The initial voltage appearing across the front end of the line is now

$$v(0, 0) = \frac{50}{50 + 25} \times 60 \text{ V} = 40 \text{ V} \tag{2–34}$$

This is the value of the first wave and is shown on the bounce diagram of Fig. 2–12(b).

Since $\Gamma_2 = 0$, there is no reflected voltage at the load, and the bounce diagram reduces to a single forward wave again. The voltages at the two ends of the line are determined in the same manner as in Example 2–5 and are shown in Figs. 2–12(c) and (d).

Comparing these results with those of Example 2–5, it is clear that there is no reflected wave in either case. Thus, when $\Gamma_2 = 0$, there is no reflected wave from the load even when the source is mismatched. The effect of the source mismatch is to change the value of the voltage on the line from that obtained with a match, but the form is the same.

EXAMPLE 2-7

Assume that the circuit of Example 2–5 is modified by changing the load resistance to 250 Ω as shown in Fig. 2–13(a). Repeat the analysis of Example 2–5.

Solution

As in Example 2–5, the generator reflection coefficient is $\Gamma_1 = 0$. However, the load reflection coefficient is now

$$\Gamma_2 = \frac{R_2 - R_0}{R_2 + R_0} = \frac{250 - 50}{250 + 50} = \frac{2}{3} \tag{2–35}$$

The initial voltage at the input of the line is the same as in Example 2–5; i.e., $v(0, 0) = 30$ V. This voltage is shown as the first wave on the bounce diagram of Fig. 2–13(b).

The first wave encounters the mismatch at the load at $t = 1$ μs, and a reflected wave of value $(\frac{2}{3}) \times 30$ V $= 20$ V is created. At $t = 2$ μs, this reverse wave reaches the generator end, but since $\Gamma_1 = 0$, there is no further reflection.

From the bounce diagram of Fig. 2–13(b), the voltages at the ends may be constructed as shown in Figs. 2–13(c) and (d). At the generator end, the reverse wave adds to the forward wave at $t = 2$ μs as shown, and that voltage remains at 50 V thereafter. At the load end, the voltage is zero until $t = 1$ μs. At that time, the 30-V forward wave reaches the load, but the 20-V reverse wave is generated at the same time. Thus, the voltages add together instantaneously to produce 50 V as shown.

As an additional check, a dc steady-state analysis for the voltage everywhere on the line would yield

$$v(x, \infty) = \frac{250}{250 + 50} \times 60 \text{ V} = \frac{5}{6} \times 60 = 50 \text{ V} \tag{2–36}$$

Comparing this example with the previous two, it is apparent that when the load end is matched, there is only one forward wave irrespective of whether the generator is matched or not. However, when the generator end is matched but the load end is mismatched, there is one forward wave and one reverse wave. In this case, the transient interval for the line to reach steady-state conditions is $2t_1$.

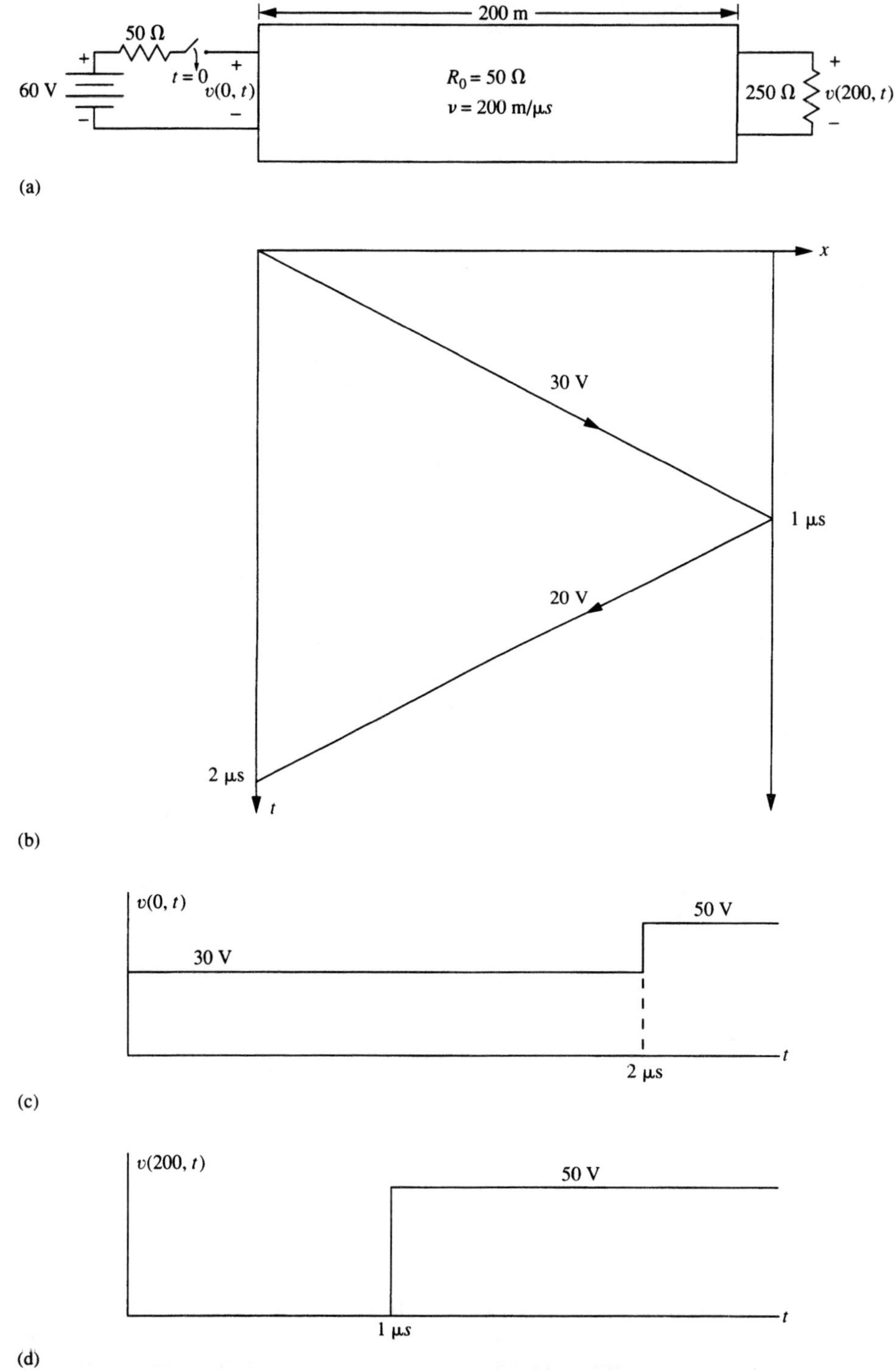

FIGURE 2–13 Circuit, bounce diagram, and waveforms of Example 2–7.

EXAMPLE 2-8

The circuit of Fig. 2–14 represents the situation considered in Examples 2–1 through 2–4 with $d = 60$ m and $E = 64$ V. Draw the voltage bounce diagram and use it to plot the voltage at the generator end of the line for at least four round trips.

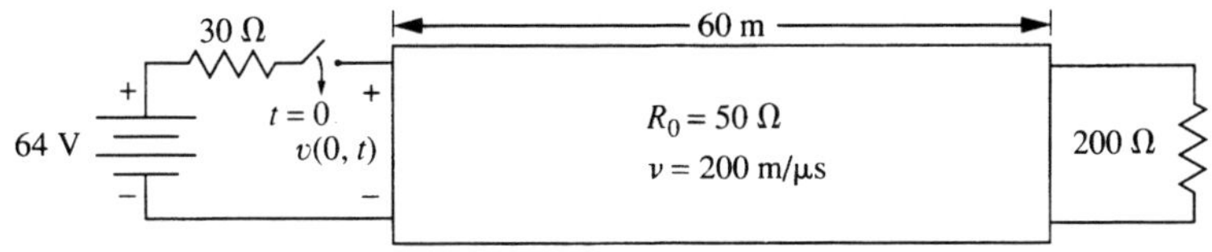

FIGURE 2–14 Circuit of Example 2–8.

Solution

Much of the preliminary analysis for this problem was performed in Examples 2–1 through 2–3, so those results can be readily employed. The two voltage reflection coefficients were determined in Example 2–1 to be $\Gamma_1 = -0.25$ and $\Gamma_2 = 0.6$. The one-way delay time is

$$t_1 = \frac{d}{v} = \frac{60 \text{ m}}{200 \text{ m}/\mu\text{s}} = 0.3\ \mu\text{s} \tag{2–37}$$

The value of the first forward voltage wave was determined in Example 2–3 to be 40 V, and the first reverse voltage wave was determined in Example 2–2 to be 24 V. These two values are labeled on the top two waves of the voltage bounce diagram shown in Fig. 2–15.

The value of the second forward voltage wave is determined by multiplying the first reverse voltage wave by Γ_1. This value is $-0.25 \times 24\text{ V} = -6$ V. The value of the second reverse voltage wave is determined by multiplying the second forward voltage wave by Γ_2, and this value is $0.6 \times (-6\text{ V}) = -3.6$ V. This process is repeated for several more round trips as shown on the voltage bounce diagram of Fig. 2–15.

The voltage $v(0, t)$, which is shown in Fig. 2–16, is constructed by moving down the left-hand side of the voltage bounce diagram and adding the various waves as they are generated. Starting at $t = 0$, the first forward wave of value 40 V is included. At $t = 0.6\ \mu$s, a reverse wave of 24 V returns, and a new forward wave of value -6 V is generated. The net voltage immediately after 0.6 μs is thus $40 + 24 - 6 = 58$ V.

The preceding process is repeated at as many multiples of 0.3 μs as desired. Note from Fig. 2–16 that the changes become smaller as subsequent waves are encountered. After $t = 2.4\ \mu$s, the voltage rounded to four significant digits is 55.64 V. From the steady-state model of Example 2–4, the final voltage everywhere on the line was determined to be 55.65 V to four significant digits. Thus, for all practical purposes, the voltage has reached its limiting value in this circuit after four round trips.

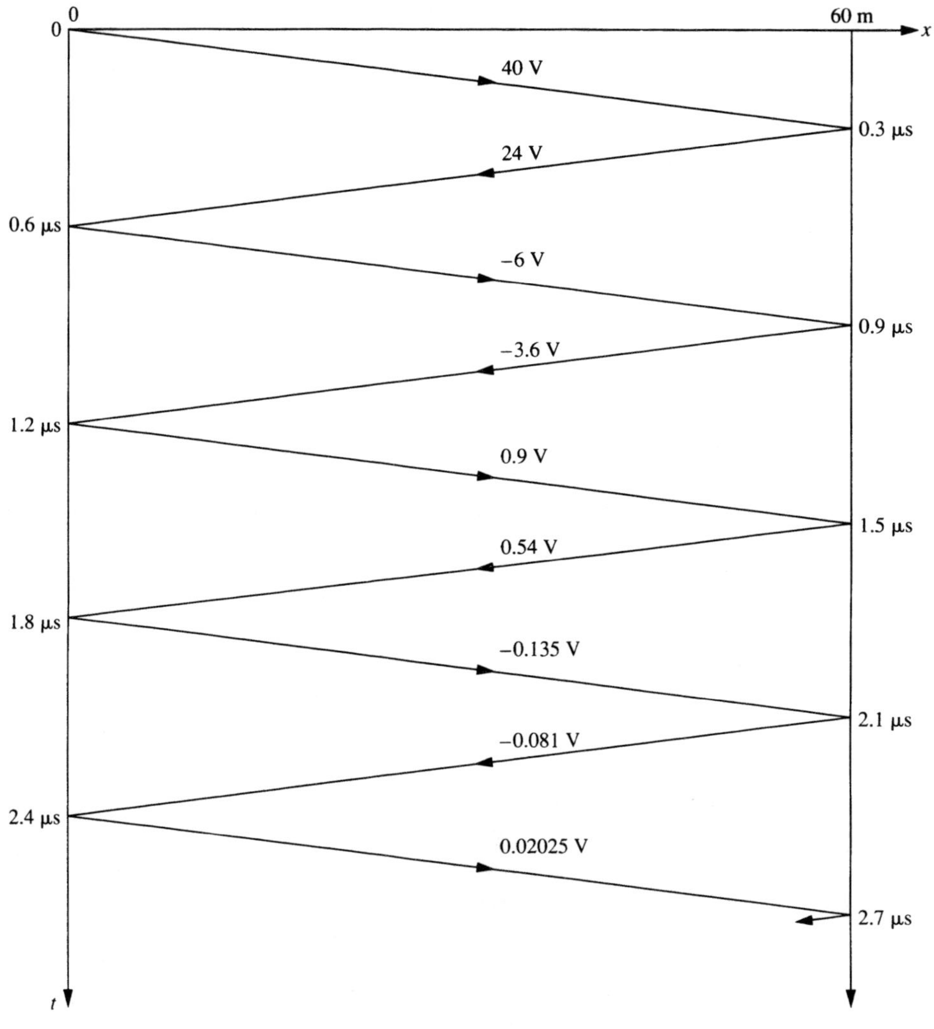

FIGURE 2–15 Voltage bounce diagram of Example 2–8.

EXAMPLE 2–9

For the circuit of Example 2–8, draw the current bounce diagram, and use it to plot the current at the generator end of the line for at least four round trips.

Solution

The two current reflection coefficients were determined in Example 2–1 to be $-\Gamma_1 = 0.25$ and $-\Gamma_2 = -0.6$.

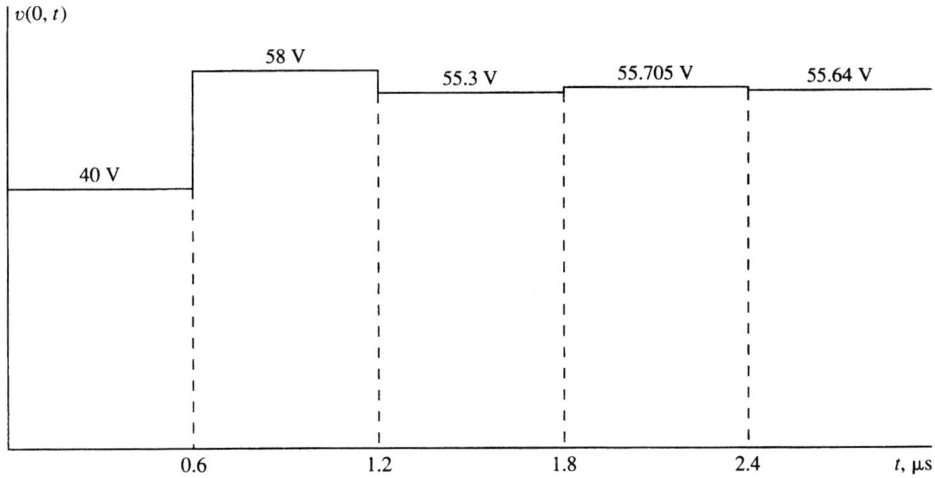

FIGURE 2–16 Voltage waveform of Example 2–8.

The value of the first forward current wave was determined in Example 2–3 to be 0.8 A, and the value of the first reverse current wave was determined in Example 2–2 to be −0.48 A. These two values are labeled on the top two waves of the current bounce diagram shown in Fig. 2–17.

The value of the second forward current wave is determined by multiplying the first reverse current wave by $-\Gamma_1$. This value is $0.25 \times (-0.48 \text{ A}) = -0.12$ A. The value of the second reverse current wave is determined by multiplying the second forward current wave by $-\Gamma_2$, and its value is $-0.6 \times (-0.12 \text{ A}) = 0.072$ A. This process is repeated for several more round trips as shown on the current bounce diagram of Fig. 2–17.

The current $i(0, t)$, which is shown in Fig. 2–18, is constructed in the same manner as the voltage waveform of Example 2–8, except, of course, that the current bounce diagram is now used. Starting at $t = 0$, the first forward wave of 0.8 A is included. At $t = 0.6\ \mu$s, a reverse wave of value −0.48 A returns, and a new forward wave of value −0.12 A is generated. The net current immediately after 2 μs is thus $0.8 - 0.48 - 0.12 = 0.2$ A.

The preceding process is repeated for several more round trips, and the results are shown in Fig. 2–18. After $t = 2.4\ \mu$s, the current rounded to four significant digits is 0.2785 A. From the steady-state model of Example 2–4, the final current everywhere on the line was determined to be 0.2783 A, which is very close to the value reached after 2.4 μs.

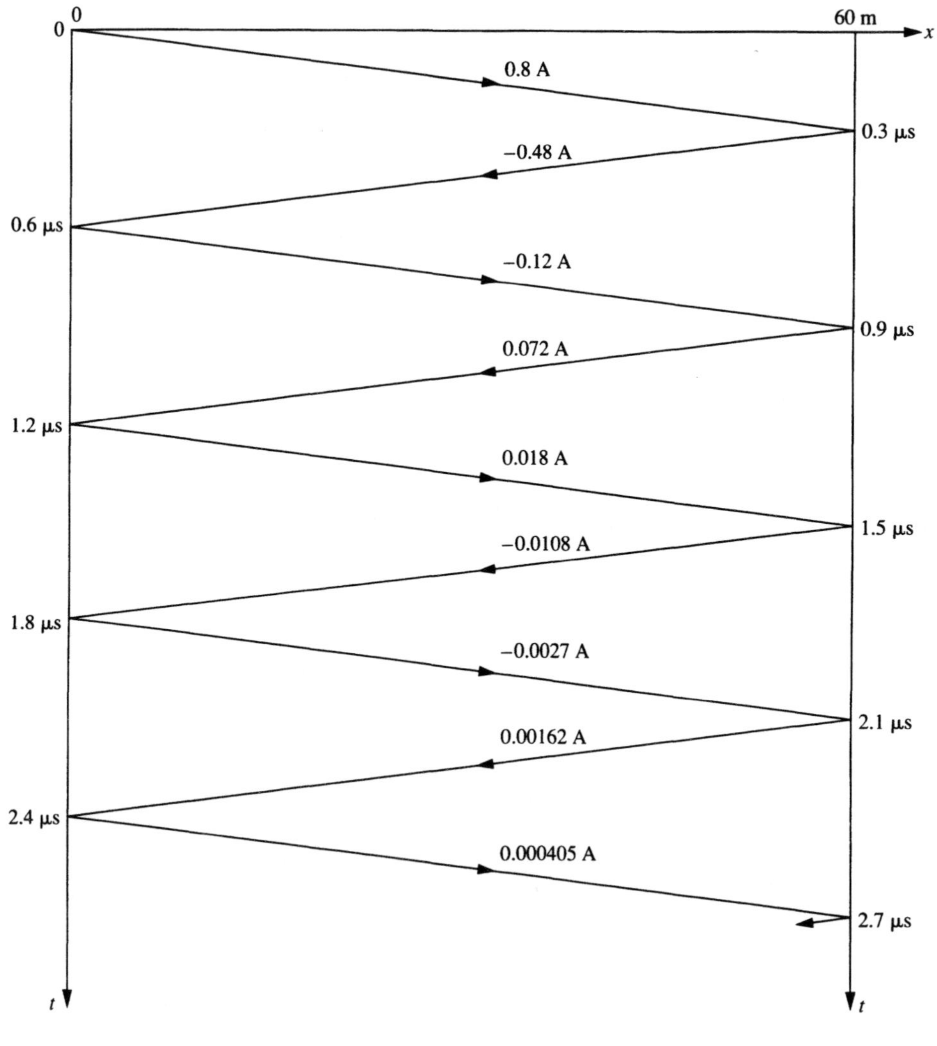

FIGURE 2–17 Current bounce diagram of Example 2–9.

2–6 ■ PULSE RESPONSE ON TRANSMISSION LINE

Another special transient case of interest in lossless lines is the response due to a finite-width pulse. For the purpose of this development, the symbol used in Fig. 2–19(a) will be used. The values on the left of the source represent the value E of the pulse and its width T. The form of the pulse is shown in Fig. 2–19(b). For the pulse waveform, it is unnecessary to show a switch since the pulse is always turned on and is later turned off. Unless otherwise indicated, it will be assumed that the pulse begins at $t = 0$.

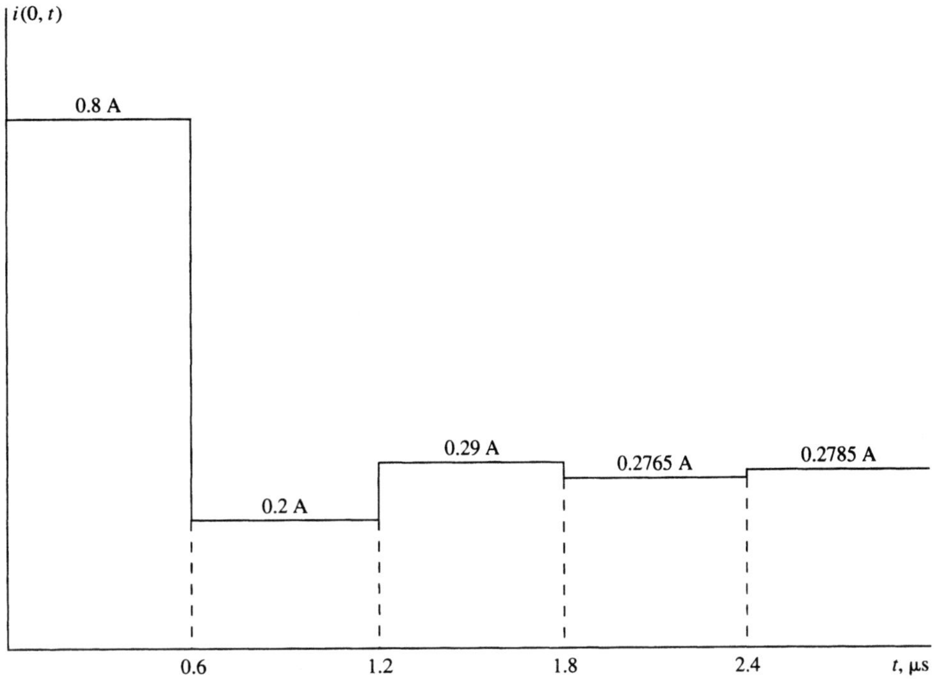

FIGURE 2–18 Current waveform of Example 2–9.

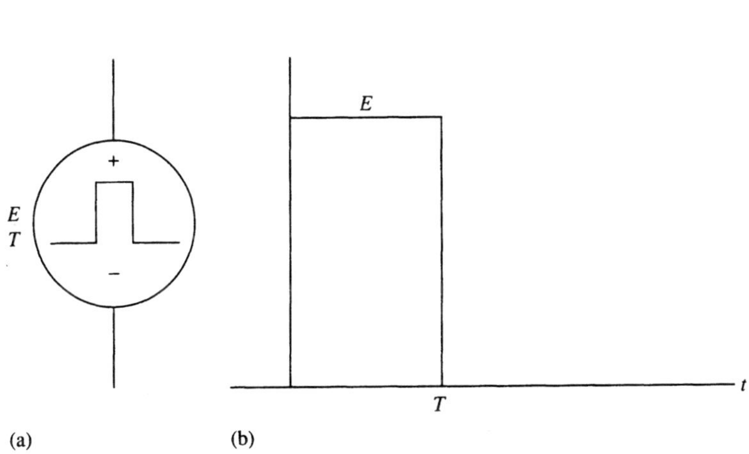

FIGURE 2–19 Symbol for finite-width pulse and the associated waveform.

The bounce diagram may be used for constructing a graph of the response at any point on the line by the use of a "window" technique. The concept is illustrated in Fig. 2–20. First, the bounce diagram is constructed in the same manner as it would be for a dc input using the value of the pulse E. Next, an imaginary "window" of width T is placed just above the bounce diagram as shown in Fig. 2–20. In this illustration, the voltage $v(0, t)$ is being determined, and $T < 2t_1$; but the concept may be applied at any point and for any width of the pulse.

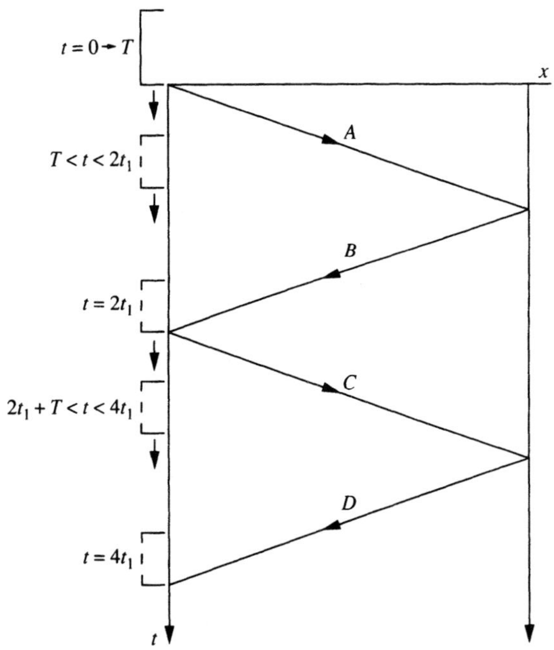

FIGURE 2–20 Use of sliding window on bounce diagram for finite-width pulse.

Starting at $t = 0$, the window is slid downward at a constant rate along the t-axis. (A piece of paper corresponding to length T may be used if desired.) At any time t, the voltage is determined by summing all waves that *appear within the window*. Thus, when the end of a window moves past a certain wave, the wave is subtracted from the sum existing prior to that time. The result is that only a finite number of waves will ever be used in determining the voltage at a given time.

In Fig. 2–20, for the first T seconds, wave A will be the only component contained in the voltage. For $T < t < 2t_1$, the window moves past the first wave, and the voltage is zero for a period of time. At $t = 2t_1$, the window picks up waves B and C simultaneously, and the voltage at that time is the sum of the two. That voltage remains for a duration of T. For $2t_1 + T < t < 4t_1$, the window is past any waves, and the voltage becomes zero again. The pattern continues to repeat itself.

EXAMPLE 2–10

The circuit of Fig. 2–21(a) is the same as that of Example 2–6 except for one difference: The dc voltage of 60 V has been replaced by a 60-V pulse of width 0.2 μs starting at $t = 0$. Plot the voltages at the two ends of the line.

Solution

The bounce diagram, which consists of only one wave in this case, is shown again in Fig. 2–21(b). A window of width 0.2 μs is first placed on the left. As the window is moved along the t-axis, the wave is included for 0.2 μs, and it is then removed. This leads to the voltage waveform shown in Fig. 2–21(c).

Next, a window of width 0.2 μs is placed on the right-hand side. As the window is moved along, nothing is included for an interval of 1 μs. At that time, the 40-V wave is included in the window for 0.2 μs and is then removed. This leads to the voltage in Fig. 2–21(d).

Obviously, neither the bounce diagram nor the window was required in this simple case, but the process should help to illustrate the concept as a basis for more complex examples.

EXAMPLE 2–11

The circuit of Fig. 2–22(a) is the same as that of Example 2–7 except that the 60-V dc voltage has been replaced by a 60-V pulse of width 0.2 μs. Plot the voltages at the two ends of the line.

Solution

The bounce diagram was drawn in Fig. 2–13(b) and is repeated in Fig. 2–22(b) for convenience. A window of width 0.2 μs is first placed on the left. As the window is moved, the forward wave of 30 V is included for 0.2 μs and is then removed. At $t = 2$ μs, the reverse wave of 20 V is included within the window, and it remains there for 0.2 μs. The resulting voltage $v(0, t)$ is shown in Fig. 2–22(c).

A window of width 0.2 μs is next placed on the load end. No wave is included for $t < 1$ μs. However, when t reaches the value of 1 μs, both the 30-V forward wave and the 20-V reverse wave are included simultaneously to produce a level of 50 V. This value remains for a duration of 0.2 μs, and it then becomes zero. The voltage $v(200, t)$ is shown in Fig. 2–22(d).

Note that in this case, the voltage at the generator end consists of two separate pulses, but the load end consists of a single pulse of value 50 V. As soon as the 30-V pulse arrives, an additional reflected pulse of value 20 V is instantaneously created, so the net effect is that of a single 50-V pulse.

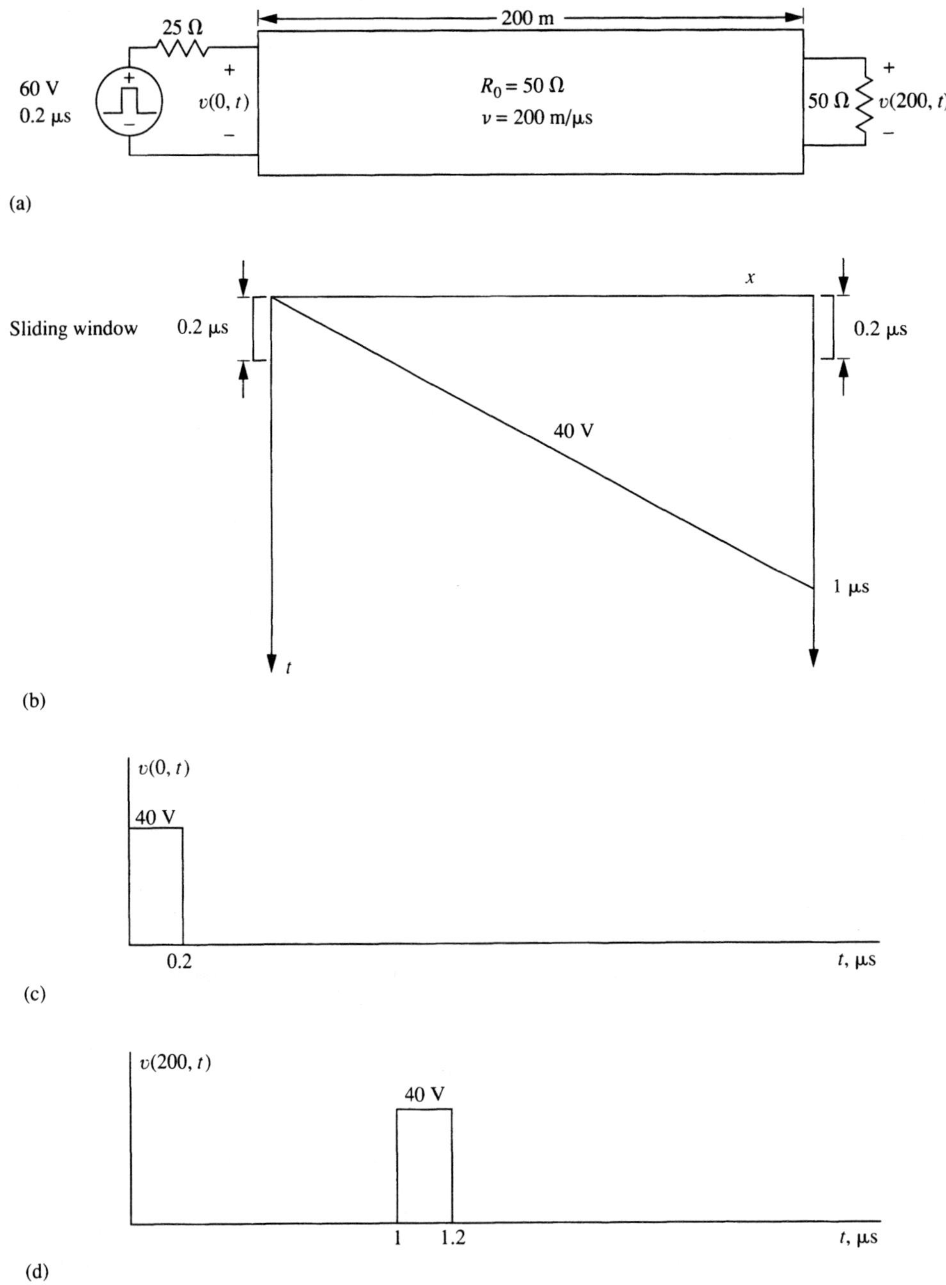

FIGURE 2–21 Circuit, bounce diagram, and waveforms of Example 2–10.

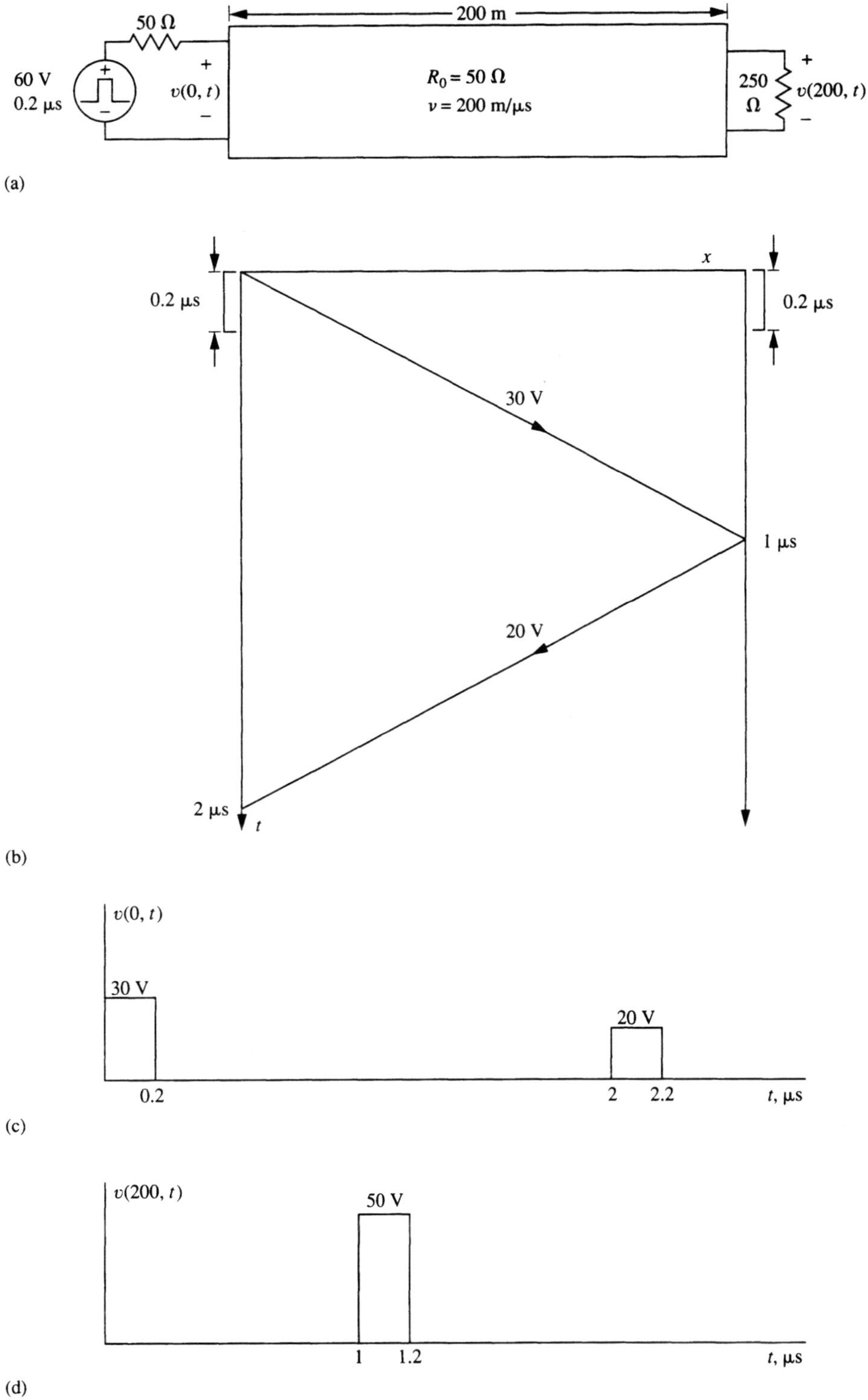

FIGURE 2–22 Circuit, bounce diagram, and waveforms of Example 2–11.

EXAMPLE 2–12

The circuit of Fig. 2–23 is the same as that of Example 2–8 except that the 64-V dc source has been replaced by a 64-V pulse with width 0.1 μs. Plot the voltage waveforms at the two ends of the line.

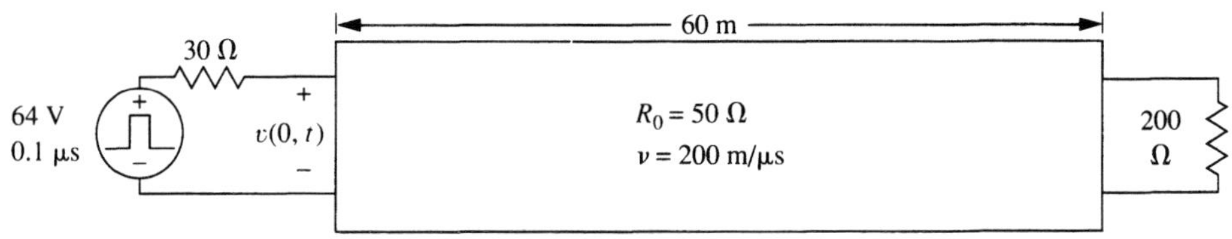

FIGURE 2–23 Circuit of Example 2–12.

Solution

The bounce diagram was constructed in Fig. 2–15 but is repeated in Fig. 2–24 for convenience. A window of width 0.1 μs is first slid along the left-hand side of the diagram. The 40-V wave is included for 0.1 μs, at which time the wave is dropped. At $t = 0.6$ μs, the 24-V wave and the −6-V wave appear within the window for the next 0.1 μs, so this produces an effective pulse of amplitude 18 V. This process is repeated at several other multiples of 0.6 μs, and the resulting voltage $v(0, t)$ is shown in Fig. 2–25(a).

At the load end, no wave appears until $t = 0.3$ μs, at which time both the 40-V forward wave and the 24-V reverse wave combine to yield a net pulse of 64 V, which lasts until the window passes by at $t = 0.4$ μs. This process is repeated at $t = 0.9$ μs, at which time −6-V and −3.6-V waves combine to produce a −9.6-V pulse. The pattern continues, and $v(60, t)$ is shown in Fig. 2–25(b).

Note that in this completely mismatched case, multiple reflections, consisting of pulses, continue for some time until their magnitudes diminish to insignificant values.

2–7 ■ SHORTED AND OPEN LINES

Two situations of special interest in the transient response of lines occur when the load end is (a) a **short circuit** and (b) an **open circuit.** The cases discussed here will be limited to those in which the line is matched at the generator end, but several problems at the end of the chapter (Problems 2–19 through 2–22) will consider shorted and open lines with a mismatch also at the generator.

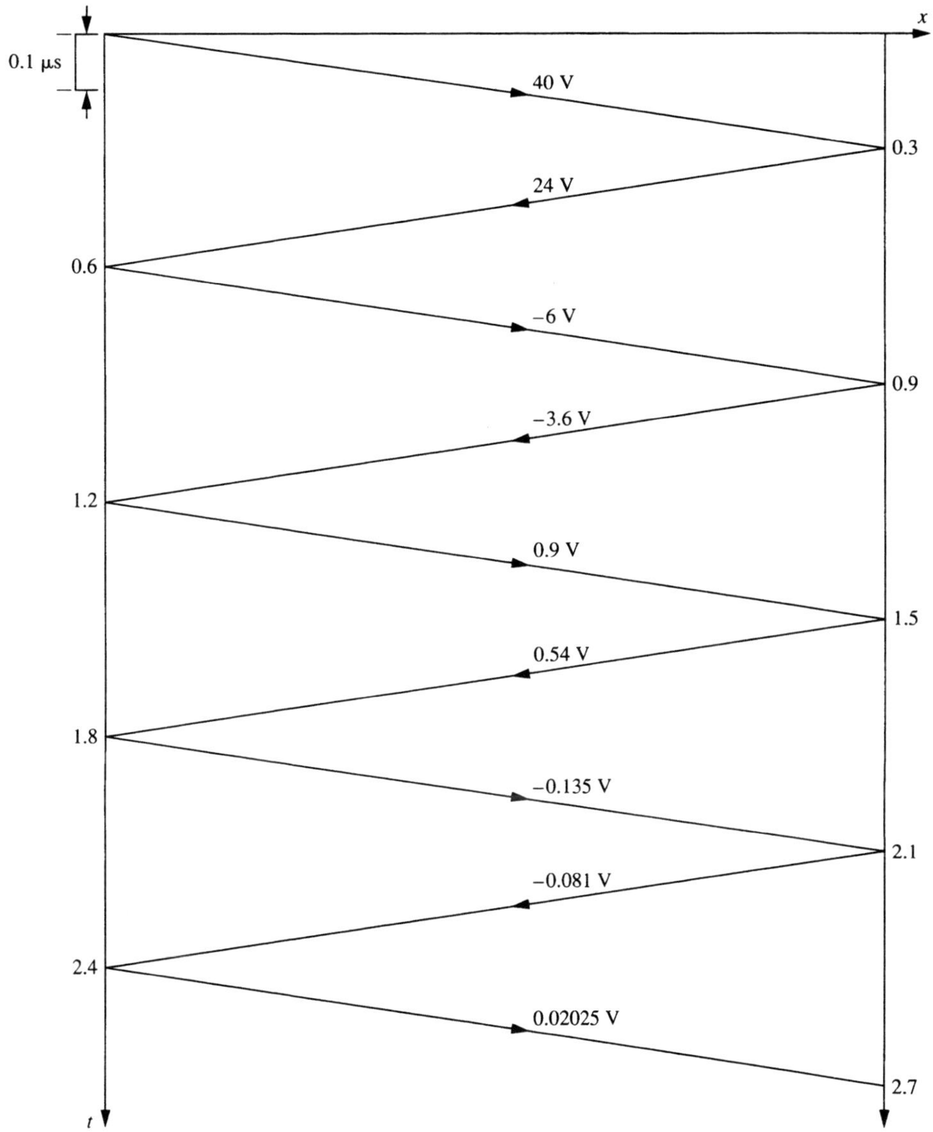

FIGURE 2–24 Voltage bounce diagram of Example 2–12.

Short-Circuited Line

Consider the system of Fig. 2–26(a), in which the line is shorted at the load end but in which the line is matched at the generator end. The generator is a dc voltage of value E. The load resistance may be considered as $R_2 = 0$, and

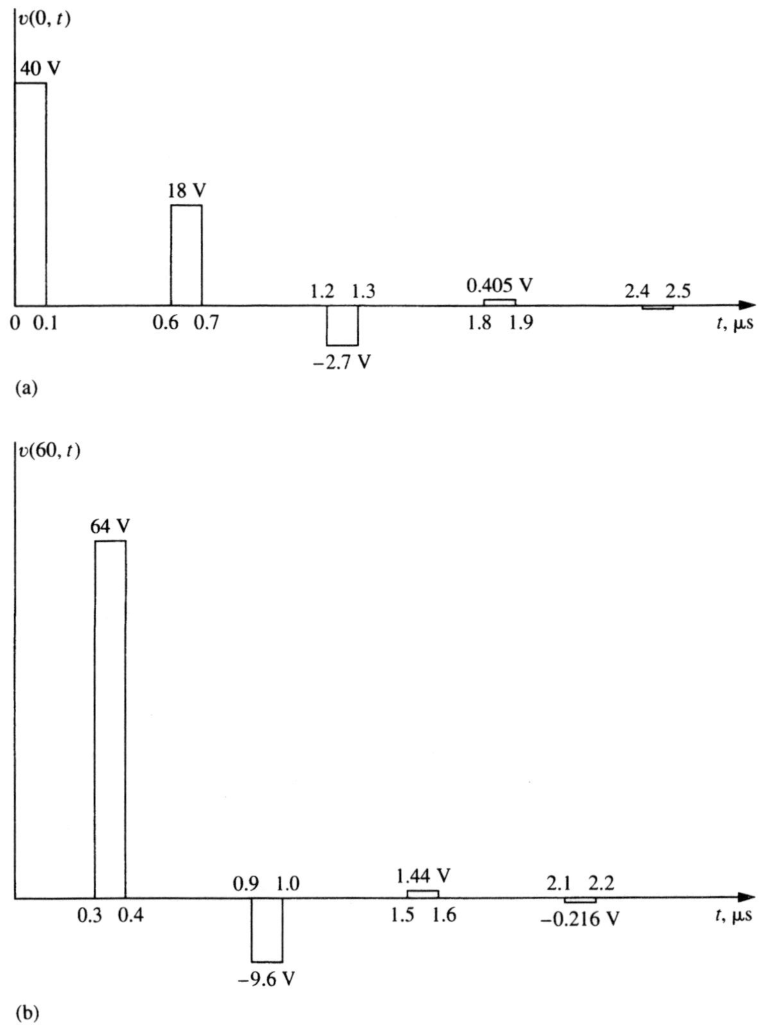

FIGURE 2–25 Voltage waveforms of Example 2–12.

the corresponding voltage reflection coefficient is

$$\Gamma_2 = \frac{R_2 - R_0}{R_2 + R_0} = \frac{0 - R_0}{0 + R_0} = -1 \tag{2–38}$$

Refer now to the bounce diagram of Fig. 2–26(b). The forward wave is $E/2$ due to the equal division of the generator voltage between the internal resistance R_0 and the characteristic impedance of the line. At time $t = t_1$, the forward wave reaches the load, and a reverse wave of value $-1 \times E/2 = -E/2$ is generated.

One way to visualize the preceding result is to remember that the voltage at the load end must always be zero since there is a short circuit. Thus, when

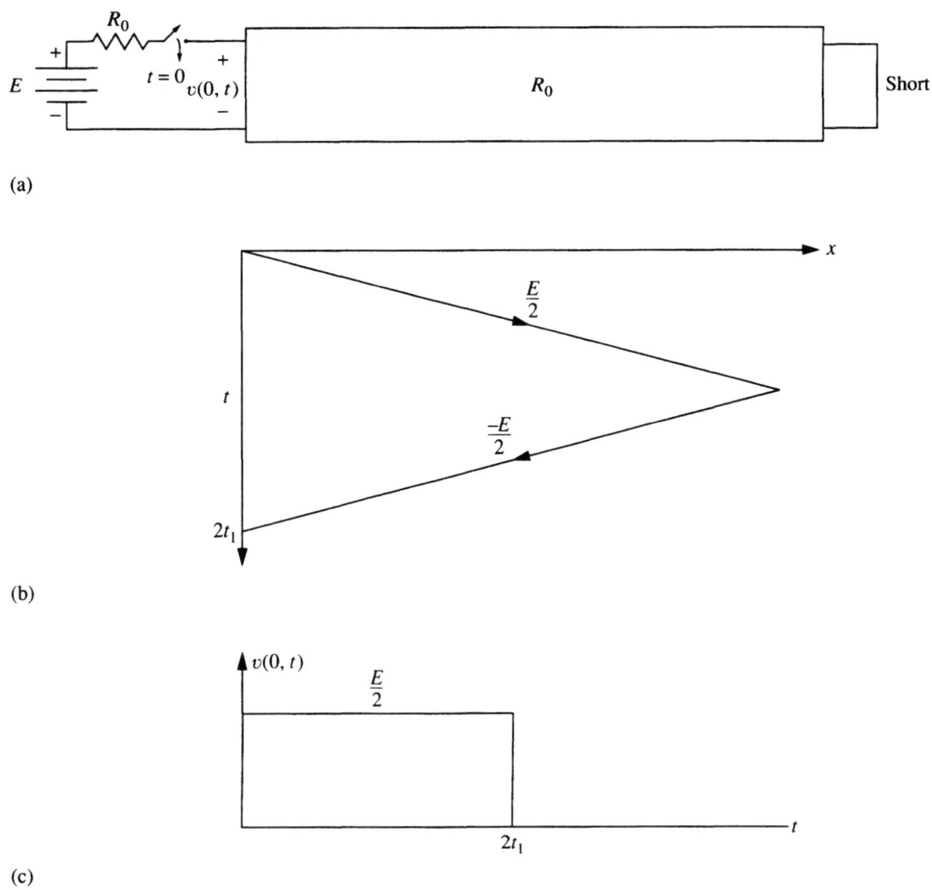

FIGURE 2–26 Voltage response of *shorted* line with dc input.

the forward voltage arrives, a reverse voltage wave of equal magnitude and opposite sign is generated to force the net voltage at the load to be zero. At time $t = 2t_1$, the reverse wave reaches the generator, and since that end is matched, no further reflection occurs.

The voltage $v(0, t)$ at the generator end is shown in Fig. 2–26(c). The result is a pulse of amplitude $E/2$ and width $2t_1$. This concept has been used to generate pulses in some applications. For $t > 2t_1$, the presence of the short is felt at the generator end, and the voltage across the input to the line is zero. In a sense, it takes one complete round trip or a time $2t_1$ for the short to establish its effect at the input.

Consider next the system of Fig. 2–27(a). The only difference between this system and that of Fig. 2–26(a) is that the dc voltage has been replaced by a pulse generator, which produces a pulse of amplitude E and width t_p. It is assumed that $t_p < 2t_1$.

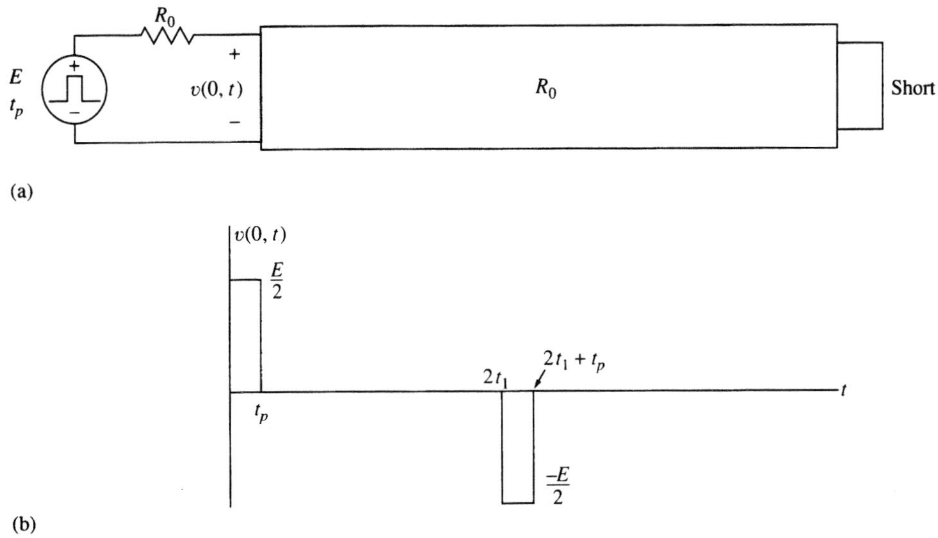

FIGURE 2–27 Voltage response of *shorted* line with pulse input.

The bounce diagram is the same as that of Fig. 2–26(b) except that a window of width t_p is used as discussed earlier in the chapter. The resulting waveform is shown in Fig. 2–27(b). A positive pulse of amplitude $E/2$ and width t_p appears early in the response. At time $t = 2t_1$, an *inverted* pulse of amplitude $-E/2$ and width t_p appears across the input.

Open Line

Consider the system of Fig. 2–28(a), in which the line is open at the load end, but in which the line is matched at the generator end. The load resistance may be considered as $R_2 = \infty$, and the corresponding voltage reflection coefficient is

$$\Gamma_2 = \lim_{R_2 \to \infty} \frac{R_2 - R_0}{R_2 + R_0} = \lim_{R_2 \to \infty} \frac{1 - \dfrac{R_0}{R_2}}{1 + \dfrac{R_0}{R_2}} = 1 \tag{2–39}$$

where "lim" refers to the mathematical process of taking the limit as R_2 approaches infinity, which is a rigorous procedure for performing this evaluation. (An alternate, practical way is to assume some value such as $R_2 = 10^8\ \Omega$ and $R_0 = 50\ \Omega$ and evaluate Γ_2 with a calculator.)

Refer now to the bounce diagram of Fig. 2–28(b). The forward wave is again $E/2$, but the reverse wave is now $1 \times E/2 = E/2$. The voltage $v(0, t)$ is shown in Fig. 2–28(c). For $t > 2t_1$, the presence of the open is felt at the generator end, and the voltage across the input to the line is E.

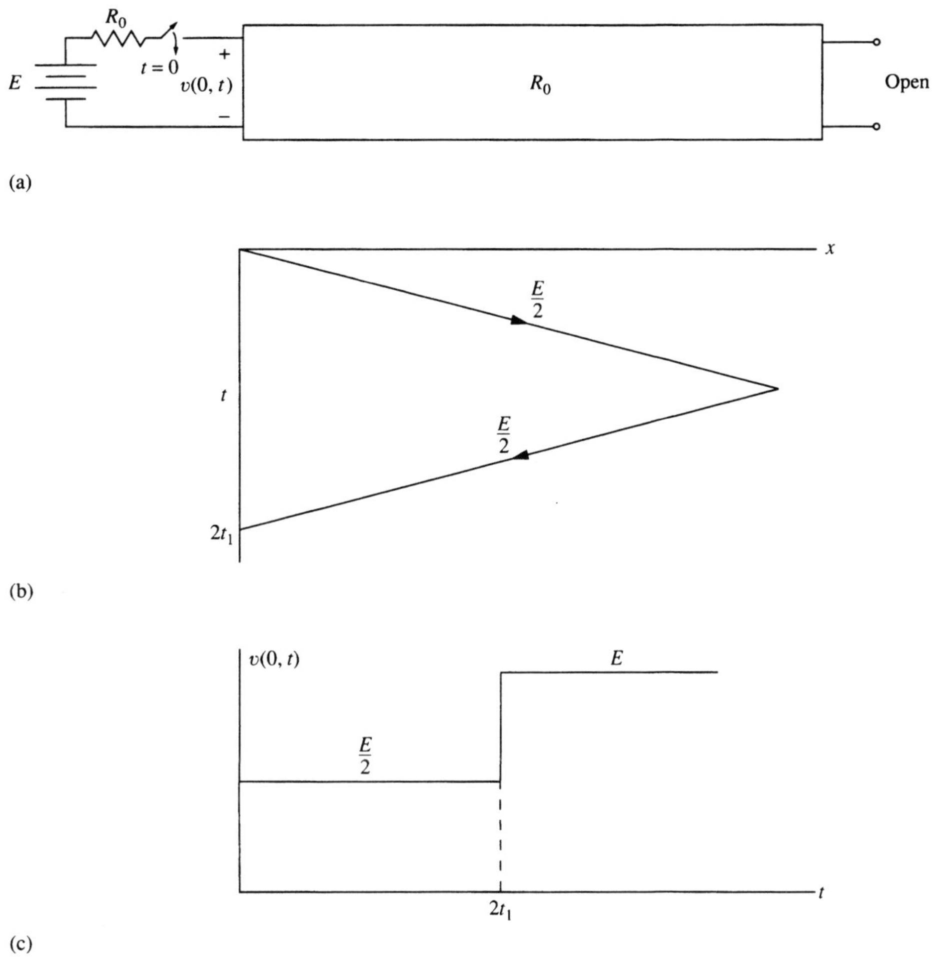

FIGURE 2–28 Voltage response of *open* line with dc input.

Finally, consider the system of Fig. 2–29(a), in which the dc voltage at the generator is replaced by a pulse generator. As in the case of the shorted line, it is assumed that $t_p < 2t_1$. The resulting waveform is shown in Fig. 2–29(b). The difference between this case and the case of the shorted line is that the return pulse for the open has the *same* polarity as that of the initial pulse.

Time-Domain Reflectometry

Short and open circuits constitute extreme cases of limiting values of loads. If the line is assumed to be lossless, the return pulse in each case will have the same magnitude as the initial pulse, but it will have the opposite sign for a

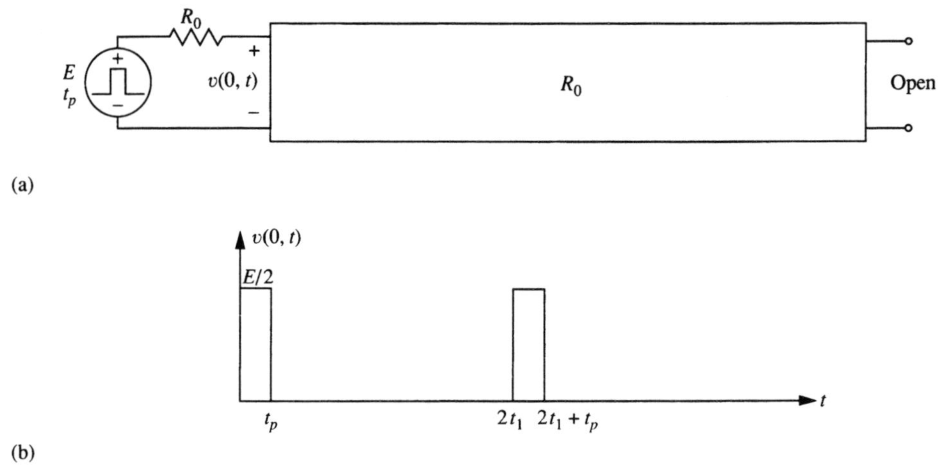

FIGURE 2–29 Voltage response of *open* line with pulse input.

short and the same sign for an open. In general, for a resistive load and with a pulse width $t_p < 2t_1$, a return voltage pulse from a load will have an arbitrary magnitude less than or equal to $E/2$, and it will have either the same sign as or the opposite sign of the initial pulse. The following pattern may be deduced from the nature of the reflection coefficient:

1. If the return voltage pulse has the opposite sign of the initial pulse, the load resistance is less than the characteristic impedance; i.e., $R_2 < R_0$.
2. If the return voltage pulse has the same sign as the initial pulse, the load resistance is greater than the characteristic impedance; i.e., $R_2 > R_0$.

The preceding concepts form the basis of **time-domain reflectometry,** an important process widely used to determine the locations of faults in transmission lines and cables. Commercial instruments, called **time-domain reflectometers,** are designed specifically for this application. The presence of a fault will be observed by the appearance of a return pulse. By measuring the round-trip time, one may determine the location of the fault. By observing the amplitude and polarity of the return pulse, one may also deduce the nature of the fault.

EXAMPLE 2–13

A transmission line with negligible losses was originally matched at both the generator and the load. However, some trouble has been experienced, and a time-domain reflectometer system is used to attempt to determine the location and nature of the fault. When the line is excited by the pulse waveform on the left in Fig. 2–30, the return pulse shown on the right is observed. The velocity of propagation on the line is 2×10^8 m/s. Determine (a) the distance from the generator end to the fault and (b) the net equivalent load resistance at the fault location. Assume that $R_0 = 50\ \Omega$.

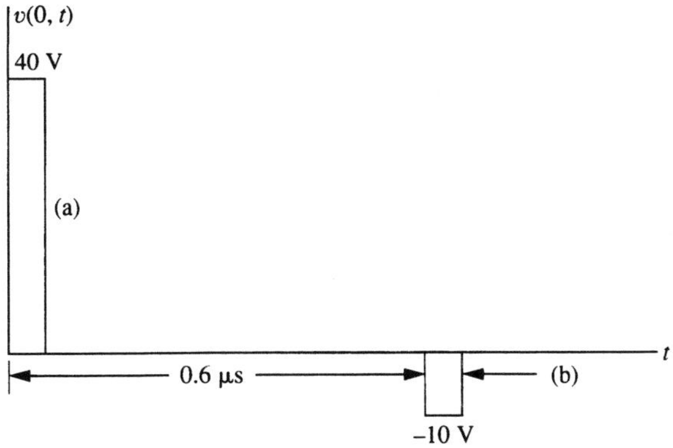

FIGURE 2–30 Waveforms of Example 2–13.

Solution

(a) Since the two-way delay time is $2t_1 = 0.6\ \mu s$, the one-way propagation time is $t_1 = 0.3\ \mu s$. The distance d to the fault is then

$$d = vt_1 = (2 \times 10^8) \times (0.3 \times 10^{-6}) = 60\ \text{m} \qquad \textbf{(2–40)}$$

(b) The return pulse is inverted, but its magnitude is less than the forward pulse. Thus, assuming that the line is lossless, the fault does not represent a short circuit. The reflection coefficient at the load is readily calculated as

$$\Gamma_2 = \frac{-10\ \text{V}}{40\ \text{V}} = -0.25 \qquad \textbf{(2–41)}$$

The relationship between Γ_2 and R_0 is

$$\Gamma_2 = \frac{R_2 - R_0}{R_2 + R_0} \qquad \textbf{(2–42)}$$

We must solve for R_2 in terms of R_0 and Γ_2 in (2–42). The result is

$$R_2 = \frac{(1 + \Gamma_2)}{(1 - \Gamma_2)} R_0 \qquad \textbf{(2–43)}$$

Substitution of $\Gamma_2 = -0.25$ and $R_0 = 50\ \Omega$ in (2–43) yields

$$R_2 = \frac{1 - 0.25}{1 - (-0.25)} \times 50 = \frac{0.75}{1.25} \times 50 = 30\ \Omega \qquad \textbf{(2–44)}$$

■ PROBLEMS

Note: Unless otherwise indicated, assume that all lines are initially uncharged and that switches are closed at $t = 0$. It is recommended that bounce diagrams be constructed in all cases.

2–1. (a) For the circuit of Fig. P2–1, plot the voltages at both ends of the line.
(b) Repeat (a) for the currents at both ends of the line.

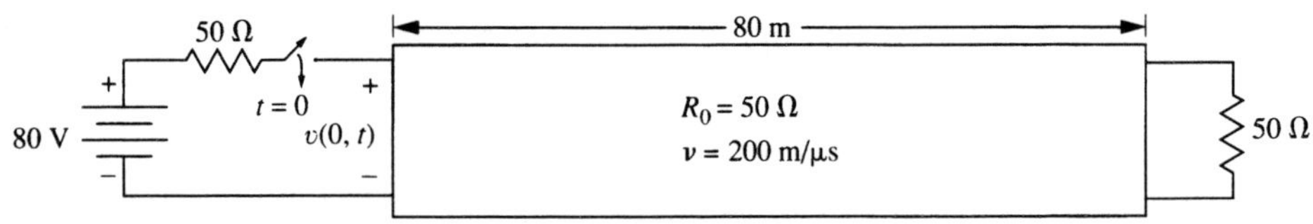

FIGURE P2–1

2–2. (a) For the circuit of Fig. P2–2, plot the voltages at both ends of the line.
(b) Repeat (a) for the currents at both ends of the line.

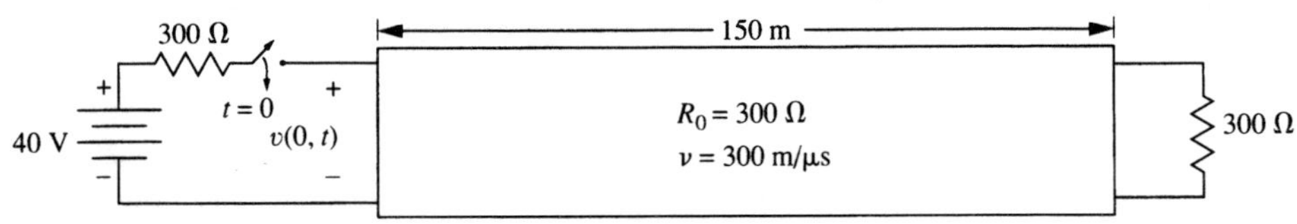

FIGURE P2–2

2–3. (a) For the circuit of Fig. P2–3, plot the voltages at both ends of the line.
(b) Repeat (a) for the currents at both ends of the line.

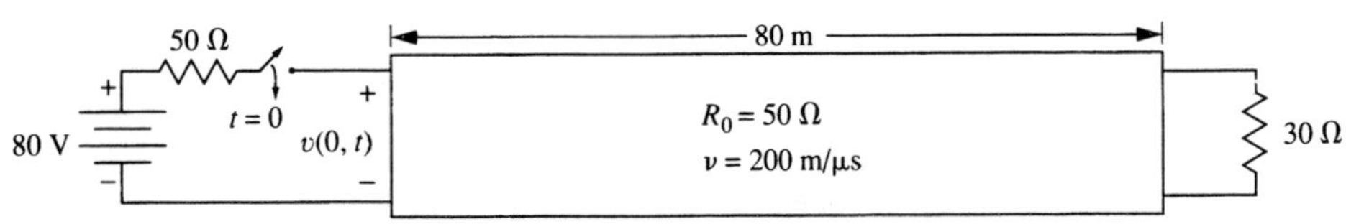

FIGURE P2–3

2–4. (a) For the circuit of Fig. P2–4, plot the voltages at both ends of the line.
(b) Repeat (a) for the currents at both ends of the line.

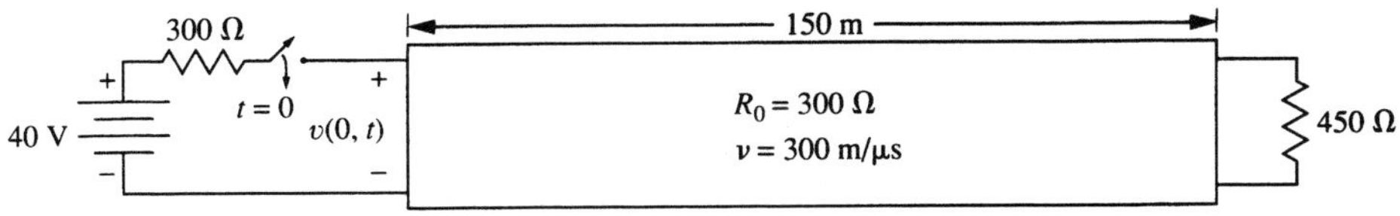

FIGURE P2–4

2–5. **(a)** For the circuit of Fig. P2–5, plot the voltages at both ends of the line.
(b) Repeat (a) for the currents at both ends of the line.

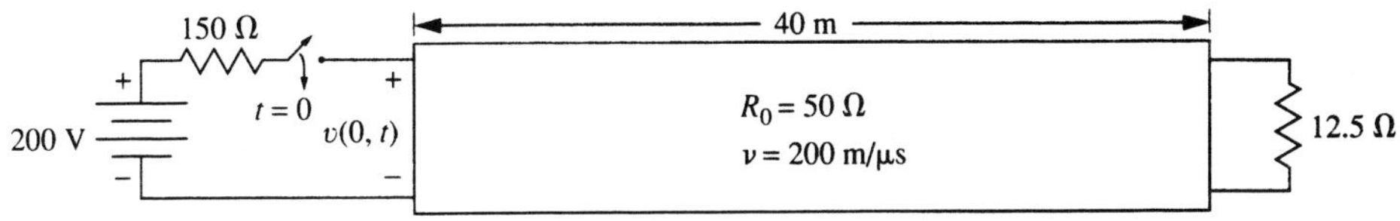

FIGURE P2–5

2–6. **(a)** For the circuit of Fig. P2–6, plot the voltages at both ends of the line.
(b) Repeat (a) for the currents at both ends of the line.

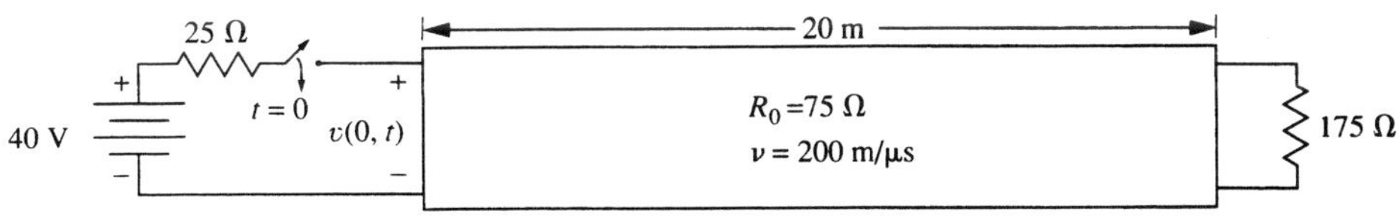

FIGURE P2–6

2–7. Assume in the circuit of Problem 2–3 (Fig. P2–3) that the 80-V dc source is replaced by an 80-V pulse of width 0.1 μs starting at $t = 0$.
(a) Plot the voltages at the two ends of the line.
(b) Plot the currents at the two ends of the line.

2–8. Assume in the circuit of Problem 2–4 (Fig. P2–4) that the 40-V dc source is replaced by a 40-V pulse of width 0.1 μs starting at $t = 0$.
(a) Plot the voltages at the two ends of the line.
(b) Plot the currents at the two ends of the line.

2–9. Assume in the circuit of Problem 2–5 (Fig. P2–5) that the 200-V dc source is replaced by a 200-V pulse of width 0.1 μs starting at $t = 0$.
(a) Plot the voltages at the two ends of the line.
(b) Plot the currents at the two ends of the line.

2–10. Assume in the circuit of Problem 2–6 (Fig. P2–6) that the 40-V dc source is replaced by a 40-V pulse of width 0.1 μs starting at $t = 0$.
(a) Plot the voltages at the two ends of the line.
(b) Plot the currents at the ends of the line.

2–11. For the circuit of Fig. P2–11, plot the voltage $v(0, t)$.

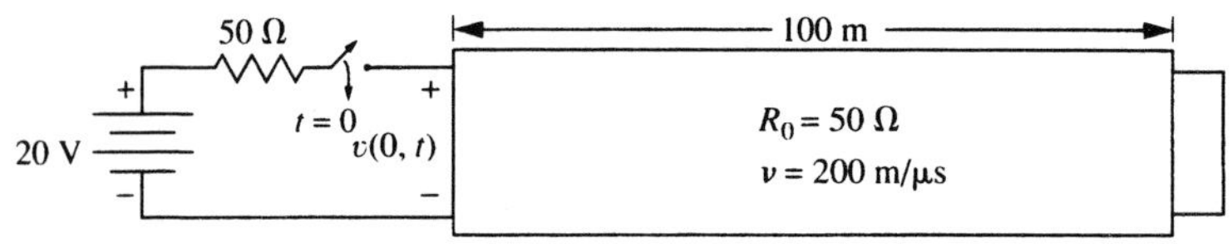

FIGURE P2–11

2–12. For the circuit of Fig. P2–12, plot the voltage $v(0, t)$.

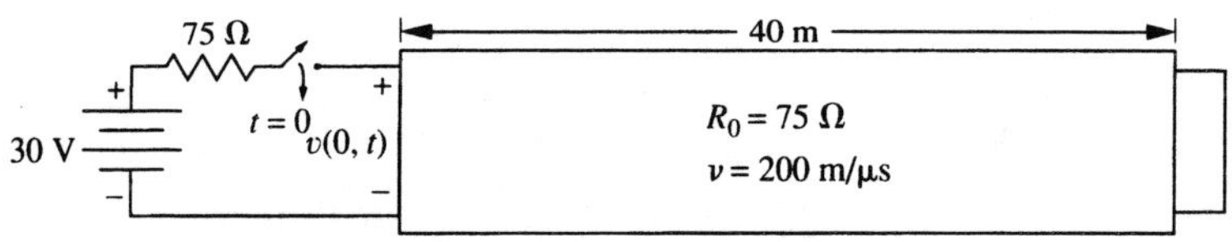

FIGURE P2–12

2–13. For the circuit of Fig. P2–13, plot the voltage $v(0, t)$.

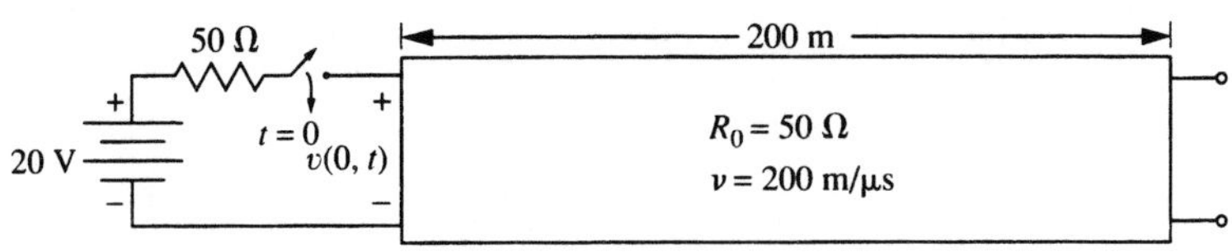

FIGURE P2–13

2–14. For the circuit of Fig. P2–14, plot the voltage $v(0, t)$.

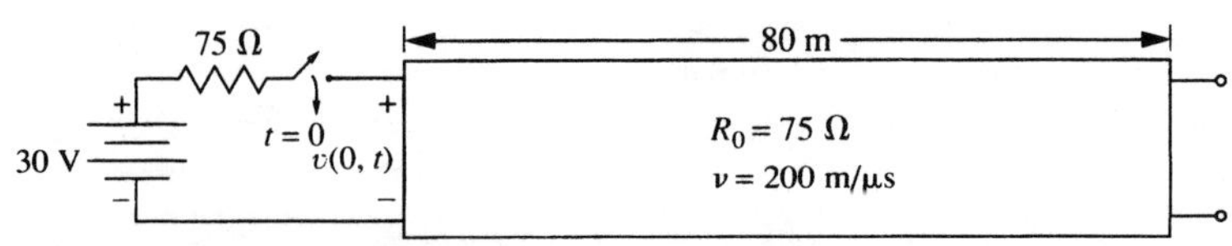

FIGURE P2–14

2–15. For the circuit of Problem 2–11 (Fig. P2–11), assume that the 20-V dc source is replaced by a 20-V pulse of width 0.1 μs starting at $t = 0$. Plot the voltage $v(0, t)$.

2–16. For the circuit of Problem 2–12 (Fig. P2–12), assume that the 30-V dc source is replaced by a 30-V pulse of width 50 ns starting at $t = 0$. Plot the voltage $v(0, t)$.

2–17. For the circuit of Problem 2–13 (Fig. P2–13), assume that the 20-V dc source is replaced by a 20-V pulse of width 0.4 μs starting at $t = 0$. Plot the voltage $v(0, t)$.

2–18. For the circuit of Problem 2–14 (Fig. P2–14), assume that the 30-V dc source is replaced by a 30-V pulse of width 20 ns starting at $t = 0$. Plot the voltage $v(0, t)$.

2–19. For the circuit of Fig. P2–19, plot the voltage $v(0, t)$.

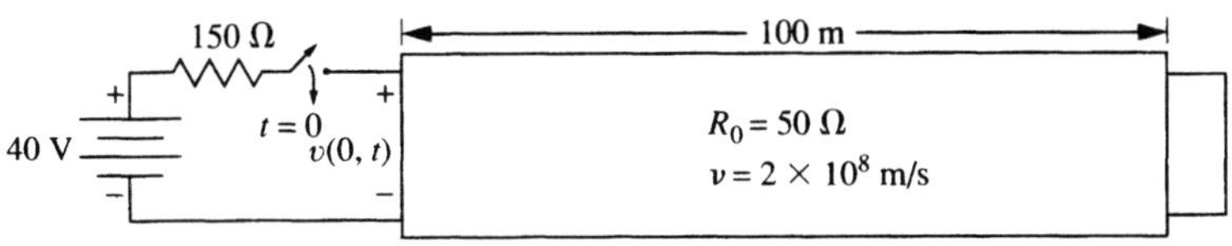

FIGURE P2–19

2–20. For the circuit of Fig. P2–20, plot the voltage $v(0, t)$.

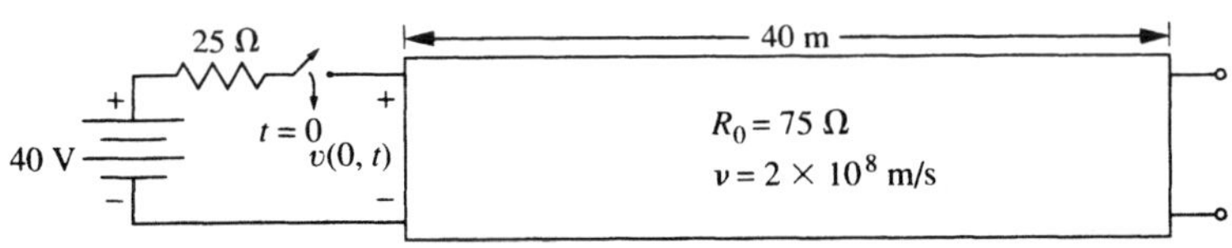

FIGURE P2–20

2–21. For the circuit of Problem 2–19 (Fig. P2–19), assume that the 40-V dc source is replaced by a 40-V pulse of width 0.2 μs starting at $t = 0$. Plot the voltage $v(0, t)$.

2–22. For the circuit of Problem 2–20 (Fig. P2–20), assume that the 40-V dc source is replaced by a 40-V pulse of width 100 ns starting at $t = 0$. Plot the voltage $v(0, t)$.

2–23. A 50-Ω line is terminated in an unknown load resistance R_2. The line is excited by a pulse generator with an output impedance of 50 Ω resistive. The waveform across the input to the line is shown in Fig. P2–23, and the velocity of propagation is 200 m/μs.

(a) Determine the length of the line.

(b) Determine the value of the load resistance.

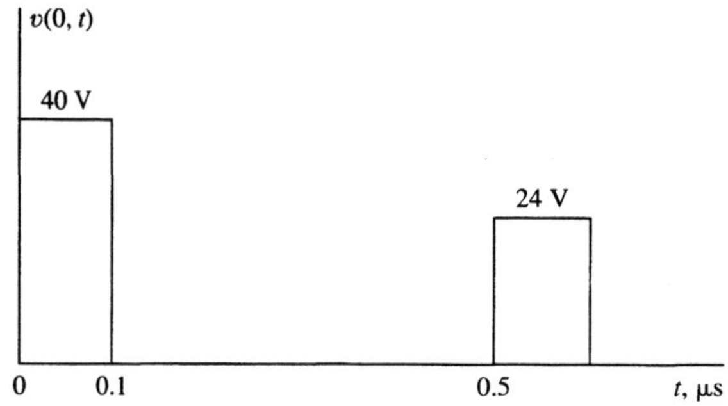

FIGURE P2–23

2–24. Repeat the analysis of Problem 2–23 if the voltage across the input is as shown in Fig. P2–24.

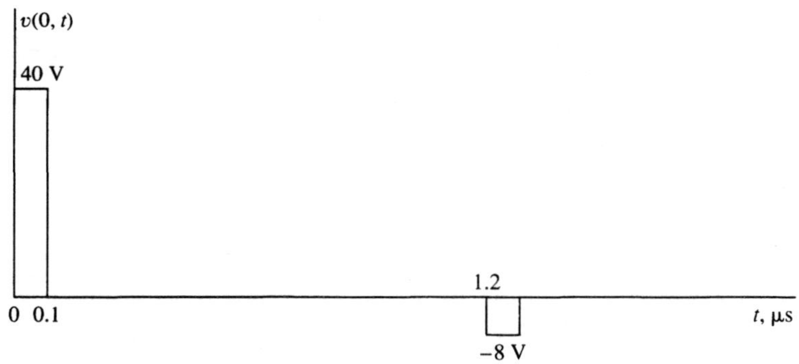

FIGURE P2–24

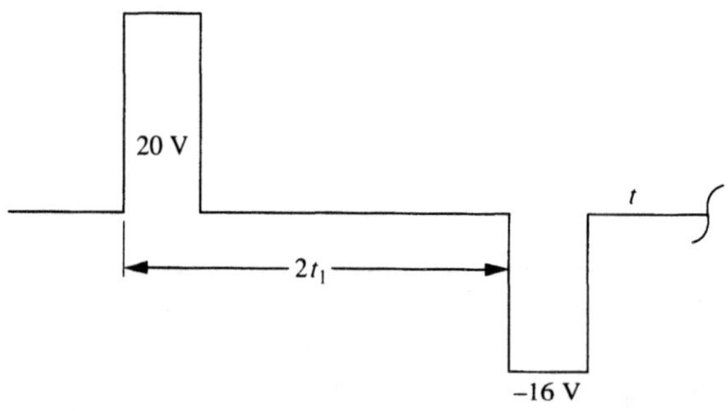

FIGURE P2–25

2–25. A pulse generator having an unknown internal resistive impedance is used to excite a 50-Ω lossless line that is known to be *shorted* at the load end. The voltage across the input is shown in Fig. P2–25.
(a) Determine the output resistance of the generator.
(b) Determine the value of the return pulse at $t = 4t_1$.

2–26. Repeat the analysis of Problem 2–25 if the input voltage waveform has the form of Fig. P2–26.

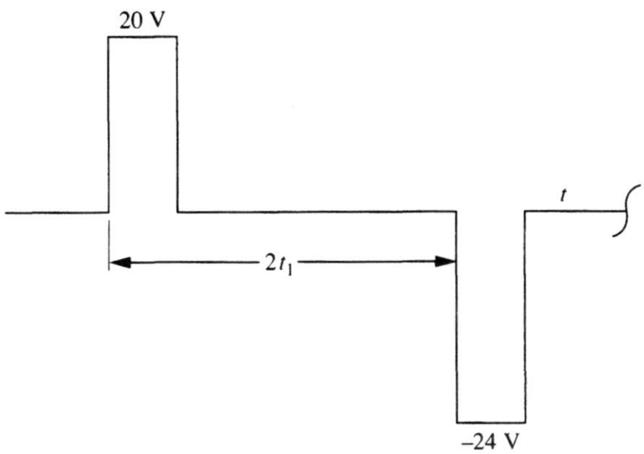

FIGURE P2–26

2–27. A dc voltage connected through an unknown source resistance R_1 is used to excite a 50-Ω lossless line that is known to be *shorted* at the load end. The voltage across the input is shown in Fig. P2–27.

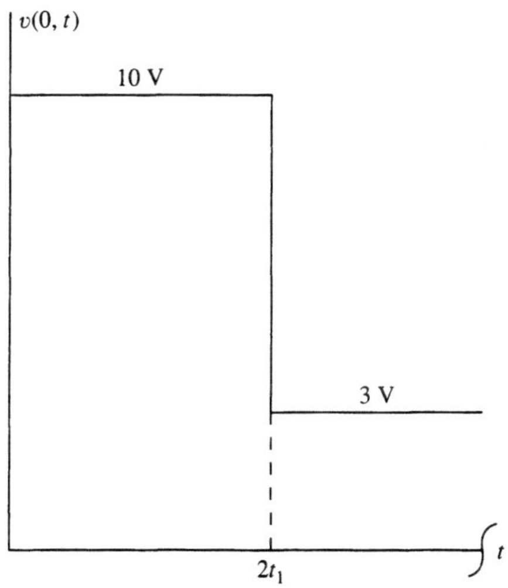

FIGURE P2–27

(a) Determine the value of R_1.
(b) Determine the value of the voltage just after $t = 4t_1$.

2–28. Repeat the analysis of Problem 2–27 if the voltage across the input is as shown in Fig. P2–28.

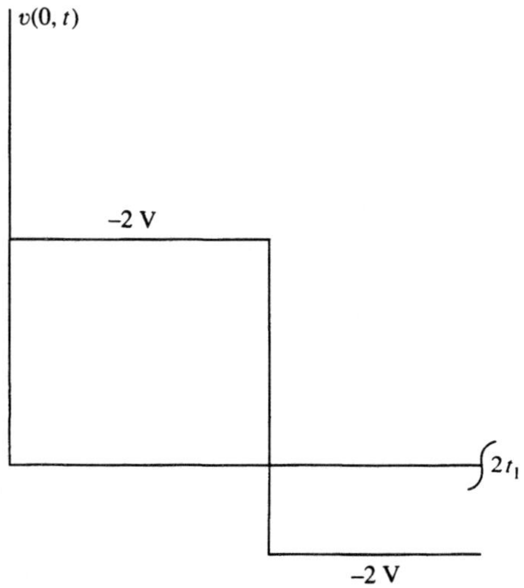

FIGURE P2–28

3

Steady-State ac Transmission Lines

OBJECTIVES

After completing this chapter, the reader should be able to:

1. Discuss quantitatively and qualitatively the concept of a sinusoidal traveling wave.
2. Discuss the concepts of time-varying and distance-varying waves.
3. Define **cyclic frequency, radian frequency, period, phase constant,** and **wavelength.**
4. Demonstrate physically and mathematically the differences between a positive traveling wave and a negative traveling wave.
5. Represent a traveling sinusoidal wave in terms of its phasor form.
6. Explain the difference between a peak phasor and an rms phasor.
7. Express the general form of the steady-state ac solution for both a lossless line and a lossy line, and discuss the significance of the different terms.
8. Define **propagation constant** and **attenuation constant.**
9. For a lossy line, determine **propagation constant, attenuation constant, phase constant, wavelength, velocity of propagation,** and **characteristic impedance.**
10. Define **decibels** and **nepers,** and convert between them.
11. Determine the complete solution for voltage, current, and power for a lossy line matched at the load.

3–1 ■ SINUSOIDAL FUNCTION

Prior to discussing sinusoidal traveling waves, a few properties of the sinusoidal function need to be reviewed and strengthened. The terms **sinusoid** and **sinusoidal function** will be used to refer to the sine function or the cosine function or a combination of both.

The basic forms of the sine and cosine functions are shown in Fig. 3–1 and are labeled as sin θ and cos θ, respectively. The only difference between the two functions is the phase shift between them. The function sin θ starts with zero value for $\theta = 0$, while cos θ starts with unity value at $\theta = 0$.

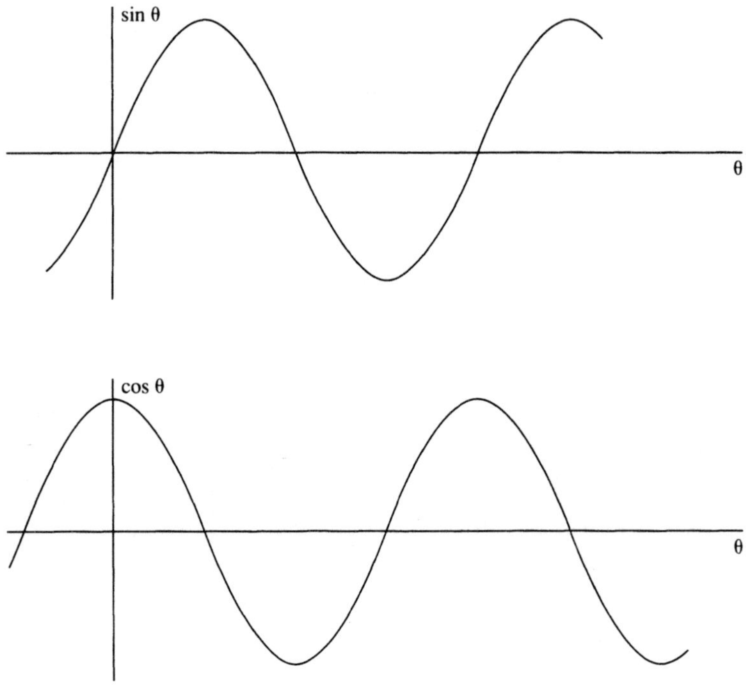

FIGURE 3–1 Sine and cosine functions.

Either the sine or the cosine function may be used as the basis of representation for sinusoidal traveling waves. For certain reasons of symmetry in the results, the choice in this book has been to use the cosine function.

The quantity θ is the **angle,** whose units will be discussed shortly. When θ involves a combination of two or more terms, as will be seen later, it is referred to as the **argument** of the sinusoidal function.

The most basic unit of angle is the **radian (rad),** and there are 2π radians in one complete cycle. The radian is a byproduct of the infinite series representation of the sinusoidal function as a solution of a differential equation. However, the units of **degrees** (°) were introduced many years ago, and most people are accustomed to working in degrees. There are 360° in one complete cycle, so 2π rad = 360°, or 1 rad = 57.296°.

For the majority of trigonometric applications, where a function of a single argument such as $\sin\theta$ or $\cos\theta$ is given, the use of degrees is perfectly fine. A problem can arise, however, when the angle involves the sum of a time-varying function and a constant unless one returns to the basic units of radians.

In basic ac circuit theory, functions of the form $\cos(\omega t + \theta)$ are encountered on a regular basis. The quantity ω is the angular frequency in radians/second (rad/s), t is the independent variable time in seconds (s), and θ is the angle. Many engineering books will frequently express θ in degrees so that expressions such as $\cos(\omega t + 45°)$ are quite common. This practice is longstanding and will no doubt continue. It is usually easier for technical people to visualize 45° than $\pi/4$ rad, probably because of our earlier training; so from an intuitive point of view, it is easy to justify this notation.

A problem arises when it is necessary to perform specific computations of the function for different values of t. The product ωt yields radians, so an argument such as $(\omega t + 45°)$ is like "adding oranges and apples." Although ωt could be altered to yield degrees, this process is a bit clumsy. The best way to resolve the dilemma is to write the preceding function as $\cos(\omega t + \pi/4) = \cos(\omega t + 0.7854)$. Most scientific calculators have radian modes, so, with a little practice, this form can be readily adapted.

Another reason for using radians is that some of the important transmission line parameters to be discussed later are determined in terms of an angle, and the angle in such cases *must be expressed in radians*. Therefore, *the general practice in this text will be to express all phase angles in sine or cosine functions in radians unless otherwise indicated.* Since it is awkward to place the units of the angle in the argument, this practice will be assumed throughout the book. Thus, in an expression of the form $\cos(\omega t + 2)$, the value 2 will be understood to be 2 radians.

3–2 ■ SINUSOIDAL TRAVELING WAVE

The concepts of dc and pulse traveling waves were established in Chapter 2. In this chapter, the concept of a sinusoidal traveling wave will be developed. A sinusoidal traveling wave serves as the basis for studying radio and television signals propagating both on transmission lines and in space.

Consider first a **time-varying** sinusoidal voltage $v(t)$ of the type considered in basic ac circuit theory with peak value V_p. If the phase angle is assumed as zero, the function may be expressed as

$$v(t) = V_p \cos\omega t \tag{3–1}$$

where

V_p = peak value in volts

ω = angular frequency in radians/seconds (rad/s)

The angular frequency is related to the cyclic frequency f in hertz (Hz) by

$$\omega = 2\pi f \tag{3–2}$$

The cyclic frequency f is related to the period T by

$$f = \frac{1}{T} \tag{3–3}$$

The time-varying voltage is shown in Fig. 3–2.

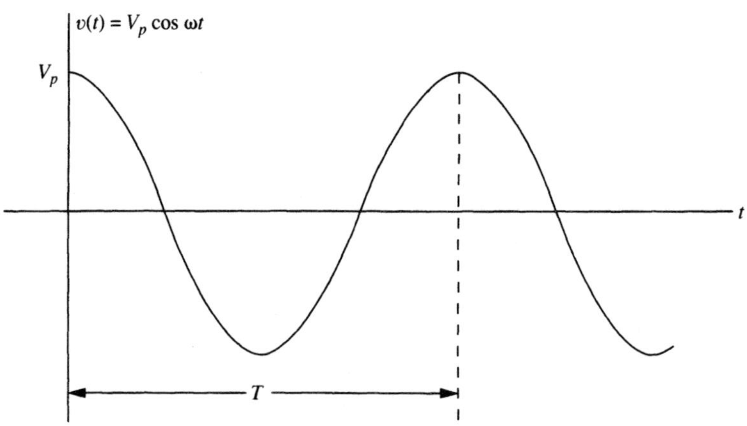

FIGURE 3–2 Time-varying sinusoid.

Now let us introduce the concept of a **distance-varying** sinusoidal voltage $v(x)$ of the form

$$v(x) = V_p \cos \beta x \tag{3–4}$$

where V_p is again the peak value and x is a displacement variable. The quantity β serves the same purpose with respect to displacement as does ω with respect to time, and it is called the **phase constant.** Its units are radians/distance, e.g., radians/meter (rad/m), with the requirement that the distance units be the same as those of x. Thus, if β is expressed in radians/meter, x must be expressed in meters.

The distance-varying voltage is shown in Fig. 3–3. Obviously, this function appears the same along a distance axis x as did the first function along a time axis t. The "period" in this case is the wavelength λ, as will be discussed shortly.

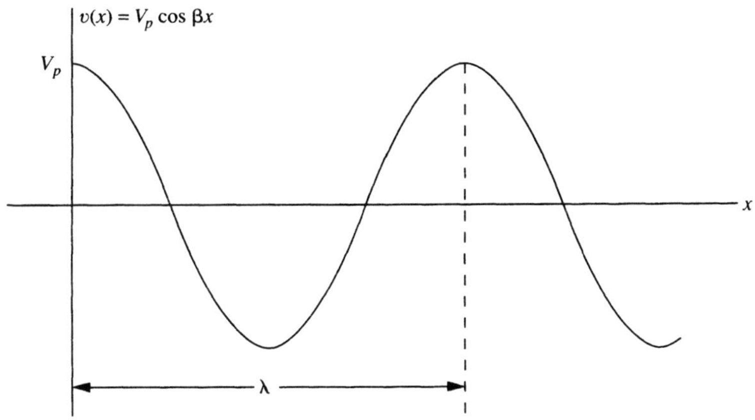

FIGURE 3–3 Distance-varying sinusoid.

A traveling sinusoidal wave involves the sum or difference between ωt and βx as the argument of the sinusoidal function. Consider first the difference expressed in the form

$$v \equiv v(x, t) = V_p \cos(\beta x - \omega t) \tag{3–5}$$

It would require a three-dimensional plot to show simultaneously the variation of v as a function of both x and t. Instead, either x or t may be fixed, and v can then be plotted as a function of the other variable.

Consider the process of showing Eq. (3–5) as a function of x for several fixed values of t, as will be illustrated in Fig. 3–4. First assume that $t = 0$, in which case

$$v = V_p \cos \beta x \tag{3–6}$$

This function is shown in Fig. 3–4(a), and it is obviously of the same form as in Fig. 3–3.

Next consider a positive value of time t_1, in which case

$$v = V_p \cos(\beta x - \omega t_1) \tag{3–7}$$

The quantity ωt_1 represents a phase shift, which, as a result of the minus sign, means that x must assume a more positive value to reach any original value of the argument. A distance plot of the new function is of the form shown in Fig. 3–4(b). Note that the point P at the crest of wave (a) has moved to the *right* as shown in (b).

At a still later time t_2 ($t_2 > t_1$), the function is

$$v = V_p \cos(\beta x - \omega t_2) \tag{3–8}$$

A still greater shift has occurred as shown in Fig. 3–4(c).

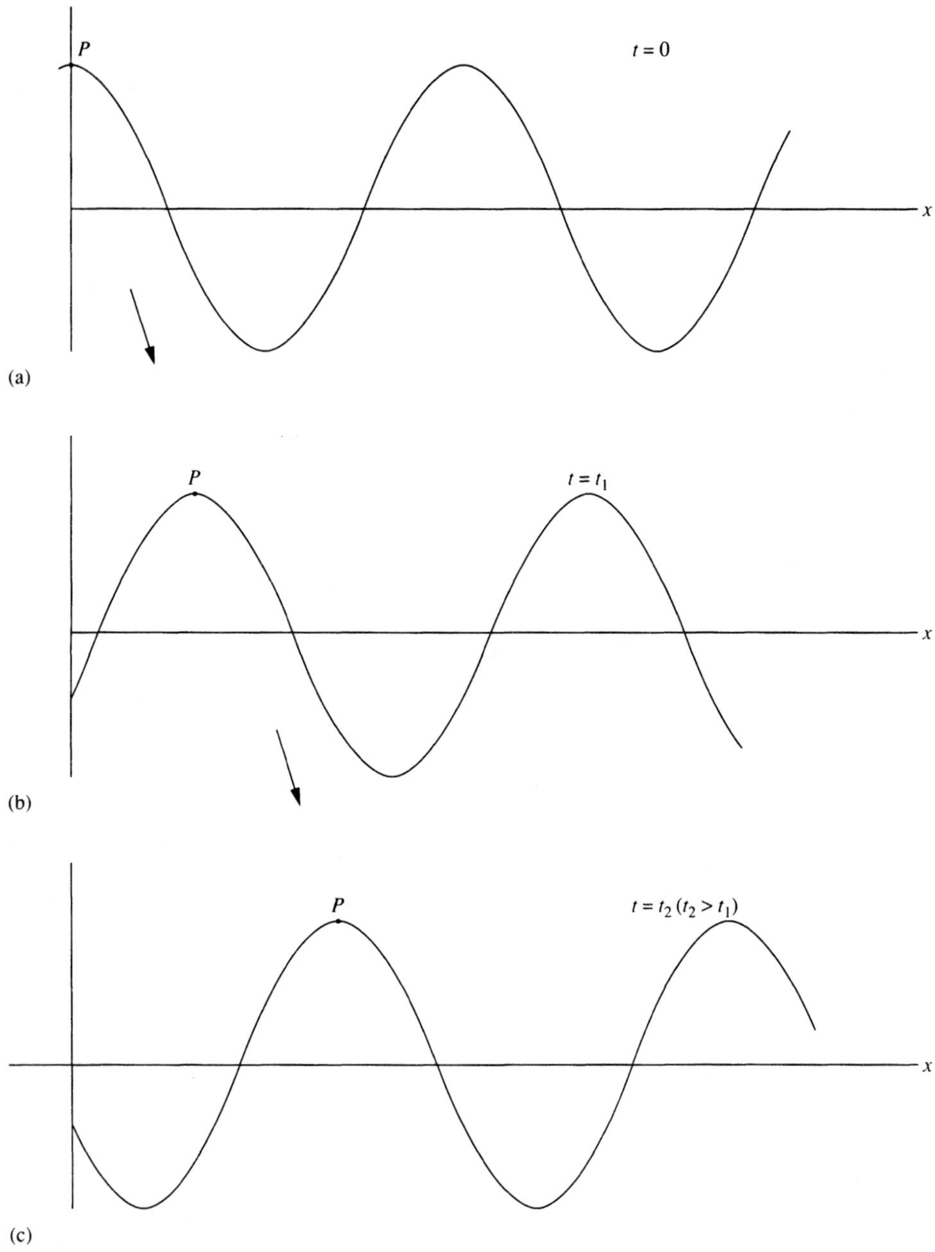

FIGURE 3–4 Sinusoidal function propagating in positive x-direction.

What is happening here is that the wave is shifting successively to the right as time increases. Thus, the wave is traveling in the direction of positive x. We conclude, therefore, that *a function of the form $V_p \cos(\beta x - \omega t)$ describes a sinusoidal voltage of peak value V_p traveling in the positive x-direction.* At a given value of x, say, x_1, the voltage as a function of time is described by

$$v = V_p \cos(\beta x_1 - \omega t) = V_p \cos(\omega t - \beta x_1) \tag{3–9}$$

where the property $\cos(-\theta) = \cos\theta$ was used. The quantity βx_1 is simply a constant phase shift. Thus, *at a fixed value of x, the traveling wave appears to be simply a time-varying sinusoid which assumes all possible values of the function over a time cycle. However, at a given value of time, the voltage is a sinusoid spread out over one or more cycles in a distance or spatial sense.*

We will now determine some essential properties of the traveling wave function. First assume that x is fixed at a value x_1, and consider the argument in the form $\omega t - \beta x_1$. A complete cycle is defined as the period T, and this corresponds to a change in the angle of 2π rad. Setting $t = T$, we have

$$\omega T = 2\pi \tag{3–10}$$

which yields

$$T = \frac{2\pi}{\omega} = \frac{2\pi}{2\pi f} = \frac{1}{f} \tag{3–11}$$

where f is the frequency in hertz. We already knew this, of course, but the process of working it out helps to establish a similar procedure for displacement.

Next assume that t is fixed at a value t_1, and consider the argument in the form $\beta x - \omega t_1$. A complete cycle along the x-axis corresponds to the wavelength λ. Setting $x = \lambda$, a complete cycle corresponds to

$$\beta\lambda = 2\pi \tag{3–12}$$

which yields

$$\lambda = \frac{2\pi}{\beta} \tag{3–13}$$

Thus, the wavelength may be determined in terms of the phase constant by (3–13). However, *it is essential that the angle units of β be in radians.* The units of λ will be the same as the per-unit distance in β. For example, if β is expressed in radians/meter (rad/m), λ will be expressed in meters.

Note the similarities between time and displacement as they concern period and wavelength. The following analogies are useful:

Time		**Distance**
ω	$\longleftrightarrow$	β
T	$\longleftrightarrow$	λ

The result of (3–13) may be modified so that β can be expressed in terms of frequency f. First multiply numerator and denominator of the right-hand side of (3–13) by f, and substitute $\omega = 2\pi f$. Next substitute $\nu = f\lambda$. The result can be rearranged as

$$\beta = \frac{\omega}{\nu} \tag{3–14}$$

The result of (3–13) will be inverted so that β is expressed in terms of λ and is grouped with (3–14) to provide two alternate relationships as follows:

$$\beta = \frac{2\pi}{\lambda} = \frac{\omega}{\nu} \tag{3–15}$$

The two forms of (3–15) should be memorized for reference throughout the book.

Negative Traveling Wave

Thus far, the function considered has employed the difference in the argument of the sinusoidal function, and this has resulted in a positive traveling wave. Consider now the function

$$v = V_p \cos(\beta x + \omega t) \tag{3–16}$$

First set $t = 0$, in which case we have

$$v = V_p \cos \beta x \tag{3–17}$$

This function is shown in Fig. 3–5(a).

Next consider a positive value of time t_1, in which case

$$v = V_p \cos(\beta x + \omega t_1) \tag{3–18}$$

The phase shift ωt_1 now adds a positive value to the argument. Thus, the argument will be the same as before for a smaller value of x. A distance plot of the new function would have the form of Fig. 3–5(b). Note that the point P at the crest of waveform (a) has moved to the *left* as shown in (b).

At a still later time t_2 ($t_2 > t_1$), the function is

$$v = V_p \cos(\beta x + \omega t_2) \tag{3–19}$$

A still greater shift to the left has occurred as shown in Fig. 3–5(c).

From the preceding development, we conclude that *a function of the form $V_p \cos(\beta x + \omega t)$ describes a sinusoidal voltage of peak value V_p traveling in the negative x-direction.*

EXAMPLE 3–1

A traveling wave of voltage (in volts) is given by

$$v = 20 \cos(10^6 t - 5 \times 10^{-3} x) \tag{3–20}$$

with t in seconds and x in meters. Determine (a) the direction of propagation, (b) the peak value, (c) the angular frequency, (d) the phase constant, (e) the cyclic frequency, (f) the period, (g) the wavelength, and (h) the velocity of propagation.

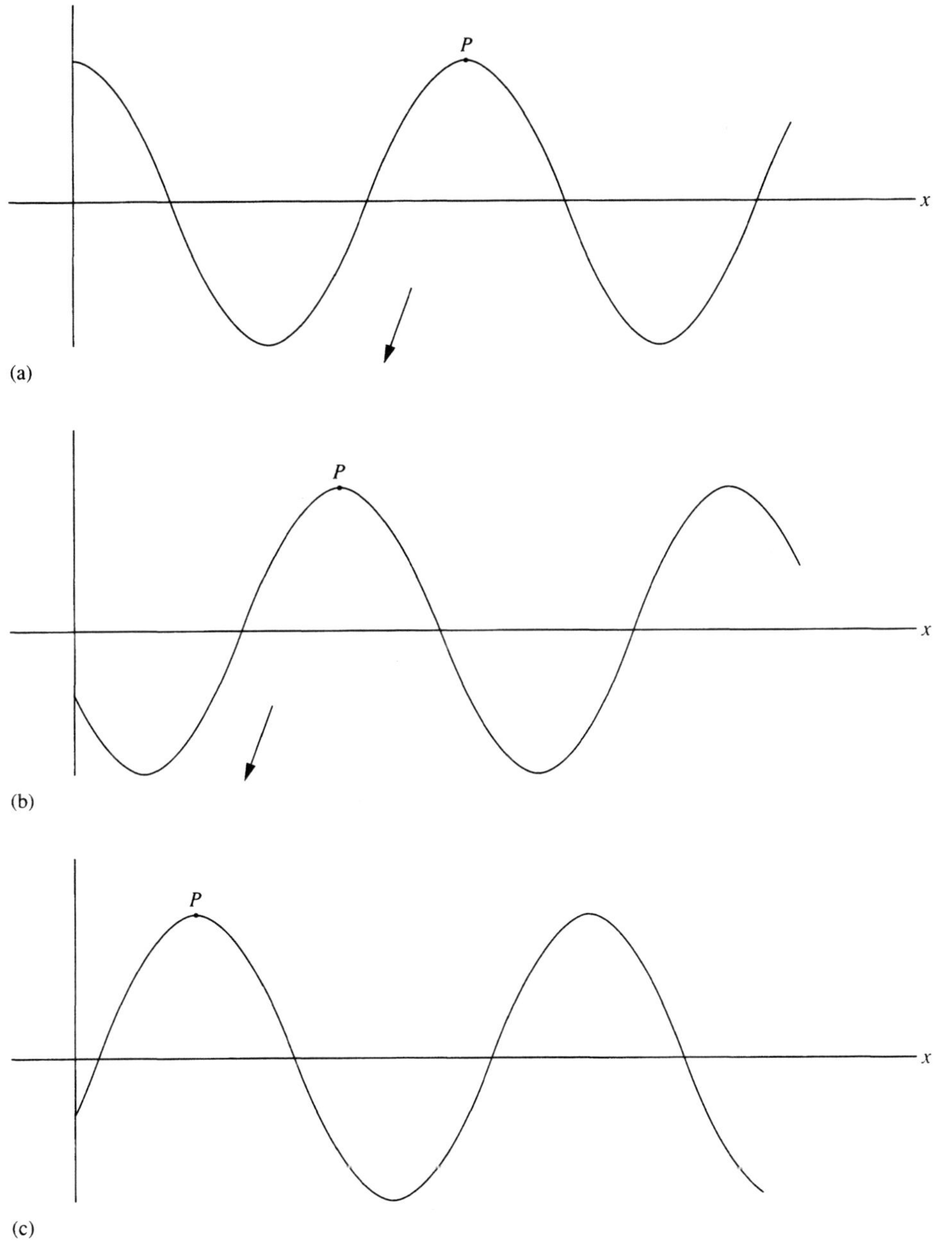

FIGURE 3–5 Sinusoidal function propagating in negative x-direction.

Solution

(a) Since the argument of the sinusoidal function has the *difference* between the time and displacement terms, the wave is traveling in the *positive* x-direction.

(b) The peak value is

$$V_p = 20 \text{ V} \tag{3–21}$$

(c) Comparing the general form of the argument $\omega t - \beta x$ with $10^6 t - (5 \times 10^{-3})x$, we see that

$$\omega = 10^6 \text{ rad/s} \tag{3–22}$$

(d) Continuing with the comparison of (c), we see that

$$\beta = 5 \times 10^{-3} \text{ rad/m} \tag{3–23}$$

Note that since x is understood to be in meters, β must be in radians/meter.

(e) The cyclic frequency is

$$f = \frac{\omega}{2\pi} = \frac{10^6}{2\pi} = 159.2 \text{ kHz} \tag{3–24}$$

(f) The period is

$$T = \frac{1}{f} = \frac{1}{159.2 \times 10^3} = 6.281\ \mu\text{s} \tag{3–25}$$

(g) The wavelength is

$$\lambda = \frac{2\pi}{\beta} = \frac{2\pi}{5 \times 10^{-3}} = 1257 \text{ m} \tag{3–26}$$

(h) The velocity of propagation is

$$\nu = f\lambda = (159.2 \times 10^3) \times (1.257 \times 10^3) = 2 \times 10^8 \text{ m/s} \tag{3–27}$$

Alternatively,

$$\nu = \frac{\omega}{\beta} = \frac{10^6}{5 \times 10^{-3}} = 2 \times 10^8 \text{ m/s} \tag{3–28}$$

EXAMPLE 3–2

A sinusoidal voltage with a peak value of 6 V and a frequency of 1 MHz is traveling in the negative x-direction at a velocity of 2.5×10^8 m/s. Determine (a) the period, (b) the angular frequency, (c) the phase constant, and (d) the wavelength. (e) Also write an equation for the voltage.

Solution

(a) The period is simply

$$T = \frac{1}{f} = \frac{1}{10^6} = 1\ \mu\text{s} \tag{3–29}$$

(b) The angular frequency is

$$\omega = 2\pi f = 2\pi(1 \times 10^6) = 6.283 \times 10^6 \text{ rad/s} \quad \textbf{(3–30)}$$

(c) The phase constant is

$$\beta = \frac{\omega}{\nu} = \frac{2\pi(1 \times 10^6)}{2.5 \times 10^8} = 25.13 \times 10^{-3} \text{ rad/m} \quad \textbf{(3–31)}$$

(d) The wavelength is

$$\lambda = \frac{2\pi}{\beta} = \frac{2\pi}{25.13 \times 10^{-3}} = 250 \text{ m} \quad \textbf{(3–32)}$$

Alternately,

$$\lambda = \frac{\nu}{f} = \frac{2.5 \times 10^8}{1 \times 10^6} = 250 \text{ m} \quad \textbf{(3–33)}$$

(e) Combining all the preceding results, we can express the waveform as

$$v = 6\cos(\omega t + \beta x) = 6\cos[(6.283 \times 10^6)t + 0.02513x] \quad \textbf{(3–34)}$$

3–3 ■ PHASOR FORM OF TRAVELING SINUSOID

In basic ac circuit theory, a sinusoidal function is converted to a complex phasor form. This conversion permits the determination of steady-state responses with the use of complex algebra utilizing the j-operator ($j = \sqrt{-1}$). The computation is possible through the use of **Euler's formula,** which reads

$$e^{j\omega t} = \cos \omega t + j \sin \omega t \quad \textbf{(3–35)}$$

where $e = 2.718$ (to four significant figures) is the base of the natural system of logarithms. The function $\cos \omega t$ is the **real part** of $e^{j\omega t}$, which is denoted as $\text{Re}[e^{j\omega t}]$, and $\sin \omega t$ is the **imaginary part** of $e^{j\omega t}$, which is denoted as $\text{Im}[e^{j\omega t}]$. Either the real or the imaginary part may be used as the basis for the development, and both approaches appear in the literature. For certain reasons of symmetry, the cosine will be used here.

Seldom do beginners understand the full significance of the development of ac phasor analysis, but the application of the results can be readily learned. To reinforce the concepts of basic circuit theory, a short development will follow. First assume a positive traveling voltage wave of the form

$$v(x, t) = V_p \cos(\omega t - \beta x + \theta) \quad \textbf{(3–36)}$$

where an arbitrary additional constant phase shift of value θ has been added to the argument. This voltage may be expressed as

$$v(x, t) = \text{Re}[V_p e^{j(\omega t - \beta x + \theta)}] \quad \textbf{(3–37a)}$$

$$v(x, t) = \text{Re}[V_p e^{j\theta} e^{-j\beta x} e^{j\omega t}] \quad \textbf{(3–37b)}$$

Define $v(x, t)$ as

$$v(x, t) \equiv V_p e^{j\theta} e^{-j\beta x} e^{j\omega t} \tag{3–37c}$$

In (3–37c), the real part designation is dropped for convenience, but it is understood.

Next a **fixed phasor** $\overline{V}^+$ representing the complex form of the positive traveling wave will be defined. It is given by

$$\overline{V}^+ \equiv V_p e^{j\theta} = V_p \angle \theta \tag{3–38}$$

The exponential form of the positive traveling wave can be expressed as

$$v(x, t) = \overline{V}^+ e^{-j\beta x} e^{j\omega t} \tag{3–39}$$

Thus, the positive traveling wave can be expressed as the product of a phasor and two phase factors, one corresponding to variation in distance and one corresponding to variation in time. As is customary in ac phasor analysis, the $e^{j\omega t}$ factor will be dropped shortly.

Next assume a negative traveling wave of the form

$$v(x, t) = V_p \cos(\omega t + \beta x + \phi) \tag{3–40}$$

where ϕ is an arbitrary constant phase shift. We will leave as an exercise for the analytically oriented reader to show that when the preceding process is repeated, the following form results:

$$v(x, t) = \overline{V}^- e^{j\beta x} e^{j\omega t} \tag{3–41}$$

where $\overline{V}^-$ is the *fixed* phasor for the negative traveling wave as given by

$$\overline{V}^- \equiv V_p e^{j\phi} = V_p \angle \phi \tag{3–42}$$

A major simplification arises when phasor forms are introduced, because the processes of differentiation and integration are replaced by multiplication and division by $j\omega$. This is a result of the property that

$$\frac{de^{j\omega t}}{dt} = j\omega e^{j\omega t} \tag{3–43}$$

$$\int e^{j\omega t}\, dt = \frac{e^{j\omega t}}{j\omega} \tag{3–44}$$

where any constant of integration in (3–44) has been ignored. Once all derivatives and integrals have been computed, the common $e^{j\omega t}$ factor can be momentarily suppressed or removed, and the ac phasor circuit will remain.

The preceding leads to the concept of **impedance**. The impedance Z is defined as

$$Z = \frac{\text{phasor voltage}}{\text{phasor current}} \tag{3–45}$$

Let us return briefly to the concept of lumped inductance L (in henries) and lumped capacitance C (in farads). The steady-state ac impedances for

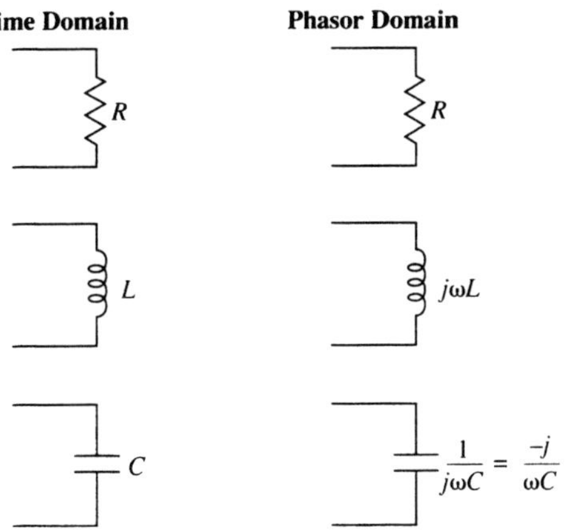

FIGURE 3–6 Basic circuit elements and their phasor-domain impedances.

inductance and capacitance are $j\omega L$ and $1/j\omega C = -j/\omega C$, respectively, as illustated in Fig. 3–6.

The connection to basic circuit theory can now be deduced. An inductance of value L appears as a purely imaginary impedance of $j\omega L$ to either traveling wave, and a capacitance of value C appears as a purely imaginary impedance of $1/j\omega C = -j/\omega C$ to either wave. The primary difference here is that the phasor voltage and current are functions of the location on the line due to the $e^{-j\beta x}$ and $e^{j\beta x}$ factors, respectively.

When the $e^{j\omega t}$ factor is removed, a positive traveling phasor is now a function of x only and may be expressed as

$$\overline{V}(x) = \overline{V}^{+}e^{-j\beta x} \tag{3–46}$$

Similarly, a negative traveling phasor may be expressed as

$$\overline{V}(x) = \overline{V}^{-}e^{j\beta x} \tag{3–47}$$

where $\overline{V}^{+} = V_p\angle\theta$ and $\overline{V}^{-} = V_p\angle\phi$ are fixed phasors. When the $e^{\pm j\beta x}$ factors are included as in (3–46) and (3–47), the phasors will be referred to as **distance-varying phasors.**

rms Phasor

The **root-mean-square (rms)** value V_{rms} of a sinusoidal voltage of peak value V_p is given by

$$V_{\text{rms}} = \frac{V_p}{\sqrt{2}} = 0.7071V_p \tag{3–48}$$

Similarly, the rms value I_{rms} of a sinusoidal current of peak value I_p is given by

$$I_{\text{rms}} = \frac{I_p}{\sqrt{2}} = 0.7071 I_p \tag{3–49}$$

The rms value (also called the **effective value**) is the equivalent power-producing measure of a waveform. Let P represent the average power produced by the sinusoidal waveform. For a resistance R, the average power may be expressed as either

$$P = I_{\text{rms}}^2 R \tag{3–50}$$

or

$$P = \frac{V_{\text{rms}}^2}{R} \tag{3–51}$$

where V_{rms} is measured across the resistance and I_{rms} is measured through the resistance.

When there is a major interest in power calculations, it is common in the electrical and electronics field to express the magnitudes of phasors in terms of their rms values. For example, a voltage phasor expressed in peak volts as $\overline{V} = 100\angle 2$ V could be redefined in rms volts as $\overline{V} = 70.71\angle 2$ V. Similarly, a current phasor expressed in peak amperes as $\overline{I} = 4\angle -1.5$ A could be redefined in rms amperes as $\overline{I} = 2.828\angle -1.5$ A. Thus, when phasor forms are given, it is necessary to specify whether the magnitudes are given in peak units or in rms units.

EXAMPLE 3–3

Consider the voltage traveling wave of Example 3–1, which was

$$v(x, t) = 20 \cos(10^6 t - 5 \times 10^{-3} x) \tag{3–52}$$

Determine (a) the fixed phasor $\overline{V}^+$ in peak units, (b) the distance-varying phasor $\overline{V}(x)$ in peak units, and (c) the value of the distance-varying phasor at $x = 400$ m.

Solution

(a) Since there is no other phase angle added to the time and distance variations, and since the function is given as a cosine, the fixed phasor is

$$\overline{V}^+ = V_p e^{j0} = 20\angle 0 \text{ V} \tag{3–53}$$

(b) The distance-varying phasor is

$$\overline{V}(x) = \overline{V}^+ e^{-j\beta x} = (20e^{j0})(e^{-j5\times 10^{-3}x}) \tag{3–54a}$$

$$= (20\angle 0)(1\angle -5 \times 10^{-3}x) = (20\angle -5 \times 10^{-3}x) \text{ V} \tag{3–54b}$$

(c) At a value of $x = 400$ m, the net angle is $-5 \times 10^{-3} \times 400 = -2$ rad. Thus,

$$\overline{V}(400) = 20\underline{/-2}\text{ V} \qquad \textbf{(3–55)}$$

EXAMPLE 3–4

A current traveling wave has the value

$$i(x, t) = 6\cos[(5 \times 10^6 t) + 0.02x + 1.2] \qquad \textbf{(3–56)}$$

Determine (a) the fixed phasor in peak units, (b) the distance-varying phasor $\overline{I}(x)$ in peak units, and (c) the value of the distance-varying phasor at $x = 100$ m.

Solution

(a) Since the ωt and βx terms have the same sign, the wave is traveling in the negative x-direction. Thus, the fixed phasor will be denoted as $\overline{I}^-$. Its value is

$$\overline{I}^- = 6e^{j1.2} = 6\underline{/1.2}\text{ A} \qquad \textbf{(3–57)}$$

where the angle is understood to be 1.2 rad.

(b) The distance-varying phasor function is

$$\overline{I}(x) = \overline{I}^- e^{j\beta x} = (6e^{j1.2})(e^{j0.02x}) = (6\underline{/1.2})(1\underline{/0.02x}) \qquad \textbf{(3–58a)}$$

$$= 6\underline{/1.2 + 0.02x}\text{ A} \qquad \textbf{(3–58b)}$$

(c) At $x = 100$ m, the preceding phasor is

$$\overline{I}(100) = 6\underline{/1.2 + 0.02(100)} = 6\underline{/3.2}\text{ A} \qquad \textbf{(3–59)}$$

EXAMPLE 3–5

Redefine the fixed phasor of Example 3–3 so that the phasor magnitude is expressed in rms units, and determine the average power dissipated in a 50-Ω resistance.

Solution

The rms value corresponding to a peak sinusoidal voltage of 20 V is simply $20/\sqrt{2} = 14.142$ V. Using the same notation, i.e., $\overline{V}^+$, but redefining the magnitude as rms volts, we have

$$\overline{V}^+ = 14.142\underline{/0}\text{ V} \qquad \textbf{(3–60)}$$

The average power dissipated in a 50-Ω resistance is

$$P = \frac{(14.142)^2}{50} = 4\text{ W} \qquad \textbf{(3–61)}$$

EXAMPLE 3–6

Redefine the fixed phasor of Example 3–4 so that the phasor magnitude is expressed in rms units, and determine the average power dissipated in a 12-Ω resistance.

Solution

The rms value corresponding to a peak sinusoidal current of 6 A is $6/\sqrt{2} =$ 4.243 A. Using the same notation, i.e., $\overline{I}^-$, but redefining the magnitude as rms amperes, we have

$$\overline{I}^- = 4.243\underline{/1.2}\text{ A} \tag{3–62}$$

The average power dissipated in a 12-Ω resistance is

$$P = (4.243)^2 \times 12 = 216\text{ W} \tag{3–63}$$

3–4 ■ STEADY-STATE ac LOSSLESS SOLUTION

The incremental steady-state ac model of a lossless transmission line is shown in Fig. 3–7. Since we are now working with steady-state ac phasors, the series impedance of the section is $j\omega L\Delta x$, and the shunt admittance of the section is $j\omega C\Delta x$. The voltage phasor to the left of the section is denoted as $\overline{V}(x)$, and the voltage to the right is $\overline{V}(x) + \Delta\overline{V}$ where $\Delta\overline{V}$ is the change in voltage across the series impedance. Similarly, the current phasor to the left of the section is denoted as $\overline{I}(x)$, and the current to the right is $\overline{I}(x) + \Delta\overline{I}$, where $\Delta\overline{I}$ is the change in current resulting from the shunt admittance.

The change in voltage is

$$\Delta\overline{V} = -(j\omega L\Delta x)\overline{I}(x) \tag{3–64}$$

The change in current is

$$\Delta\overline{I} = -(j\omega C\Delta x)\overline{V}(x + \Delta x) \tag{3–65}$$

The minus signs are required for both equations because $\Delta\overline{V}$ and $\Delta\overline{I}$ were defined in a positive sense, but the changes actually represent reductions in the voltage and current.

If both sides of (3–64) and (3–65) are divided by Δx, there results

$$\frac{\Delta\overline{V}}{\Delta x} = -j\omega L\overline{I}(x) \tag{3–66}$$

$$\frac{\Delta\overline{I}}{\Delta x} = -j\omega C\overline{V}(x + \Delta x) \tag{3–67}$$

Next, let Δx become infinitesimally small. From differential calculus, we have

$$\lim_{\Delta x \to 0} \frac{\Delta\overline{V}}{\Delta x} = \frac{d\overline{V}}{dx} \tag{3–68}$$

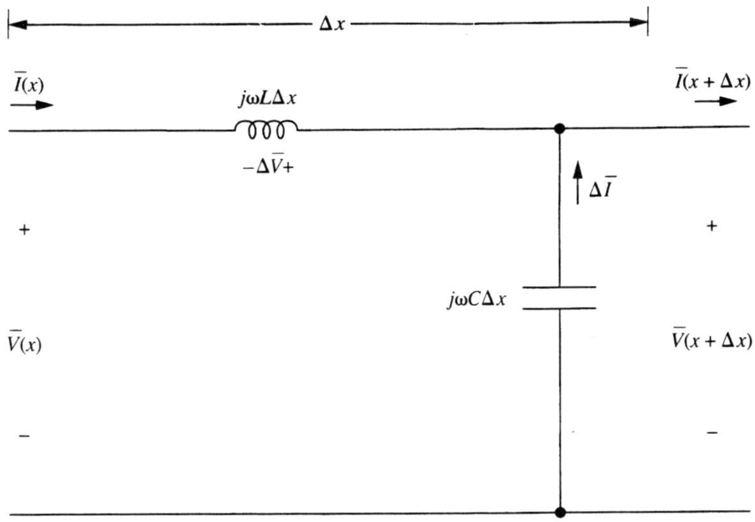

FIGURE 3–7 Incremental model of lossless line.

$$\lim_{\Delta x \to 0} \frac{\Delta \overline{I}}{\Delta x} = \frac{d\overline{I}}{dx} \tag{3–69}$$

When these limits are taken, the additional Δx term on the right-hand side of (3–67) becomes insignificant, and both $\overline{V}$ and $\overline{I}$ become functions of x. The result is

$$\frac{d\overline{V}}{dx} = -j\omega L\overline{I} \tag{3–70}$$

and

$$\frac{d\overline{I}}{dx} = -j\omega C\overline{V} \tag{3–71}$$

where $\overline{V} = \overline{V}(x)$ and $\overline{I} = \overline{I}(x)$

Equations (3–70) and (3–71) are *the steady-state ac transmission line equations for a lossless line.* In order to solve them, it is necessary to first eliminate $\overline{I}$ between the two equations and solve for $\overline{V}$, with the process then reversed to solve for $\overline{I}$.

The quantity $\overline{I}$ may be eliminated by first differentiating both sides of (3–70) with respect to x, which yields

$$\frac{d^2\overline{V}}{dx^2} = -j\omega L\frac{d\overline{I}}{dx} \tag{3–72}$$

Next, (3–71) is substituted in (3–72), which yields

$$\frac{d^2\overline{V}}{dx^2} = -\omega^2 LC\overline{V} \tag{3–73}$$

Equation (3–73) is a special case of a more general equation known in electromagnetic theory as the **wave equation.** Its solution is well known from the theory of differential equations and can be expressed initially as

$$\overline{V} = \overline{K}_1 e^{j\omega\sqrt{LC}x} + \overline{K}_2 e^{-j\omega\sqrt{LC}x} \tag{3–74}$$

where $\overline{K}_1$ and $\overline{K}_2$ are complex constants that must be determined from conditions at the ends of the lines (**boundary conditions**).

At this point, an interesting observation can be readily made. The first term in (3–74) is multiplied by $e^{j\beta x}$, which was shown in the last section to be a phasor representation of a sinusoidal wave propagating in the negative x-direction. Similarly, the second term on the right-hand side of (3–74) has the form $e^{-j\beta x}$, which was shown to be a phasor representation of a sinusoidal wave propagating in the positive x-direction. Thus, the constant $\overline{K}_2$ will be redefined as $\overline{V}^+$, and the constant $\overline{K}_1$ will be redefined as $\overline{V}^-$. Reversing the order of the terms so that the positive traveling wave is listed first, we have

$$\overline{V} = \overline{V}^+ e^{-j\beta x} + \overline{V}^- e^{j\beta x} \tag{3–75}$$

A similar procedure for current can be performed, and the solution will be of the form

$$\overline{I} = \overline{I}^+ e^{-j\beta x} + \overline{I}^- e^{j\beta x} \tag{3–76}$$

Thus, *the steady-state ac voltage or current on a transmission line may be expressed as the sum of a sinusoidal wave traveling in the positive x-direction and a sinusoidal wave traveling in the negative x-direction.*

In the transient analysis of lines in Chapter 2, it was observed that for a line mismatched at both ends, there could exist many forward waves and many reverse waves. However, in steady-state ac lines, all transients have already disappeared, and the final sinusoidal voltage and current waves can each be expressed in terms of a **single forward wave** and a **single reverse wave.**

Comparing (3–75) and (3–76) with (3–74), we can readily determine the value of β to be

$$\beta = \omega\sqrt{LC} \tag{3–77}$$

It has already been established that $\beta = \omega/v$, so when these quantities are equated, we have $\omega\sqrt{LC} = \omega/v$, or

$$v = \frac{1}{\sqrt{LC}} \tag{3–78}$$

which was stated in Chapter 1 without proof.

Next, relationships between the individual components of the waves of voltage and current will be established. This can be accomplished by first forming $d\overline{V}/dx$. Using the form of (3–75), we have

$$\frac{d\overline{V}}{dx} = -j\beta\overline{V}^+ e^{-j\beta x} + j\beta\overline{V}^- e^{j\beta x} \tag{3–79}$$

Substituting this expression along with (3–76) in (3–70), we obtain

$$-j\beta\overline{V}^+e^{-j\beta x} + j\beta\overline{V}^-e^{j\beta x} = -j\omega L(\overline{I}^+e^{-j\beta x} + \overline{I}^-e^{j\beta x}) \tag{3–80}$$

In order for this equation to be true, the coefficients of the individual exponential factors on both sides of the equation must be the same. Thus, we have two separate equations:

$$-j\beta\overline{V}^+ = -j\omega L\overline{I}^+ \tag{3–81}$$

and

$$j\beta\overline{V}^- = -j\omega L\overline{I}^- \tag{3–82}$$

After common factors are canceled and the terms are rearranged, we have

$$\frac{\overline{V}^+}{\overline{I}^+} = \frac{\omega L}{\beta} = \frac{\omega L}{\omega\sqrt{LC}} = \sqrt{\frac{L}{C}} \tag{3–83}$$

Let

$$\frac{\overline{V}^-}{\overline{I}^-} = \frac{-\omega L}{\beta} = \frac{-\omega L}{\omega\sqrt{LC}} = -\sqrt{\frac{L}{C}} \tag{3–84}$$

in which case

$$R_0 = \sqrt{\frac{L}{C}} \tag{3–85}$$

$$\frac{\overline{V}^+}{\overline{I}^+} = R_0 \tag{3–86}$$

$$\frac{\overline{V}^-}{\overline{I}^-} = -R_0 \tag{3–87}$$

We recognize, of course, that $R_0 = \sqrt{L/C}$ is the characteristic impedance of a lossless transmission line, which was stated without justification in Chapter 1. What we have shown here is that the quantity arises through the natural process of solving the differential equation governing the transmission line.

We should carefully note the difference in signs between (3–86) and (3–87). For the positive traveling wave, the ratio of voltage to current is R_0, while for the negative traveling wave, the ratio is $-R_0$. The minus sign never arose in Chapter 1 because the ratio was used only for the first wave, which was a positive wave.

EXAMPLE 3–7

Under steady-state ac conditions, the forward voltage wave on a lossless 50-Ω line is

$$\overline{V}^+ = 80\underline{/2}\text{ V} \tag{3–88}$$

Determine the current forward wave.

Solution

The current forward wave is

$$\bar{I}^+ = \frac{\bar{V}^+}{R_0} = \frac{80\underline{/2}}{50} = 1.6\underline{/2}\text{ A} \tag{3–89}$$

EXAMPLE 3–8

Under steady-state ac conditions, the reverse current wave on a lossless 300-Ω line is

$$\bar{I}^- = 4\underline{/-1.5}\text{ mA} \tag{3–90}$$

Determine the reverse voltage wave.

Solution

The reverse voltage wave is

$$\bar{V}^- = -R_0\bar{I}^- = (-300)(4 \times 10^{-3}\underline{/-1.5}) = -1.2\underline{/-1.5} \tag{3–91a}$$

$$= 1.2\underline{/-1.5 + \pi} = 1.2\underline{/1.642}\text{ V} \tag{3–91b}$$

3–5 ■ LINES WITH LOSSES

We will now introduce losses into the incremental transmission line model. Dissipative losses occur as a result of the resistance of the conductors and as a result of dielectric losses. The resulting incremental model considering these effects is shown in Fig. 3–8.

The quantity R represents the series resistance/unit length as referred to one side of the line in a manner similar to that of inductance. Thus, the net series resistance of the incremental model section is $R\Delta x$.

The quantity G represents the shunt conductance per unit length resulting from the dielectric losses. The net shunt conductance of the section is $G\Delta x$. By popular convention, neither R nor G has any subscripts, but it should be understood that there is no relationship between the two quantities; i.e., $G \neq 1/R$.

It is now convenient to introduce the concept of **series impedance per unit length,** which will be denoted by Z, and **shunt admittance per unit length,** which will be denoted by Y. These quantities are

$$Z = R + j\omega L \tag{3–92}$$

and

$$Y = G + j\omega C \tag{3–93}$$

The net series impedance for the incremental model section is $Z\Delta x$, and the net shunt admittance for the section is $Y\Delta x$.

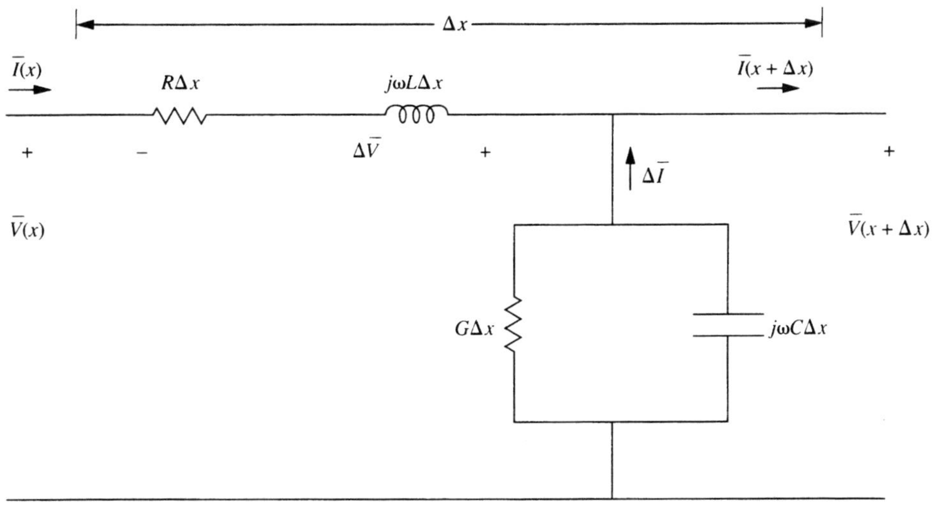

FIGURE 3–8 Incremental model of lossy line.

The differential equation for the lossy section may now be developed by a process similar to that for the lossless section by following the procedure given in Eqs. (3–64) through (3–71). Without showing all the steps, the results are

$$\frac{d\overline{V}}{dx} = -Z\overline{I} \tag{3–94}$$

$$\frac{d\overline{I}}{dx} = -Y\overline{V} \tag{3–95}$$

Equations (3–94) and (3–95) are the *steady-state ac transmission line equations for a lossy line*. The phasor $\overline{I}$ may be eliminated by differentiating both sides of (3–94) and substituting in (3–95). This results in

$$\frac{d^2\overline{V}}{dx^2} = ZY\overline{V} \tag{3–96}$$

which is another form of the **wave equation.**

The solution of (3–96) is well known from the theory of differential equations and can be initially expressed as

$$\overline{V} = \overline{K}_1 e^{\sqrt{ZY}x} + \overline{K}_2 e^{-\sqrt{ZY}x} \tag{3–97}$$

If (3–97) is compared with (3–74), it is seen that $j\omega\sqrt{LC}$ for the lossless line is replaced by $\sqrt{ZY}$ for the lossy line.

Let us now investigate the nature of $\sqrt{ZY}$. In general, both Z and Y are complex numbers. The square root of a complex number is also a complex number. For reasons that will be clear shortly, this result will be defined as the **complex propagation constant,** for which the symbol γ will be used. In general, γ can be expressed as

$$\gamma = \alpha + j\beta \tag{3–98}$$

The imaginary part of γ is the **phase constant** β which was discussed in detail in previous sections. In the case of the lossless line, the propagation constant is simply $\gamma = j\beta$ as previously noted, and the wave encounters only phase shift down the line. However, the effect of the real part α is to introduce the possible growth or decay of a propagating wave.

The quantity α is called the **attenuation constant,** and its units are **nepers/unit length.** The significance of **nepers** will be discussed shortly. Utilizing the definition of *propagation constant* and recognizing the significance of the signs of βx and $-\beta x$ in the two terms, we can express (3–97) as

$$\overline{V} = \overline{V}^{+}e^{-\gamma x} + \overline{V}^{-}e^{\gamma x} = \overline{V}^{+}e^{-\alpha x}e^{-j\beta x} + \overline{V}^{-}e^{\alpha x}e^{j\beta x} \tag{3–99}$$

Comparing (3–99) with (3–75) for the lossless line, we see that $\pm j\beta x$ for the lossless line is replaced by $\pm\gamma x$ for the lossy line. When γ is expanded, the positive traveling wave is multiplied by the factor $e^{-\alpha x}$, which results in a diminishing or attenuation of the positive traveling wave as x increases. Conversely, the negative traveling wave is multiplied by $e^{\alpha x}$, which results in an increase in the level of that wave as x increases. This does not mean that this wave is growing as it travels. Remember, it is considered to be traveling from right to left. As x is increased, an "earlier" point on the wave is reached, so its magnitude is greater. Thus, both waves are actually being attenuated.

Illustrations of the corresponding time-domain behavior of the two waves are shown in Fig. 3–9.

The characteristic impedance of the lossy line may be derived by determining the relationship between $\overline{V}^{+}$ and $\overline{I}^{+}$ along with the corresponding relationship between $\overline{V}^{-}$ and $\overline{I}^{-}$. This procedure follows the same pattern as for the lossless case as given in Eqs. (3–79) through (3–87). The results are

$$\frac{\overline{V}^{+}}{\overline{I}^{+}} = \sqrt{\frac{Z}{Y}} = Z_0 \tag{3–100}$$

and

$$\frac{\overline{V}^{-}}{\overline{I}^{-}} = -\sqrt{\frac{Z}{Y}} = -Z_0 \tag{3–101}$$

where Z_0 is the **complex characteristic impedance** as given by

$$Z_0 = \sqrt{\frac{Z}{Y}} \tag{3–102}$$

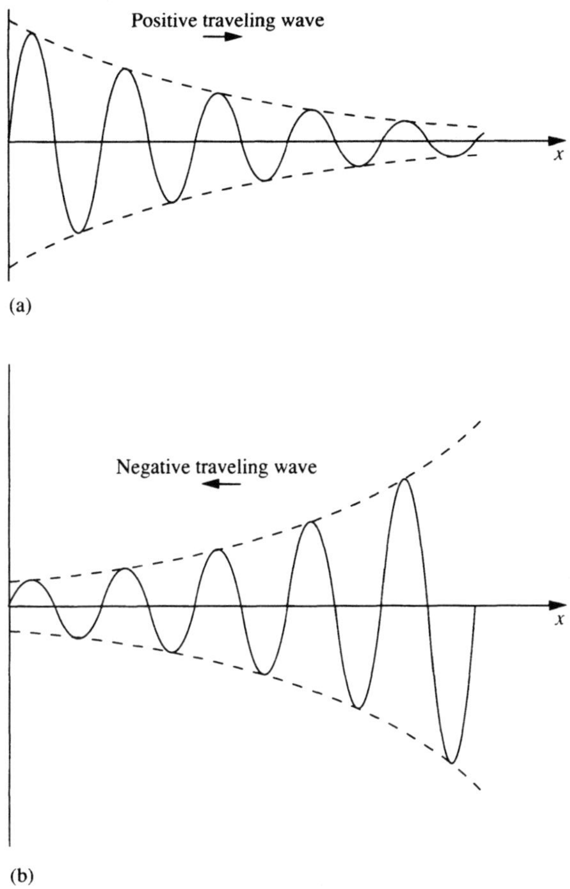

FIGURE 3–9 Positive and negative traveling waves on lossy line.

It should be noted that when $Z = j\omega L$ and $Y = j\omega C$, then

$$Z_0 = \sqrt{\frac{j\omega L}{j\omega C}} = \sqrt{\frac{L}{C}} = R_0 \tag{3–103}$$

and

$$\gamma = \sqrt{ZY} = \sqrt{(j\omega L)(j\omega C)} = j\omega\sqrt{LC} = j\beta \tag{3–104}$$

Thus, for the lossless case, Z_0 reduces to R_0, and γ reduces to $j\beta$, as expected. For the general case when $R \neq 0$ and/or $G \neq 0$, Z_0 will have both a real part and an imaginary part; i.e.,

$$Z_0 = R_0 + jX_0 \tag{3–105}$$

where R_0 is the real part of Z_0 and X_0 is the imaginary part.

3-6 ■ SQUARE ROOTS OF COMPLEX NUMBERS

There are some peculiarities involved in manipulating the complex quantities required to calculate the propagation constant and the characteristic impedance of a lossy line. A fundamental operation is that of taking the square root of a complex number. In general, there are two square roots of any complex number, but the usual root of interest is the **principal value.**

Before taking the square root of a complex number, one should express it in polar form. Let $Z = M\underline{/\theta}$ represent any complex number for which M is the magnitude and θ is the angle. The square root of the number is given by

$$\sqrt{Z} = \sqrt{M\underline{/\theta}} = \sqrt{M}\underline{/(\theta/2)} \tag{3–106}$$

Thus, to take the square root of a complex number, *we take the square root of the magnitude and divide the angle by two.* The result may be then converted back to rectangular form if necessary, such as when evaluating γ.

EXAMPLE 3-9

Determine the square root (principal value) of the following complex numbers: (a) 9, (b) $j16$, (c) $25\underline{/2.4}$, and (d) $3 - j4$. Express each answer in both polar and rectangular forms.

Solution

(a) The number 9 is simply a positive real number, which can be thought of as a complex number with an angle of 0 rad. Thus,

$$\sqrt{9} = \sqrt{9\underline{/0}} = 3\underline{/0} = 3 + j0 \tag{3–107}$$

(b) The quantity $j16$ can be written as $16\underline{/\pi/2}$. The principal value of the square root is

$$\sqrt{16\underline{/\pi/2}} = 4\underline{/\pi/4} = 2.828 + j2.828 \tag{3–108}$$

(c) The square root of $25\underline{/2.4}$ is

$$\sqrt{25\underline{/2.4}} = 5\underline{/1.2} = 1.812 + j4.660 \tag{3–109}$$

(d) The number $3 - j4$ must first be expressed in polar form as

$$3 - j4 = 5\underline{/-0.9273}$$

The square root is

$$\sqrt{3 - j4} = \sqrt{5\underline{/-0.9273}} = 2.236\underline{/-0.4636} = 2 - j1 \tag{3–110}$$

3–7 ■ NEPERS AND DECIBELS

In Section 3–5, the real part α of the propagation constant was identified as the attenuation constant, and its units were identified as **nepers/unit length.** The term *neper* (Np) is derived from the natural or Naperian logarithmic base $e = 2.718$. As will be shown shortly, neper measurement is directly proportional to the more commonly used decibel measurement.

Let us briefly review the concept of decibel measurement. If P_1 is the input power to a network and P_2 is the output power, the **decibel loss** or **attenuation** in **decibels (dB)** will be denoted as L_{dB}, defined as

$$L_{\text{dB}} = 10 \log_{10}\frac{P_1}{P_2} \tag{3–111}$$

Assume that P_1 and P_2 are each based on a constant reference resistance level R. In terms of the corresponding input and output rms voltage magnitudes, (3–111) may be expressed as

$$L_{\text{dB}} = 10 \log_{10}\frac{V_1^2/R}{V_2^2/R} = 20 \log_{10}\frac{V_1}{V_2} \tag{3–112}$$

A similar development could be made to express L_{dB} in terms of the corresponding rms currents I_1 and I_2, and this would result in

$$L_{\text{dB}} = 20 \log\frac{I_1}{I_2} \tag{3–113}$$

Thus, 10 is the constant for a power ratio, and 20 is the constant for either a voltage or a current ratio.

Next assume that the ratio V_2/V_1 is determined from (3–112). The result is

$$\frac{V_2}{V_1} = 10^{-L_{\text{dB}}/20} \tag{3–114}$$

However, for the exponential form of a wave traveling in the positive x-direction, the ratio of output voltage to input voltage for a distance d can be expressed as

$$\frac{V_2}{V_1} = e^{-L} = e^{-\alpha d} \tag{3–115}$$

where L is defined as the total loss in nepers as given by

$$L = \alpha d \tag{3–116}$$

Equating (3–114) and (3–115), we have

$$10^{-L_{\text{dB}}/20} = e^{-\alpha d} \tag{3–117}$$

This equation may be solved by taking the logarithm to the base 10 of both sides. We have

$$\frac{-L_{\text{dB}}}{20} = -\alpha d \log_{10} e \tag{3–118}$$

or

$$L_{\text{dB}} = 8.686\alpha d = 8.686L \tag{3–119}$$

In some cases, it is convenient to define an attenuation factor in decibels/unit length, which will be denoted as α_{dB}. This quantity is defined as

$$\alpha_{\text{dB}} = 8.686\alpha \tag{3–120}$$

The net loss in decibels is then simply given by

$$L_{\text{dB}} = \alpha_{\text{dB}} d \tag{3–121}$$

Thus, whether it is loss per unit length or total loss, the decibel loss is 8.686 times the neper loss.

EXAMPLE 3–10

The propagation constant of a lossy line is $\alpha = 0.02$ Np/m. For a line of length $d = 250$ m, determine (a) the attenuation factor in decibels/meter, (b) the total attenuation in nepers, (c) the total attenuation in decibels, and (d) the V_2/V_1 ratio using both nepers and decibels for a single wave.

Solution

(a) The attenuation factor in dB/m is

$$\alpha_{\text{dB}} = 8.686\alpha = 8.686 \times 0.02 = 0.1737 \text{ dB/m} \tag{3–122}$$

(b) The total attenuation in nepers is

$$L = \alpha d = 0.02 \times 250 = 5 \text{ Np} \tag{3–123}$$

(c) The total attenuation in decibels is

$$L_{\text{dB}} = \alpha_{\text{dB}} d = 0.1737 \times 250 = 43.43 \text{ dB} \tag{3–124}$$

(d) In terms of nepers, the V_2/V_1 ratio can be expressed as

$$\frac{V_2}{V_1} = e^{-L} = e^{-5} = 6.738 \times 10^{-3} \tag{3–125}$$

In terms of decibels, this ratio may be expressed as

$$\frac{V_2}{V_1} = 10^{-L_{\text{dB}}/20} = 10^{\;43.43/20} = 10^{\;2.1715} = 6.738 \times 10^{-3} \tag{3–126}$$

which obviously agrees with (3–125).

3–8 ■ COMPUTING PROPAGATION CONSTANT AND CHARACTERISTIC IMPEDANCE

The formulas and techniques for computing the complex propagation constant and the characteristic impedance have now been established. Assume that for a given line, the parameters L, C, R, and G are given. Based on a specified frequency, the procedure may be outlined as follows:

1. Compute $Z = R + j\omega L$ and $Y = G + j\omega C$, and express these quantities in polar forms.
2. Compute $\gamma = \sqrt{ZY}$ using the formula for taking the square root in polar form, and express the quantity in rectangular form. The real part is α and the imaginary part is β.
3. Compute $Z_0 = \sqrt{Z/Y}$. Depending on the application, Z_0 may be left in polar form or it may be converted to rectangular form.

The examples that follow will illustrate the procedure.

EXAMPLE 3–11

A transmission line is characterized by the following parameters: $L = 200$ nH/m, $R = 0.25\ \Omega$/m, $C = 100$ pF/m, and $G = 1\ \mu$S/m. At a frequency of 1 MHz, determine (a) Z; (b) Y; (c) γ, α, and β; (d) the attenuation in dB/m; (e) ν; and (f) Z_0.

Solution

(a) The series impedance per unit length is

$$Z = R + j\omega L = 0.25 + j[(2\pi \times 10^6) \times (200 \times 10^{-9})] \quad \textbf{(3–127a)}$$

$$= 0.25 + j1.2566\ \Omega/\text{m} \quad \textbf{(3–127b)}$$

$$= 1.2812\underline{/1.3744}\ \Omega/\text{m} \quad \textbf{(3–127c)}$$

(b) The shunt admittance per unit length is

$$Y = G + j\omega C = 1 \times 10^{-6} + j[(2\pi \times 10^6) \times (100 \times 10^{-12})] \quad \textbf{(3–128a)}$$

$$= 1 \times 10^{-6} + j6.2832 \times 10^{-4}\ \text{S/m} \quad \textbf{(3–128b)}$$

$$= 6.2832 \times 10^{-4}\underline{/1.5692}\ \text{S/m} \quad \textbf{(3–128c)}$$

(c) The complex propagation constant γ is

$$\gamma = \sqrt{ZY} = \sqrt{(1.2812\underline{/1.3744})(6.2832 \times 10^{-4}\underline{/1.5692})} \quad \textbf{(3–129a)}$$

$$= \sqrt{8.0500 \times 10^{-4}\underline{/2.9436}} = 2.8373 \times 10^{-2}\underline{/1.4718} \quad \textbf{(3–129b)}$$

$$= 2.804 \times 10^{-3} + j2.8234 \times 10^{-2} \quad \textbf{(3–129c)}$$

From the real and imaginary parts, we see that

$$\alpha = 2.804 \times 10^{-3}\ \text{Np/m} \quad \textbf{(3–130)}$$

and

$$\beta = 2.8234 \times 10^{-2} \text{ rad/m} \tag{3–131}$$

(d) The attenuation α_{dB} in dB/m is

$$\alpha_{dB} = 8.686 \times (2.804 \times 10^{-3}) = 0.02436 \text{ dB/m} \tag{3–132}$$

(e) The velocity of propagation is

$$\nu = \frac{\omega}{\beta} = \frac{2\pi \times 10^6}{2.8234 \times 10^{-2}} = 2.225 \times 10^8 \text{ m/s} \tag{3–133}$$

As an interesting observation, suppose we had attempted to calculate ν as $1/\sqrt{LC}$. The result would be

$$\nu = \frac{1}{\sqrt{(200 \times 10^{-9})(100 \times 10^{-12})}} = 2.236 \times 10^8 \text{ m/s} \tag{3–134}$$

which differs from the true velocity by less than 0.5%. Strictly speaking, the formula $1/\sqrt{LC}$ is true only for a lossless line. However, when $R << \omega L$ and $G << \omega C$, as is true in this example, the velocity is very close to $1/\sqrt{LC}$.

(f) The characteristic impedance is

$$Z_0 = \sqrt{\frac{Z}{Y}} = \sqrt{\frac{1.2812\underline{/1.3744}}{6.283 \times 10^{-4}\underline{/1.5692}}} = \sqrt{2039\underline{/-0.1948}} \tag{3–135a}$$

$$= 45.16\underline{/-0.0974}\ \Omega = 44.95 - j4.392\ \Omega \tag{3–135b}$$

As another interesting observation, suppose we had attempted to calculate the characteristic impedance as $\sqrt{L/C}$. The result would be

$$\sqrt{200 \times 10^{-9}/100 \times 10^{-12}} = 44.72\ \Omega$$

which is a real number, of course. However, its value is not significantly different from the real part of Z_0, a result again of the inequalities of (e). The next example, however, will tell a different story.

EXAMPLE 3–12

A telephone line has parameters given on a per-mile (/mi) basis as follows:

$$L = 4 \text{ mH/mi}$$
$$R = 30\ \Omega\text{/mi}$$
$$C = 10 \text{ nF/mi}$$
$$G = 1\ \mu\text{S/mi}$$

At a frequency of 1 kHz, determine (a) Z; (b) Y; (c) γ, α, and β; (d) the attenuation in dB/mi; (e) ν; and (f) Z_0.

Solution

(a) The series impedance per unit length is

$$Z = R + j\omega L = 30 + j[(2\pi \times 10^3) \times (4 \times 10^{-3})] \quad \textbf{(3–136a)}$$

$$= 30 + j25.13\ \Omega = 39.13\underline{/0.6973}\ \Omega \quad \textbf{(3–136b)}$$

(b) The shunt admittance per unit length is

$$Y = G + j\omega C = 1 \times 10^{-6} + j[(2\pi \times 10^3) \times (10 \times 10^{-9})] \quad \textbf{(3–137a)}$$

$$= 1 \times 10^{-6} + j62.83 \times 10^{-6}\ \text{S} \quad \textbf{(3–137b)}$$

$$= 62.84 \times 10^{-6}\underline{/1.5549}\ \text{S} \quad \textbf{(3–137c)}$$

Note that because the imaginary part of Y is so much larger than the real part, the magnitude is almost exactly equal to the imaginary part, and the angle 1.5549 rad corresponds to 89.1°, which is very close to 90°.

(c) The complex propagation constant γ is

$$\gamma = \sqrt{ZY} = \sqrt{(39.13\underline{/0.6973})(62.84 \times 10^{-6}\underline{/1.5549})} \quad \textbf{(3–138a)}$$

$$= \sqrt{2.4589 \times 10^{-3}\underline{/2.2522}} = 0.04959\underline{/1.1261} \quad \textbf{(3–138b)}$$

$$= 0.02133 + j0.04477 \quad \textbf{(3–138c)}$$

The attenuation constant is

$$\alpha = 0.02133\ \text{Np/mi} \quad \textbf{(3–139)}$$

The phase constant is

$$\beta = 0.04477\ \text{rad/mi} \quad \textbf{(3–140)}$$

(d) The attenuation in dB/mi is

$$\alpha_{\text{dB}} = 0.02133 \times 8.686 = 0.1853\ \text{dB/mi} \quad \textbf{(3–141)}$$

(e) The velocity of propagation ν is

$$\nu = \frac{\omega}{\beta} = \frac{2\pi \times 10^3}{0.04477} = 140{,}300\ \text{mi/s} \quad \textbf{(3–142)}$$

Now let us try $1/\sqrt{LC}$ to see what it predicts. We obtain

$$1/\sqrt{(4 \times 10^{-3})(10 \times 10^{-9})} = 158{,}114\ \text{mi/s}$$

which differs substantially from the true velocity.

(f) The characteristic impedance is

$$Z_0 = \sqrt{\frac{Z}{Y}} = \sqrt{\frac{39.13\underline{/0.6973}}{62.84 \times 10^{-6}\underline{/1.5549}}} = \sqrt{6.2269 \times 10^5\underline{/-0.8576}} \quad \textbf{(3–143a)}$$

$$= 789.1\underline{/-0.4288}\ \Omega = 717.7 - j328.1\ \Omega \quad \textbf{(3–143b)}$$

Again, let us see what $\sqrt{L/C}$ gives. The result is

$$\sqrt{(4 \times 10^{-3})/(10 \times 10^{-9})} = 632.5\ \Omega$$

which is quite different from the real part of Z_0.

EXAMPLE 3–13

An open-wire line has the following parameters at a frequency of 500 Hz:

series resistance = 24 Ω/mi
series reactance = 10 Ω/mi
shunt conductance = 3 μS/mi
shunt susceptance = 32 μS/mi

Determine (a) γ, α, and β; (b) the attenuation in dB/mi; (c) ν; and (d) Z_0.

Solution

This example differs from the previous two in that the series reactance and shunt susceptance are directly given in this case. Thus, Z and Y may be immediately expressed as

$$Z = 24 + j10\ \Omega/\text{mi} = 26\underline{/0.3948}\ \Omega/\text{mi} \tag{3–144}$$

$$Y = (3 \times 10^{-6}) + j(32 \times 10^{-6})\ \text{S/mi} = 32.14 \times 10^{-6}\underline{/1.477}\ \text{S/mi} \tag{3–145}$$

(a) The propagation constant is

$$\gamma = \sqrt{ZY} = \sqrt{(26\underline{/0.3948})(32.14 \times 10^{-6}\underline{/1.477})} \tag{3–146a}$$

$$= \sqrt{8.356 \times 10^{-4}\underline{/1.872}} = 0.02891\underline{/0.9360} \tag{3–146b}$$

$$= 0.01714 + j0.02328 \tag{3–146c}$$

The attenuation factor is

$$\alpha = 0.01714\ \text{Np/mi} \tag{3–147}$$

The phase factor is

$$\beta = 0.02328\ \text{rad/mi} \tag{3–148}$$

(b) The attenuation in dB/mi is

$$\alpha_{\text{dB}} = 8.686 \times 0.01714 = 0.1489\ \text{dB/mi} \tag{3–149}$$

(c) The velocity of propagation is

$$\nu = \frac{\omega}{\beta} = \frac{2\pi \times 500}{0.02328} = 134{,}900\ \text{mi/s} \tag{3–150}$$

(d) The characteristic impedance is

$$Z_0 = \frac{Z}{Y} = \sqrt{\frac{26\angle 0.3948}{32.14 \times 10^{-6}\angle 1.477}} = \sqrt{8.090 \times 10^5\angle -1.082} \quad \textbf{(3–151a)}$$

$$= 899.4\angle -0.5410\ \Omega = 771.0 - j463.2\ \Omega \quad \textbf{(3–151b)}$$

3–9 ■ LOSSY LINE MATCHED AT LOAD

In general, the computation of the voltage and/or current at any point on a lossy line is a very difficult process to accomplish by hand. It is necessary to evaluate both the forward and reverse waves from the boundary conditions, and this process is quite tedious. Fortunately, computer techniques are available to assist in the process.

There is one special case of a lossy line in which a complete solution may be achieved through a straightforward circuit model. Because of its practical importance and its educational value, this special case will be developed. The case of interest is that of a line *matched at the load end.*

This case is based on the assumption that the line is terminated in its characteristic impedance; i.e., $Z_2 = Z_0$. In this case, there is no reflected wave, and hence $\overline{V}^- = 0$ and $\overline{I}^- = 0$. The distance-varying voltage and current phasor may be expressed as

$$\overline{V}(x) = \overline{V}^+ e^{-\gamma x} = \overline{V}^+ e^{-\alpha x} e^{-j\beta x} \quad \textbf{(3–152)}$$

and

$$\overline{I}(x) = \overline{I}^+ e^{-\gamma x} = \overline{I}^+ e^{-\alpha x} e^{-j\beta x} \quad \textbf{(3–153)}$$

Assuming that α and β are known, it is necessary to determine only the fixed phasor $\overline{V}^+$ and $\overline{I}^+$ to obtain a complete solution for voltage and current everywhere on the line.

The key to the solution is based on the fact that when there is no reflected wave, i.e., when the line is terminated in its characteristic impedance, the line acts like an impedance of value Z_0 at all points. Thus, the line can be modeled by the simple circuit of Fig. 3–10 at the generator end, i.e., at $x = 0$.

From Fig. 3–10, the current $\overline{I}_1$ and voltage $\overline{V}_1$ at the front end of the line are readily determined from the circuit model to be

$$\overline{I}_1 = \frac{\overline{E}}{Z_0 + Z_1} \quad \textbf{(3–154)}$$

and

$$\overline{V}_1 = \frac{Z_0\overline{E}}{Z_0 + Z_1} \quad \textbf{(3–155)}$$

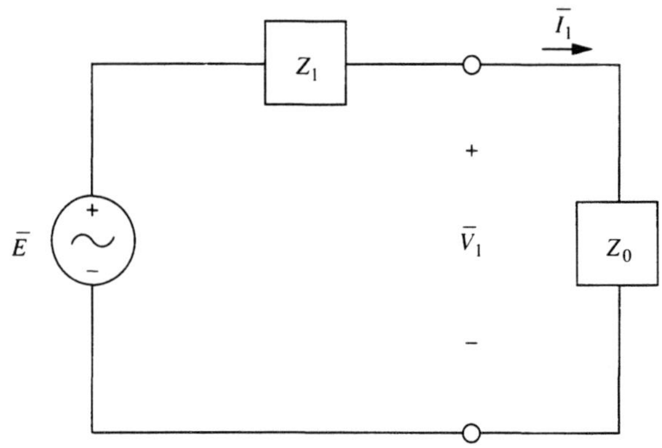

FIGURE 3–10 Circuit model of transmission line matched to a load referred to front end.

When the voltage and current from (3–152) and (3–153) are evaluated at $x = 0$ and are equated to the corresponding boundary conditions of (3–154) and (3–155), the results are

$$\bar{I}(0) = \bar{I}^+ = \bar{I}_1 \tag{3–156}$$

and

$$\bar{V}(0) = \bar{V}^+ = \bar{V}_1 \tag{3–157}$$

The meaning of the preceding results is that the current and voltage values at the front end of the line are equal to the fixed phasor values that appear in front of the $e^{-\gamma x}$ factors for the propagating wave. Thus, the current and voltage anywhere on the line can be expressed as

$$\bar{I}(x) = \bar{I}_1 e^{-\gamma x} = \bar{I}_1 e^{-\alpha x} e^{-j\beta x} \tag{3–158}$$

$$\bar{V}(x) = \bar{V}_1 e^{-\gamma x} = \bar{V}_1 e^{-\alpha x} e^{-j\beta x} \tag{3–159}$$

Let $\bar{I}_2$ and $\bar{V}_2$ represent the current and voltage at the load end of the line, i.e., at $x = d$. Substituting $x = d$ in (3–158) and (3–159), we have

$$\bar{I}_2 = \bar{I}(d) = \bar{I}_1 e^{-\gamma d} = \bar{I}_1 e^{-\alpha d} e^{-j\beta d} \tag{3–160}$$

and

$$\bar{V}_2 = \bar{V}(d) = \bar{V}_1 e^{-\gamma d} = \bar{V}_1 e^{-\alpha d} e^{-j\beta d} \tag{3–161}$$

It should be emphasized again that these simple relationships between input and output phasors are true only for a line matched at the load. In general, when there is a mismatch at the load, the relationship involves both a forward and a reverse wave and is much more complex.

For the remainder of this section, assume that

$$V_1 = |\bar{V}_1| \qquad I_1 = |\bar{I}_1| \qquad V_2 = |\bar{V}_2| \qquad I_2 = |\bar{I}_2| \tag{3–162}$$

where V_1, I_1, V_2, and I_2 represent the rms magnitude of the input and output voltages and currents. The magnitude ratios of output to input current and voltage may be determined from (3–160) and (3–161) to be

$$\frac{V_2}{V_1} = e^{-\alpha d} \quad \text{and} \quad \frac{I_2}{I_1} = e^{-\alpha d} \tag{3–163}$$

Two other quantities of interest are the input and output power levels to the line. Let $Z_0 = R_0 + jX_0$, where R_0 is the real part of the characteristic impedance and X_0 is the imaginary part. Assume for this purpose that the magnitude of the phasor source voltage $\overline{E}$ is expressed in rms volts so that the input and output voltage and current for the line will also be expressed in rms units. Let P_1 represent the input power to the line, which is

$$P_1 = I_1^2 R_0 \tag{3–164}$$

Let P_2 represent the output power to the load, which is

$$P_2 = I_2^2 R_0 \tag{3–165}$$

The presence of R_0 in both (3–164) and (3–165) is a result of the fact that the input impedance is equal to the characteristic impedance, which is also equal to the load impedance for the case at hand.

The ratio of input power to output power is given by

$$\frac{P_1}{P_2} = \frac{I_1^2}{I_2^2} = \frac{V_1^2}{V_2^2} = e^{2\alpha d} \tag{3–166}$$

The corresponding decibel loss L_{dB} is

$$L_{\text{dB}} = 20 \log_{10}\frac{I_1}{I_2} = 20 \log_{10}\frac{V_1}{V_2} = 8.686\alpha d \tag{3–167}$$

EXAMPLE 3–14

A lossy transmission line is characterized by the following parameters:

$$\alpha = 0.004 \text{ Np/m}$$
$$\beta = 0.003 \text{ rad/m}$$
$$Z_0 = 75 + j5\ \Omega = 75.17\underline{/0.06657}\ \Omega$$

A line of length 500 m is connected in the circuit of Fig. 3–11(a). Determine (a) the input current, (b) the input voltage, (c) the input power, (d) the load current, (e) the load voltage, (f) the load power, and (g) the loss in decibels.

Solution

The transmission line is obviously terminated in its characteristic impedance, and the procedure of this section may be applied. A circuit model referred to the input is shown in Fig. 3–11(b), in which the input impedance of the line

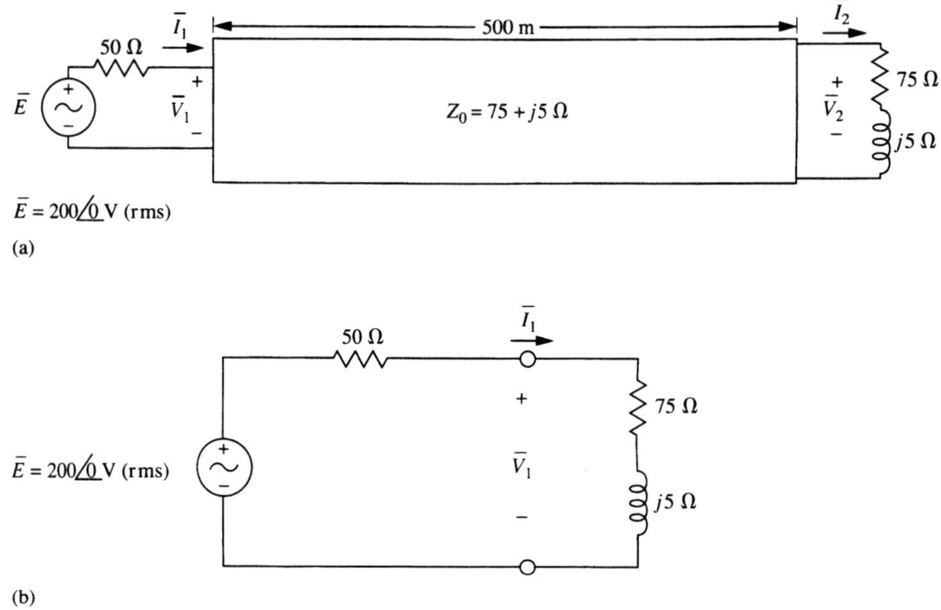

FIGURE 3–11 Circuit of Example 3–14 and equivalent circuit model at input.

is shown as $Z_{in} = 75 + j5$ Ω. Note that the line is not terminated in its characteristic impedance at the generator end.

(a) The input current is

$$\overline{I}_1 = \frac{\overline{E}}{Z_1 + Z_0} = \frac{200\underline{/0}}{50 + (75 + j5)} = \frac{200\underline{/0}}{125 + j5} \quad \textbf{(3–168a)}$$

$$= \frac{200\underline{/0}}{125.1\underline{/0.04}} = 1.599\underline{/-0.04}\ \text{A} \quad \textbf{(3–168b)}$$

(b) The input voltage is

$$\overline{V}_1 = Z_0\overline{I}_1 = (75.17\underline{/0.06657})(1.599\underline{/-0.04}) \quad \textbf{(3–169a)}$$

$$= 120.2\underline{/0.02657}\ \text{V} \quad \textbf{(3–169b)}$$

(c) The input power is

$$P_1 = I_1^2 R_0 = (1.599)^2 \times 75 = 191.8\ \text{W} \quad \textbf{(3–170)}$$

(d) The net attenuation for a line length of 500 m is $\alpha d = 0.004 \times 500 = 2$ Np, and the net phase shift is $\beta d = 0.003 \times 500 = 1.5$ rad. The load current is

$$\overline{I}_2 = \overline{I}_1 e^{-\gamma d} = \overline{I}_1 e^{-\alpha d} e^{-j\beta d} \quad \textbf{(3–171a)}$$

$$= (1.599\underline{/-0.04})(e^{-2})(1\underline{/-1.5}) = 0.2164\underline{/-1.54}\ \text{A} \quad \textbf{(3–171b)}$$

(e) The load voltage is

$$\overline{V}_2 = \overline{V}_1 e^{-\gamma d} = \overline{V}_1 e^{-\alpha d} e^{-j\beta d} \quad \textbf{(3–172a)}$$

$$= (120.2\underline{/0.02657})(e^{-2})(1\underline{/-1.5}) = 16.27\underline{/-1.473}\text{ V} \quad \textbf{(3–172b)}$$

Alternately, the load voltage could have been determined from the current in (d) by Ohm's law as follows:

$$\overline{V}_2 = Z_0 \overline{I}_2 = (75.17\underline{/0.06657})(0.2164\underline{/-1.54}) \quad \textbf{(3–173a)}$$

$$= 16.27\underline{/-1.473}\text{ V} \quad \textbf{(3–173b)}$$

which agrees with (3–172).

(f) The load power P_2 is

$$P_2 = I_2^2 R_0 = (0.2164)^2 \times 75 = 3.512\text{ W} \quad \textbf{(3–174)}$$

(g) The loss in decibels can be determined as

$$L_{\text{dB}} = 10 \log_{10} \frac{P_1}{P_2} = 10 \log_{10} \frac{191.8}{3.512} = 17.37\text{ dB} \quad \textbf{(3–175)}$$

Alternately, L_{dB} can be expressed as

$$L_{\text{dB}} = 8.686 \times \alpha d = 8.686 \times 2 = 17.37\text{ dB} \quad \textbf{(3–176)}$$

PROBLEMS

3–1. A traveling wave of current in milliamperes is given by

$$i = 8 \cos(2\pi \times 10^6 t - 0.025x)$$

with t in seconds and x in meters. Determine the following:

(a) direction of propagation
(b) peak value
(c) angular frequency
(d) phase constant
(e) cyclic frequency
(f) period
(g) wavelength
(h) velocity of propagation

3–2. A traveling wave of voltage in volts in given by

$$v = 15 \cos(10^8 t + 0.35x)$$

with t in seconds and x in meters. Determine the following:

(a) direction of propagation
(b) peak value
(c) angular frequency

(d) phase constant
(e) cyclic frequency
(f) period
(g) wavelength
(h) velocity of propagation

3–3. A sinusoidal current with a peak value of 2 A and a frequency of 50 MHz is traveling in the positive x-direction with a velocity of 2×10^8 m/s. Determine the following:
(a) period
(b) angular frequency
(c) phase constant
(d) wavelength
(e) an equation for the current

3–4. A sinusoidal voltage with a peak value of 25 V and a radian frequency of 20 Mrad/s is traveling in the negative x-direction with a velocity of 3×10^8 m/s. Determine the following:
(a) cyclic frequency
(b) period
(c) phase constant
(d) wavelength
(e) an equation for the voltage

3–5. Consider the current traveling wave of Problem 3–1. Determine the following:
(a) a fixed phasor representation in peak units as either $\overline{I}^+$ or $\overline{I}^-$ (You decide which label is appropriate.)
(b) the corresponding distance-varying phasor $\overline{I}(x)$ in peak units
(c) the value of the distance-varying phasor at $x = 100$ m

3–6. Consider the voltage traveling wave of Problem 3–2. Determine the following:
(a) a fixed phasor representation in peak units as either $\overline{V}^+$ or $\overline{V}^-$ (You decide which label is appropriate.)
(b) the corresponding distance-varying phasor $\overline{V}(x)$ in peak units
(c) the value of the distance-varying phasor at $x = 4$ m

3–7. Repeat the analysis of Problem 3–5 if the current of Problem 3–1 has a fixed phase shift such that it is described by

$$i = 8\cos(2\pi \times 10^6 t - 0.025x + 1.5)$$

3–8. Repeat the analysis of Problem 3–6 if the voltage of Problem 3–2 has a fixed phase shift such that it is described by

$$v = 15\cos(10^8 t + 0.35x - \pi/3)$$

3–9. Redefine the fixed phasor of Problem 3–5 so that the phasor magnitude is expressed in rms units, and determine the average power dissipated in a 50-Ω resistance.

3–10. Redefine the fixed phasor of Problem 3–6 so that the phasor magnitude is expressed in rms units, and determine the average power dissipated in a 75-Ω resistance.

3–11. Redefine the fixed phasor of Problem 3–7 so that the phasor magnitude is expressed in rms units. Would the power dissipated in a 50-Ω resistance be the same as in Problem 3–9?

3–12. Redefine the fixed phasor of Problem 3–8 so that the phasor magnitude is expressed in rms units. Would the power dissipated in a 75-Ω resistance be the same as in Problem 3–10?

3–13. Under steady-state ac conditions, the forward current wave on a certain lossless 50-Ω line is $\bar{I}^+ = 2\underline{/0}$ A. Determine the voltage forward wave.

3–14. Under steady-state ac conditions, the forward voltage wave in a 300-Ω lossless line is $\bar{V}^+ = 15\underline{/3}$. Determine the current forward wave.

3–15. Under steady-state ac conditions, the reverse voltage wave on a lossless 50-Ω line is $\bar{V}^- = 200\underline{/0}$ V. Determine the reverse current wave.

3–16. Under steady-state ac conditions, the reverse current wave on a lossless 75-Ω line is $\bar{I}^- = 0.5\underline{/2}$. Determine the reverse voltage wave.

3–17. A table of specifications for one version of RG–8/U 50-Ω coaxial cable indicates that the attenuation per 100 ft at 50 MHz is 1.2 dB. At this frequency, determine the following:
(a) attenuation factor in decibels per foot
(b) attenuation factor in nepers per foot
For a length of 300 ft, determine the following:
(c) total attenuation in decibels
(d) total attenuation in nepers
(e) V_2/V_1 ratio using both decibels and nepers for a single wave

3–18. A transmission line has an attenuation of 0.05 dB/m. Determine the following:
(a) attenuation factor in nepers/m
For a length of 400 m, determine the following:
(b) total attenuation in decibels
(c) total attenuation in nepers
(d) V_2/V_1 ratio using both decibels and nepers for a single wave

3–19. A single-frequency wave is propagating in one direction on a transmission line of length 200 m. With an input rms voltage of 50 V, the output rms voltage is measured as 20 V. Determine the following:
(a) total attenuation in decibels
(b) total attenuation in nepers
(c) attenuation factor in decibels/meter
(d) attenuation factor in nepers/meter

3–20. A single-frequency wave is propagating in one direction on a transmission line of length 400 m. The input power to the line is 40 W, and the output power is 12 W. Determine the following:
(a) total attenuation in decibels
(b) total attenuation in nepers
(c) attenuation constant in decibels/meters
(d) attenuation factor in nepers/meter

3–21. A transmission line has the following parameters at 50 MHz: $L = 1.2$ μH/m, $R = 15$ Ω/m, $C = 10$ pF/m, and $G = 4$ μS/m. Determine the following:

(a) Z
(b) Y
(c) γ, α, and β
(d) attenuation in dB/m
(e) ν
(f) Z_0

3–22. A lossy audio-frequency line has the following parameters at 2 kHz: $L = 0.1\ \mu\text{H/ft}$, $R = 0.2\ \Omega/\text{ft}$, $C = 2\ \text{pF/ft}$, and G is negligible. Determine the following:
(a) Z
(b) Y
(c) γ, α, and β
(d) attenuation in dB/ft
(e) ν
(f) Z_0

3–23. A coaxial cable has the following parameters at a frequency of 1 MHz:

series resistance = 0.3 Ω/m
series reactance = 2 Ω/m
shunt conductance = 0.5 μS/m
shunt susceptance = 0.6 mS/m

Determine the following:
(a) γ, α, and β
(b) attenuation in dB/m
(c) ν
(d) Z_0

3–24. For the coaxial cable of Problem 3–23, repeat the analysis at 100 MHz if the series resistance increases to 1 Ω/m, but the shunt conductance remains essentially the same. (*Note:* You must apply basic ac circuit theory to determine the new values for the reactance and susceptance.)

3–25. For the circuit of Fig. P3–25, determine the following:
(a) input current $\bar{I}_1$
(b) input voltage $\bar{V}_1$
(c) input power P_1
(d) load current $\bar{I}_2$

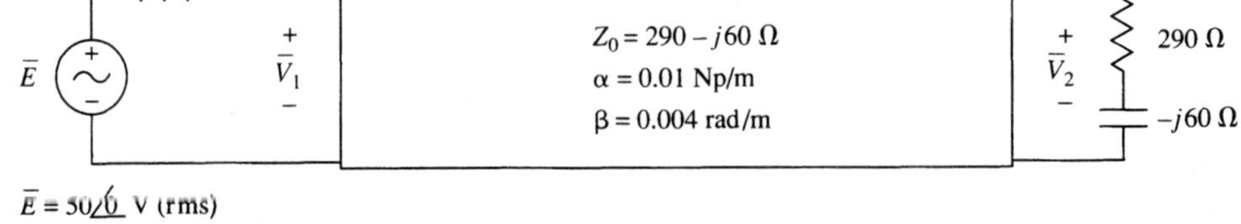

FIGURE P3–25

(e) load voltage $\overline{V}_2$
(f) load power P_2
(g) line loss in dB

3–26. For the circuit of Fig. P3–26, determine the following:
(a) input current $\overline{I}_1$
(b) input voltage $\overline{V}_1$
(c) input power P_1
(d) load current $\overline{I}_2$
(e) load voltage $\overline{V}_2$
(f) load power P_2
(g) line loss in nepers

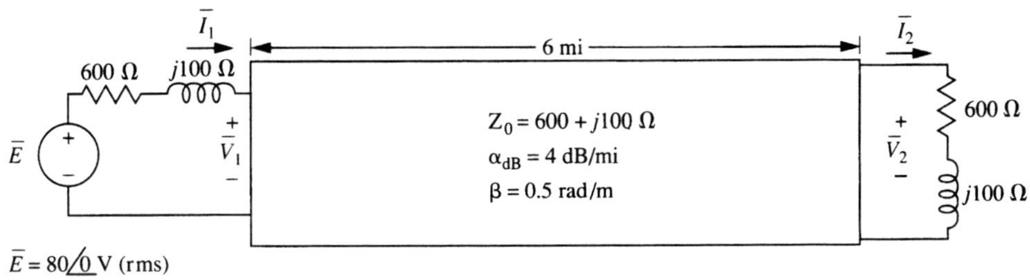

FIGURE P3–26

4

Radio-Frequency Lossless Lines

OBJECTIVES

After completing this chapter, the reader should be able to:

1. Discuss the assumptions made for a **radio-frequency (RF)** lossless line.
2. State the forms of the ac phasor voltage and current for a lossless line.
3. Define and calculate the reflection coefficient for a lossless line.
4. Define and calculate the **voltage standing-wave ratio (VSWR)** for a lossless line.
5. Discuss the significance of the reflection coefficient and the VSWR.
6. Calculate the input impedance of an arbitrary length of lossless line terminated in an arbitrary impedance.
7. Calculate the power delivered to the load for a lossless transmission line under ac steady-state conditions.
8. Discuss the significance and properties of a lossless line matched at the load and a similar line matched at both ends.
9. Discuss the properties of both a short-circuited lossless line and an open-circuited line, and sketch the forms for the voltage, current, and reactance for each.
10. Discuss the properties of both a lossless **half-wavelength line** and a lossless **quarter-wavelength line.**

4–1 ■ INTRODUCTION

Some very important applications of transmission lines are in coupling between transmitters and antennas and in coupling between receiving antennas and receivers. Such lines are loosely referred to as **radio-frequency (RF)** lines. The frequency ranges of such applications vary from well below 1 MHz to 1 GHz and above. This range includes the commercial amplitude-modulation (AM) band, the commercial frequency-modulation (FM) band, various commercial television (TV) bands, and many other communication systems.

The lengths of lines used in RF applications range from a few tenths of a wavelength in the lower frequencies to hundreds of wavelengths at the upper end of the spectrum. While no real transmission lines are truly lossless, the assumption that the losses are negligible leads to simplified methods of analysis. Such assumptions are generally valid for moderate lengths of connecting lines. As in many engineering analysis and design techniques, the assumption of lossless lines establishes certain bounds for operating conditions. Modifications may then be made to account for the actual presence of losses.

4–2 ■ REFLECTION COEFFICIENT

The concept of the reflection coefficient was introduced in Chapter 2 without proof. Although important in all aspects of transmission line theory, the reflection coefficient plays a particular and important role in RF lines. A development of the definition will be made in this section based on an assumed-lossless line under steady-state ac conditions.

Up to this point in the text, we have considered $x = 0$ as the generator end of the line. For the purpose of the reflection coefficient development, however, it is more convenient to redefine $x = 0$ as the load end of the line as shown in Fig. 4–1. The direction of positive x is not changed; i.e., it is still toward the right on the figure. Thus, the generator is now located at $x = -d$ for a line of length d, and $x > 0$ corresponds to the space to the right of the line. This seemingly odd convention avoids certain complications when dealing with the angle of the reflection coefficient. Basically, it is easier to refer all reflection coefficient calculations and measurements to the load, at which point $x = 0$.

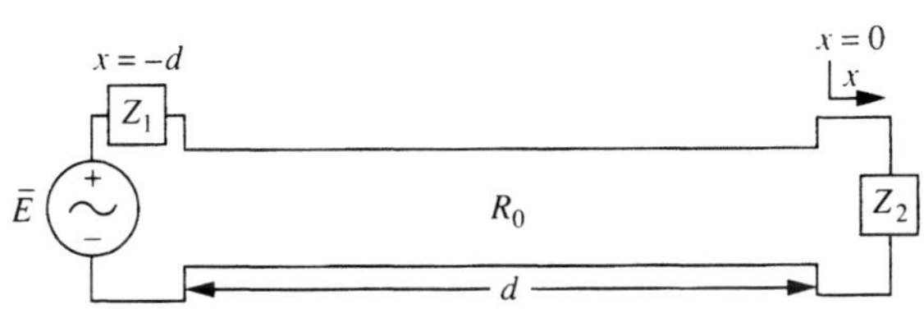

FIGURE 4–1 Coordinate system assumed for development of reflection coefficient.

Assume a lossless line with $Z_0 = R_0$ and $\gamma = j\beta$. Based on the work of Chapter 3, the voltage and current anywhere on the line may be expressed as

$$\overline{V}(x) = \overline{V}^+ e^{-j\beta x} + \overline{V}^- e^{j\beta x} \tag{4–1}$$

and

$$\overline{I}(x) = \overline{I}^+ e^{-j\beta x} + \overline{I}^- e^{j\beta x} \tag{4–2}$$

At the load, $x = 0$, and the voltage and current are denoted as $\overline{V}_2$ and $\overline{I}_2$, respectively. Substitution of these quantities in (4–1) and (4–2) yields

$$\overline{V}(0) = \overline{V}_2 = \overline{V}^+ + \overline{V}^- \tag{4–3}$$

and

$$\overline{I}(0) = \overline{I}_2 = \overline{I}^+ + \overline{I}^- \tag{4–4}$$

The two components of $\overline{I}_2$ may be expressed in terms of the two components of $\overline{V}_2$ as follows:

$$\overline{I}^+ = \frac{\overline{V}^+}{R_0} \tag{4–5}$$

and

$$\overline{I}^- = \frac{-\overline{V}^-}{R_0} \tag{4–6}$$

as proven in Chapter 3. Substituting (4–5) and (4–6) in (4–4) yields

$$\overline{I}_2 = \frac{\overline{V}^+}{R_0} - \frac{\overline{V}^-}{R_0} \tag{4–7}$$

It is further known that the load current $\overline{I}_2$ must be related to the load voltage $\overline{V}_2$ by

$$\overline{I}_2 = \frac{\overline{V}_2}{Z_2} \tag{4–8}$$

where Z_2 is the complex load impedance.

When (4–8) is substituted in (4–7), we have

$$\frac{\overline{V}_2}{Z_2} = \frac{\overline{V}^+}{R_0} - \frac{\overline{V}^-}{R_0} \tag{4–9}$$

The quantity $\overline{V}_2$ may now be eliminated between (4–3) and (4–9) by solving for $\overline{V}_2$ from either equation and substituting in the other one. After several manipulations, a relationship between $\overline{V}^-$ and $\overline{V}^+$ is obtained as follows:

$$\frac{\overline{V}^-}{\overline{V}^+} = \frac{Z_2 - R_0}{Z_2 + R_0} \tag{4–10}$$

This ratio leads to the definition of the **complex voltage reflection coefficient at the load,** which will be denoted as $\overline{\Gamma}_2$. It is given by

$$\overline{\Gamma}_2 = \frac{\overline{V}^-}{\overline{V}^+} = \frac{Z_2 - R_0}{Z_2 + R_0} \tag{4–11}$$

Note that while R_0 is assumed to be real, $\overline{\Gamma}_2$ will be complex as long as Z_2 is complex, i.e., as long as Z_2 has both resistive and reactive components.

A similar development leads to the **complex current reflection coefficient at the load,** given by

$$\frac{\overline{I}^-}{\overline{I}^+} = -\overline{\Gamma}_2 = \frac{R_0 - Z_2}{R_0 + Z_2} \tag{4–12}$$

As in Chapter 2, we will continue to assume that unless otherwise indicated, a given reflection coefficient is stated as a voltage ratio.

The reflection coefficient at any distance from the load may be expressed in terms of the coefficient at the load. Let $\overline{\Gamma}(d)$ represent the voltage reflection coefficient at a distance d back from the load, i.e., at $x = -d$. This quantity is

$$\overline{\Gamma}(d) = \frac{\text{reverse voltage wave at } x = -d}{\text{forward voltage wave at } x = -d} \tag{4–13a}$$

$$= \left[\frac{\overline{V}^- e^{j\beta x}}{\overline{V}^+ e^{-j\beta x}}\right]_{x=-d} \tag{4–13b}$$

$$= \frac{\overline{V}^- e^{-j\beta d}}{\overline{V}^+ e^{j\beta d}} = \frac{\overline{V}^-}{\overline{V}^+} e^{-j2\beta d} \tag{4–13c}$$

This result may be expressed as

$$\overline{\Gamma}(d) = \overline{\Gamma}_2 e^{-j2\beta d} = \overline{\Gamma}_2 \underline{/-2\beta d} \tag{4–14}$$

Assume that $\overline{\Gamma}_2$ may be expressed in polar form as

$$\overline{\Gamma}_2 = \Gamma_2 e^{j\theta_2} = \Gamma_2 \underline{/\theta_2} \tag{4–15}$$

where $\Gamma_2 = |\overline{\Gamma}_2|$ is the magnitude of the voltage reflection coefficient at the load and θ_2 is its angle. Substituting (4–15) in (4–14) yields

$$\overline{\Gamma}(d) = (\Gamma_2 e^{j\theta_2})(e^{-j2\beta d}) = \Gamma_2 \underline{/\theta_2 - 2\beta d} \tag{4–16}$$

The magnitude of the reflection coefficient $\Gamma(d)$ at any distance d is determined as follows:

$$\Gamma(d) = |\overline{\Gamma}(d)| = \Gamma_2 \tag{4–17}$$

Equation (4–17) is important and needs to be emphasized. *For a lossless line, the magnitude of the reflection coefficient at any point is the same.* However, the angle changes in accordance with (4–16).

Although the reflection coefficient at the load plays the most important role in the analysis of RF lines, the reflection coefficient at the generator is important in some applications. Let $\overline{\Gamma}_1$ represent the complex voltage reflection coefficient at the generator, given by

$$\overline{\Gamma}_1 = \frac{Z_1 - R_0}{Z_1 + R_0} \tag{4–18}$$

where Z_1 is the complex generator impedance.

In some measurement situations, the characteristic impedance is known, and the reflection coefficient is measured, but the load impedance is unknown. From (4–11), the load impedance may be expressed in terms of the load reflection coefficient and the characteristic impedance as

$$Z_2 = R_0 \frac{(1 + \overline{\Gamma}_2)}{(1 - \overline{\Gamma}_2)} \tag{4–19}$$

EXAMPLE 4–1

A 50-Ω lossless line is terminated in $Z_2 = R_2 = 150\ \Omega$. Determine (a) the reflection coefficient at the load and (b) the reflection coefficient at distances $\lambda/4$ and $\lambda/2$ back from the load.

Solution

(a) In this example, the load impedance is real. The reflection coefficient at the load is

$$\overline{\Gamma}_2 = \frac{Z_2 - R_0}{Z_2 + R_0} = \frac{150 - 50}{150 + 50} = 0.5 = 0.5\underline{/0} \tag{4–20}$$

Thus, the angle of the reflection coefficient is zero in this case. When the load impedance is real and the line is lossless, the angle of the reflection coefficient at the load is either 0 ($R_2 > R_0$) or π rad ($R_2 < R_0$).

(b) Before determining the reflection coefficient at any other point on the line, we should first note that the net angle $2\beta d$ may be expressed as

$$2\beta d = 2\left(\frac{2\pi d}{\lambda}\right) = \frac{4\pi d}{\lambda} \tag{4–21}$$

The reflection coefficient at a distance d from the load is determined from (4–16) as

$$\overline{\Gamma}(d) = \Gamma_2\underline{/\theta_2 - 2\beta d} = 0.5\underline{/-(4\pi d)/\lambda} \tag{4–22}$$

It should be noted that although both the load and the characteristic impedances are real, the reflection coefficient is complex and assumes all possible angles as d is varied.

At a distance $d = \lambda/4$, the reflection coefficient is

$$\overline{\Gamma}\left(\frac{\lambda}{4}\right) = 0.5\underline{/(-4\pi/\lambda) \times (\lambda/4)} = 0.5\underline{/-\pi} = -0.5 \tag{4–23}$$

At a distance $d = \lambda/2$, the reflection coefficient is

$$\bar{\Gamma}\left(\frac{\lambda}{2}\right) = 0.5\underline{/-(4\pi/\lambda) \times (\lambda/2)} = 0.5\underline{/-2\pi} = 0.5 \tag{4–24}$$

EXAMPLE 4–2

A 75-Ω lossless line is terminated in $Z_2 = 25 + j50\ \Omega$. Determine (a) the reflection coefficient at the load and (b) the reflection coefficient at a distance $\lambda/4$ back from the load.

Solution

(a) The reflection coefficient at the load is

$$\bar{\Gamma}_2 = \frac{Z_2 - R_0}{Z_2 + R_0} = \frac{(25 + j50) - 75}{(25 + j50) + 75} = \frac{-50 + j50}{100 + j50} \tag{4–25a}$$

$$= \frac{70.711\underline{/2.3562}}{111.80\underline{/0.4636}} = 0.6325\underline{/1.8926} \tag{4–25b}$$

The magnitude and angle are $\Gamma_2 = 0.6325$ and $\theta_2 = 1.8926$ rad, respectively.

(b) The reflection coefficient at a distance d from the load is determined from (4–16) to be

$$\bar{\Gamma}(d) = \Gamma_2\underline{/\theta_2 - 2\beta d} \tag{4–26a}$$

$$= 0.6325\underline{/[1.8926 - (4\pi d/\lambda)]} \tag{4–26b}$$

where the relationship of (4–21) was employed. Substituting $d = \lambda/4$ in (4–26b) yields

$$\bar{\Gamma}\left(\frac{\lambda}{4}\right) = 0.6325\underline{/[1.8926 - (4\pi/\lambda) \times (\lambda/4)]} \tag{4–27a}$$

$$= 0.6325\underline{/1.8926 - \pi} = 0.6325\underline{/-1.2490} \tag{4–27b}$$

EXAMPLE 4–3

The reflection coefficient of a load connected to a 50-Ω lossless line is known to be $\bar{\Gamma}_2 = 0.3\underline{/\pi/8}$. Determine Z_2.

Solution

The relationship of (4–19) may be employed. We have

$$Z_2 = R_0\frac{(1 + \bar{\Gamma}_2)}{(1 - \bar{\Gamma}_2)} = 50\frac{(1 + 0.3\underline{/\pi/8})}{(1 - 0.3\underline{/\pi/8})} \tag{4–28a}$$

$$= 50\frac{[1 + (0.2772 + j0.1148)]}{[1 - (0.2772 + j0.1148)]} \tag{4–28b}$$

$$= 50\frac{(1.2772 + j0.1148)}{(0.7228 - j0.1148)} \quad \text{(4–28c)}$$

$$= 50\frac{(1.2823\underline{/0.08964})}{(0.7319\underline{/-0.1575})} \quad \text{(4–28d)}$$

$$= 87.60\underline{/0.2471} = 84.94 + j21.43\ \Omega \quad \text{(4–28e)}$$

4–3 ■ STANDING-WAVE RATIO

Consider again Eqs. (4–1) and (4–2) for the distance-varying voltage and current, but factor out the first term in each expression as follows:

$$\overline{V}(x) = \overline{V}^+ e^{-j\beta x}\left(1 + \frac{\overline{V}^-}{\overline{V}^+} e^{j2\beta x}\right) \quad \text{(4–29)}$$

$$\overline{I}(x) = \overline{I}^+ e^{-j\beta x}\left(1 + \frac{\overline{I}^-}{\overline{I}^+} e^{j2\beta x}\right) \quad \text{(4–30)}$$

From the definition of *reflection coefficient,* we know that

$$\frac{\overline{V}^-}{\overline{V}^+} = \overline{\Gamma}_2 = \Gamma_2\underline{/\theta_2} \quad \text{(4–31)}$$

and

$$\frac{\overline{I}^-}{\overline{I}^+} = -\overline{\Gamma}_2 = \Gamma_2\underline{/\theta_2 + \pi} \quad \text{(4–32)}$$

Substituting (4–31) and (4–32) in (4–29) and (4–30), we obtain alternate expressions for the distance-varying voltage and current as follows:

$$\overline{V}(x) = \overline{V}^+ e^{-j\beta x}(1 + \overline{\Gamma}_2 e^{j2\beta x}) \quad \text{(4–33)}$$

$$\overline{I}(x) = \overline{I}^+ e^{-j\beta x}(1 - \overline{\Gamma}_2 e^{j2\beta x}) \quad \text{(4–34)}$$

We will now investigate the behavior of the magnitude of $\overline{V}(x)$ and $\overline{I}(x)$ as x varies. Let $V^+ = |\overline{V}^+|$ represent the magnitude of the fixed voltage phasor, and let $I^+ = |\overline{I}^+|$ represent the magnitude of the fixed current phasor. A basic operation of complex variable theory states that the magnitude of a product is the product of the magnitudes. Thus, the magnitudes $V(x) = |\overline{V}(x)|$ and $I(x) = |\overline{I}(x)|$ of the voltage and current can be expressed as

$$V(x) = V^+|(1 + \overline{\Gamma}_2 e^{j2\beta x})| \quad \text{(4–35a)}$$

$$= V^+|(1 + \Gamma_2\underline{/\theta_2 + 2\beta x})| \quad \text{(4–35b)}$$

and

$$I(x) = I^+|(1 - \overline{\Gamma}_2 e^{j2\beta x})| \quad \text{(4–36a)}$$

$$= I^+|(1 - \Gamma_2\underline{/\theta_2 + 2\beta x})| \quad \textbf{(4–36b)}$$

$$= I^+|(1 + \Gamma_2\underline{/\theta_2 + 2\beta x + \pi})| \quad \textbf{(4–36c)}$$

where the vertical bars | | indicate the magnitude of the expression in brackets and $|e^{-j\beta x}| = |1\underline{/-\beta x}| = 1$.

First let us inspect (4–35b). The quantities V^+, Γ_2, and θ_2 are all constants for a given system. However, as x varies, the quantity $2\beta x$ varies, which results in a variation of the net angle of the last term. This process is illustrated in Fig. 4–2. As $2\beta x$ varies over a range of 2π rad, the quantity $\overline{\Gamma}_2 e^{j2\beta x}$ adds vectorially to 1 at all possible angles. The maximum possible value of the voltage magnitude V_{max} occurs when $(\theta_2 + 2\beta x)$ is an integer multiple of 2π rad, and its value is

$$V_{max} = V^+(1 + \Gamma_2) \quad \textbf{(4–37)}$$

Similarly, the minimum possible value of the voltage magnitude V_{min} occurs when $(\theta_2 + 2\beta x)$ is an odd integer multiple of π rad, and its value is

$$V_{min} = V^+(1 - \Gamma_2) \quad \textbf{(4–38)}$$

A similar pattern exists for the current variation but with one important exception: Because of the minus sign in (4–36a), or equivalently the additional angle of π rad in (4–36c), *a point corresponding to a voltage maximum corre-*

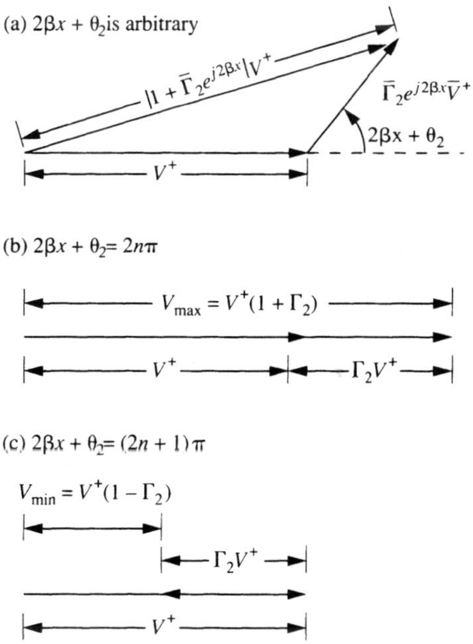

FIGURE 4–2 Phasor addition of the forward and reverse waves and maximum and minimum voltages.

sponds to a current minimum, and vice versa. The values of the maximum and minimum current magnitudes I_{max} and I_{min} are

$$I_{max} = I^{+}(1 + \Gamma_2) \tag{4–39}$$

and

$$I_{min} = I^{+}(1 - \Gamma_2) \tag{4–40}$$

However, please note again that at any point at which $V = V_{max}$, $I = I_{min}$, and at any point at which $V = V_{min}$, $I = I_{max}$.

The distance between successive voltage or current maxima or minima corresponds to a change of angle of 2π rad. Thus, the change in x (i.e., Δx) corresponds to

$$2\beta\Delta x = 2\pi$$

$$2\left(\frac{2\pi}{\lambda}\right)\Delta x = 2\pi \quad \text{or} \quad \Delta x = \frac{\lambda}{2} \tag{4–41}$$

Thus, *the voltage and current maximum and minimum magnitude values repeat at intervals of* $\lambda/2$.

Typical plots of the variation of the voltage and current magnitudes along a line are shown in Fig. 4–3. These plots are referred to as **standing-wave patterns.** Understand that at any point, the voltage or current follows a sinusoidal pattern in time as a result of the $e^{j\omega t}$ factor, which has been omitted in the phasor representative. The value of voltage or current at a given point represents the peak value of the sinusoid at that point, which varies with the location.

The variations of voltage and current in Fig. 4–3 appear to have shapes similar to that of a sinusoidal function plus a constant. However, the form is somewhat more complex as evidenced by the wider humps near the maxima as compared with those near the minima.

Assume that the peak (or rms) value of the sinusoidal voltage can be measured as a function of location along the line. Such a measurement can be used to determine the relative degree of mismatch at the load. This leads to the concept of the **standing-wave ratio,** which will be defined in terms of voltage because it is usually easier to measure. Let S represent the **voltage standing-wave ratio,** which will be referred to often in later discussions as **VSWR.** It is defined as

$$S = \frac{V_{max}}{V_{min}} \tag{4–42}$$

Substitution of (4–37) and (4–38) in (4–42) leads to

$$S = \frac{1 + \Gamma_2}{1 - \Gamma_2} \tag{4–43}$$

Thus, the VSWR is a function of the reflection coefficient at the load.

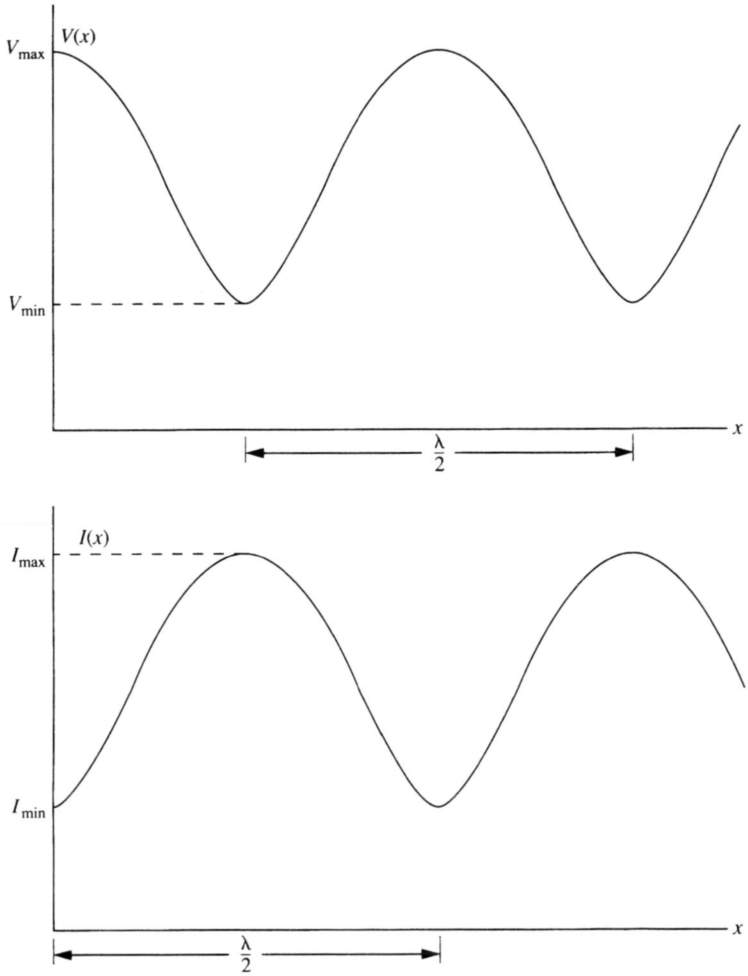

FIGURE 4–3 Typical VSWR plots of voltage and current.

The magnitude of the reflection coefficient can be determined from the VSWR in (4–43) as

$$\Gamma_2 = \frac{S-1}{S+1} \tag{4–44}$$

Note, however, that this relationship gives *only the magnitude of the reflection coefficient,* and the angle must be determined from a separate process.

The minimum value of the VSWR is unity, and there is no finite maximum. Thus, $1 \leq S < \infty$. The value $S = 1$ corresponds to a matched load, which means that $\Gamma_2 = 0$. Thus, *when a lossless line is terminated in a matched load, the peak voltage is the same everywhere on the line.* (Of course, the same also applies to the rms voltage.) As the degree of a given mismatch increases, there is a greater variation in the range of peak voltages on the line. The following corresponding values between Γ_2 and S should be noted for intuitive reasons:

Γ_2	S	Implication
0	1	Line perfectly matched at load
1	∞	Maximum possible mismatch (short or open at load); all power reflected

Resistive Load

When the load is purely resistive, the expression for the VSWR reduces to a very simple form. Let $Z_2 = R_2$ be a purely resistive load, and let $Z_0 = R_0$ be the characteristic impedance of a lossless line. It can be readily shown that the VSWR is one or the other of the following two expressions:

1. When $R_2 > R_0$,

$$S = \frac{R_2}{R_0} \tag{4–45}$$

2. When $R_2 < R_0$,

$$S = \frac{R_0}{R_2} \tag{4–46}$$

Rather than trying to remember two equations, just remember that S is the ratio greater than unity.

EXAMPLE 4–4

Determine the VSWR on the line of Example 4–1.

Solution

The value of the load reflection coefficient was determined in Example 4–1 as $|\overline{\Gamma}_2| = 0.5 = \Gamma_2$. The VSWR is then

$$S = \frac{1 + \Gamma_2}{1 - \Gamma_2} = \frac{1 + 0.5}{1 - 0.5} = 3 \tag{4–47}$$

Alternately, since the load impedance is real, S may be determined simply by forming either the ratio R_2/R_0 or R_0/R_2, depending on which is greater than unity. In this case, the first ratio is the correct one, and

$$S = \frac{R_2}{R_0} = \frac{150}{50} = 3 \tag{4–48}$$

EXAMPLE 4-5

Determine the VSWR on the line of Example 4–2.

Solution

The value of the load reflection coefficient magnitude was determined in Example 4–2 to be $\Gamma_2 = 0.6325$. The VSWR is then

$$S = \frac{1 + \Gamma_2}{1 - \Gamma_2} = \frac{1 + 0.6325}{1 - 0.6325} = 4.442 \tag{4–49}$$

EXAMPLE 4-6

The VSWR on a 50-Ω lossless line is measured as $S = 4$. Determine the magnitude of the reflection coefficient.

Solution

From (4–44), Γ_2 is determined as

$$\Gamma_2 = \frac{S - 1}{S + 1} = \frac{4 - 1}{4 + 1} = 0.6 \tag{4–50}$$

4-4 ■ INPUT IMPEDANCE

The concept of input impedance is very useful when working with transmission lines. Consider a point on a *lossless line* at a distance d from the load as illustrated in Fig. 4–4. Let $Z_{in}(d)$ represent the input impedance at this point. Since this point corresponds to $x = -d$ in the coordinate system we have employed, $Z_{in}(d)$ can be defined as

$$Z_{in}(d) = \left[\frac{\overline{V}(x)}{\overline{I}(x)}\right]_{x=-d} = \frac{\overline{V}(-d)}{\overline{I}(-d)} \tag{4–51}$$

The forms of $\overline{V}(x)$ and $\overline{I}(x)$ from (4–33) and (4–34) are substituted in (4–51), and the result is

$$Z_{in}(d) = \frac{\overline{V}^+ e^{j\beta d}(1 + \overline{\Gamma}_2 e^{-j2\beta d})}{\overline{I}^+ e^{j\beta d}(1 - \overline{\Gamma}_2 e^{-j2\beta d})} \tag{4–52}$$

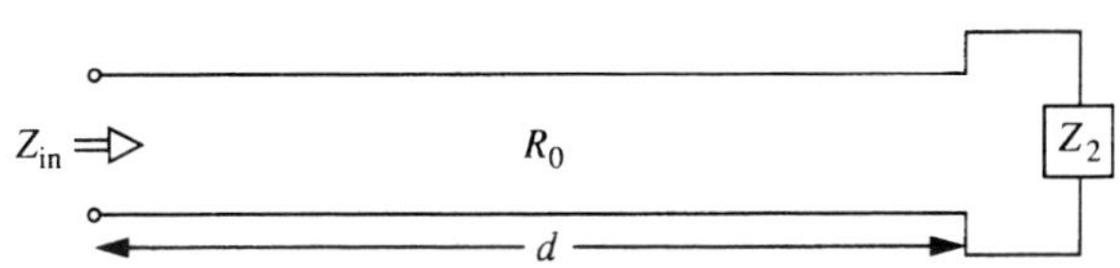

FIGURE 4–4 Concept of the input impedance of a line.

After cancellation of the common factors and substitution of $R_0 = \overline{V}^+/\overline{I}^+$, the input impedance can be expressed as

$$Z_{\text{in}}(d) = R_0 \frac{(1 + \overline{\Gamma}_2 e^{-j2\beta d})}{(1 - \overline{\Gamma}_2 e^{-j2\beta d})} \tag{4-53}$$

An alternate expression for $Z_{\text{in}}(d)$ is

$$Z_{\text{in}}(d) = R_0 \frac{(1 + \Gamma_2 \underline{/\theta_2 - 2\beta d})}{(1 - \Gamma_2 \underline{/\theta_2 - 2\beta d})} \tag{4-54}$$

where $\overline{\Gamma}_2 = \Gamma_2 \underline{/\theta_2}$ as previously defined.

As the length of the line varies for a mismatched load, the input impedance can assume a variety of values, depending on the value of $\overline{\Gamma}_2$. Irrespective of whether the load is real or complex, if the line is not matched, the input impedance will assume complex values having both inductive and capacitive reactive components for different ranges of length. The input impedance will be real only at those points at which the angle $\theta_2 - 2\beta d$ is an integer multiple of π rad.

It can be shown that the input impedance is a periodic function of the length of line satisfying the relationship

$$Z_{\text{in}}\left(d + \frac{\lambda}{2}\right) = Z_{\text{in}}(d) \tag{4-55}$$

Thus, the pattern of impedance variation established in the first half-wavelength is repeated in each subsequent half-wavelength.

Computing Load Power

There is a particular property of the input impedance of a lossless line that is very useful in determining the power delivered to the load. When the input impedance $Z_{\text{in}} = R_{\text{in}} + jX_{\text{in}}$ for a line of length d is known, an equivalent circuit of the form shown in Fig. 4–5 may be constructed at the generator end.

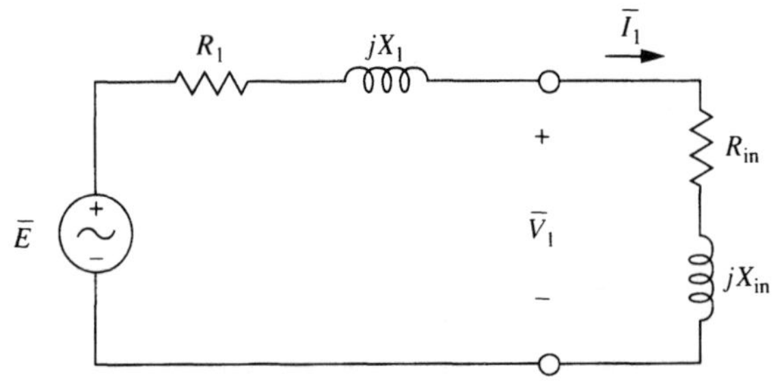

FIGURE 4–5 Equivalent circuit model at generator end based on input impedance to line.

Inductive reactances have been arbitrarily shown, but the reactive components of Z_1 or Z_2 or both could be capacitive. The power P_1 accepted by the line may be calculated from the input rms current magnitude I_1 in accordance with

$$P_1 = I_1^2 R_{in} \tag{4–56}$$

This power cannot be dissipated in the line since the line is assumed to be lossless. Since the power must be dissipated somewhere, *the logical conclusion is that it must go to the load.* Thus, the load power P_2 for a lossless line is equal to the input power to the line; i.e.,

$$P_2 = P_1 = I_1^2 R_{in} \tag{4–57}$$

The inquisitive reader may wonder if power can also be computed from a voltage-squared/resistance formula. The answer is yes, but the circuit form shown in Fig. 4–5 is not the ideal one for that purpose since the input voltage $\overline{V}_1$ is *not* the voltage across R_{in}. Instead, an alternate approach is to calculate the input admittance $Y_{in} = 1/Z_{in} = G_{in} + jB_{in}$ and draw a parallel equivalent circuit as shown in Fig. 4–6 after $\overline{V}_1$ is computed. For convenience, the quantities are labeled in terms of the Y values. Note that $G_{in} \neq 1/R_{in}$ except when $X_{in} = 0$.

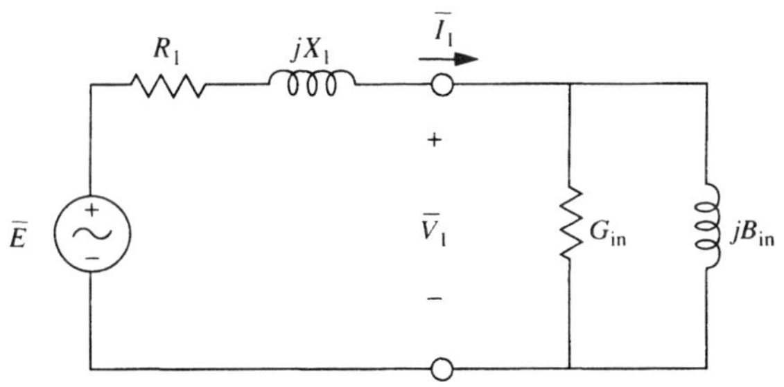

FIGURE 4–6 Equivalent circuit model at generator based on input admittance to line.

The input power may now be computed as

$$P_1 = V_1^2 G_{in} \tag{4–58}$$

where $V_1 = |\overline{V}_1|$.

EXAMPLE 4–7

In Example 4–1, the reflection coefficient was determined for a lossless 50-Ω line terminated in 150 Ω. Determine the input impedance for the following lengths: (a) 0, (b) 0.01λ, (c) 0.125λ, (d) 0.25λ, and (e) 0.5λ.

Solution

The reflection coefficient at the load was determined in Example 4–1 to be $\bar{\Gamma}_2 = 0.5\underline{/0}$. The input impedance can be determined from (4–54) as

$$Z_{in}(d) = 50\frac{(1 + 0.5\underline{/-4\pi d/\lambda})}{(1 - 0.5\underline{/-4\pi d/\lambda})} \tag{4–59}$$

where the relationship $\beta = 2\pi/\lambda$ was substituted. The input impedance can now be calculated at each of the distances.

(a) $d = 0$

This first case is trivial and amounts to no line at all! However, let us try the formula anyway. From (4–59), we have

$$Z_{in}(0) = 50\frac{(1 + 0.5\underline{/0})}{(1 - 0.5\underline{/0})} = 150\ \Omega \tag{4–60}$$

This result should certainly come as no surprise since the impedance is being measured right at the load.

(b) $d = 0.01\lambda$

$$Z_{in}(0.01\lambda) = 50\frac{(1 + 0.5\underline{/-0.04\pi})}{(1 - 0.5\underline{/-0.04\pi})} \tag{4–61a}$$

$$= 50\frac{[1 + (0.4961 - j0.0627)]}{[1 - (0.4961 - j0.0627)]} \tag{4–61b}$$

$$= 50\frac{(1.4961 - j0.0627)}{(0.5039 + j0.0627)} \tag{4–61c}$$

$$= 50\frac{(1.4974\underline{/-0.0419})}{(0.5078\underline{/0.1238})} \tag{4–61d}$$

$$= 147.4\underline{/-0.1657} = 145.4 - j24.3\ \Omega \tag{4–61e}$$

This part was designed to show that even though the length is only 1% of a wavelength, there may be a noticeable change in the input impedance when the VSWR is moderate or large. In this case, the magnitude (147.4 Ω) does not differ appreciably from the value of 150 Ω, but the striking effect is to introduce a reactive component of −24.3 Ω in this short distance.

(c) $d = 0.125\lambda$

$$Z_{in} = 50\frac{(1 + 0.5\underline{/-0.5\pi})}{(1 - 0.5\underline{/-0.5\pi})} = 50\frac{(1 - j0.5)}{(1 + j0.5)} \tag{4–62a}$$

$$= 50\frac{(1.118\underline{/-0.4636})}{(1.118\underline{/0.4636})} \tag{4–62b}$$

$$= 50\underline{/-0.9272}\ \Omega = 30 - j40\ \Omega \tag{4–62c}$$

(d) $d = \lambda/4$

$$Z_{in} = 50\frac{(1 + 0.5\underline{/-\pi})}{(1 - 0.5\underline{/-\pi})} \tag{4–63a}$$

$$= 50\frac{(1 - 0.5)}{(1 + 0.5)} = 50\left(\frac{0.5}{1.5}\right) = 16.67\ \Omega \tag{4–63b}$$

The input impedance is real at $d = \lambda/4$ but is significantly different from the load value. This property will be generalized later.

(e) $d = \lambda/2$

$$Z_{in} = 50\frac{(1 + 0.5\underline{/-2\pi})}{(1 - 0.5\underline{/-2\pi})} = 50\left(\frac{1.5}{0.5}\right) = 150\ \Omega \tag{4–64}$$

The input impedance is equal to the load impedance at $d = \lambda/2$. This property will be generalized later.

EXAMPLE 4–8

Consider the circuit of Fig. 4–7, in which the lossless transmission line and load of Example 4–7 are connected to a 250-V rms source with an internal resistive impedance of 100 Ω. Determine the power P_2 that would be delivered to the load R_2 for the following line lengths: (a) 0, (b) 0.01λ, (c) 0.125λ, (d) 0.25λ, and (e) 0.5λ.

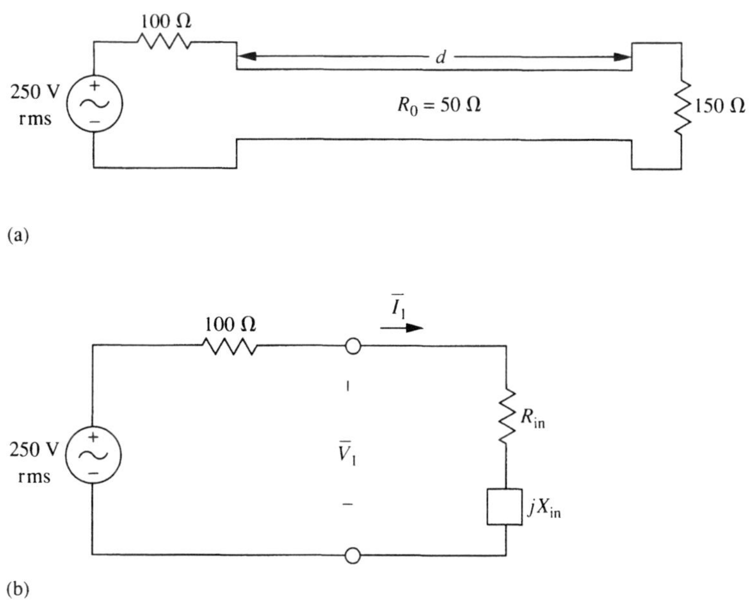

FIGURE 4–7 Circuit of Example 4–8.

Solution

For any arbitrary length d, the equivalent circuit at the input to the line is as shown in Fig. 4–7(b). By steady-state ac circuit theory, the current $\bar{I}_1$ and power P_1 may be determined for each impedance Z_{in}. The power $P_1 = P_2$ may be then determined in each case.

(a) $d = 0$

$$Z_{in} = 150\ \Omega \tag{4–65}$$

$$\bar{I}_1 = \frac{250}{(100 + 150)} = 1\underline{/0}\ \text{A} \tag{4–66}$$

$$P_2 = P_1 = I_1^2 R_{in} = (1)^2 \times 150 = 150\ \text{W} \tag{4–67}$$

It should be noted that this is not an optimum match since $R_2 \neq R_1$. The reader can verify that if the load resistance were $R_2 = R_1 = 100\ \Omega$, the power delivered to the load with $d = 0$ would be 156.25 W.

(b) $d = 0.01\lambda$

$$Z_{in} = 145.4 - j24.3\ \Omega \tag{4–68}$$

$$\bar{I}_1 = \frac{250}{[100 + (145.4 - j24.3)]} = \frac{250}{(245.4 - j24.3)} \tag{4–69a}$$

$$= \frac{250}{246.6\underline{/-0.0987}} = 1.0138\underline{/0.0987}\ \text{A} \tag{4–69b}$$

$$P_2 = P_1 = I_1^2 R_{in} = (1.0138)^2 \times 145.4 = 149.4\ \text{W} \tag{4–70}$$

We saw in Example 4–7 that the nature of the input impedance started to change somewhat with a length 0.01λ. However, there is almost no change in the power delivered to the load for this short length.

(c) $d = 0.125\lambda$

$$Z_{in} = 30 - j40\ \Omega \tag{4–71}$$

$$\bar{I}_1 = \frac{250}{[100 + (30 - j40)]} = \frac{250}{(130 - j40)} \tag{4–72a}$$

$$= \frac{250}{136.01\underline{/-0.2985}} = 1.8381\underline{/0.2985}\ \text{A} \tag{4–72b}$$

$$P_2 = P_1 = I_1^2 R_{in} = (1.8381)^2 \times 30 = 101.4\ \text{W} \tag{4–73}$$

(d) $d = \lambda/4$

$$Z_{in} = 16.67\ \Omega \tag{4–74}$$

$$\bar{I}_1 = \frac{250}{(100 + 16.67)} = 2.143\underline{/0}\ \text{A} \tag{4–75}$$

$$P_2 = P_1 = (2.143)^2 \times 16.67 = 76.56\ \text{W} \tag{4–76}$$

(e) $d = 0.5\ \lambda$

$$Z_{in} = 150\ \Omega \tag{4–77}$$

$$\bar{I}_1 = \frac{250}{100 + 150} = 1\underline{/0}\ \text{A} \tag{4–78}$$

$$P_2 = P_1 = (1)^2 \times 150 = 150\ \text{W} \tag{4–79}$$

In reviewing the various parts of this problem, we see that the power delivered to the load is a definite function of the line length even though the line is lossless. This type of situation is generally undesirable in an RF transmission system. If the system were allowed to operate in the manner given in the problem, careful adjustment of the length would be required to get maximum power to the load. The approach to this problem is impedance matching, which will be discussed in Chapter 5.

4–5 ■ LINE MATCHED AT LOAD

The concept of a line matched at the load has been considered at several points in the text. However, it is also worthwhile to consider it again in the context of a lossless line.

Consider the situation depicted in Fig. 4–8 in which the line is terminated in a load $Z_2 = R_2 = R_0$. The reflection coefficient is readily determined to be

$$\bar{\Gamma}_2 = \frac{Z_2 - R_0}{Z_2 + R_0} = \frac{R_0 - R_0}{R_0 + R_0} = 0 \tag{4–80}$$

The VSWR is given by

$$S = \frac{1 + \Gamma_2}{1 - \Gamma_2} = \frac{1 + 0}{1 - 0} = 1 \tag{4–81}$$

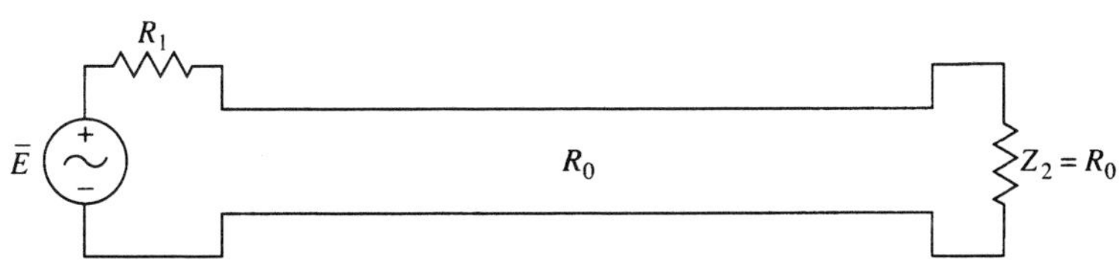

FIGURE 4–8 Line matched at load.

This result means that neither the voltage nor the current varies along the line. Thus, *the magnitudes of both the voltage and the current along a line matched at the load are constants and do not vary with distance.* The forms of the voltage and current magnitudes are shown in (a) and (b) of Fig. 4–9.

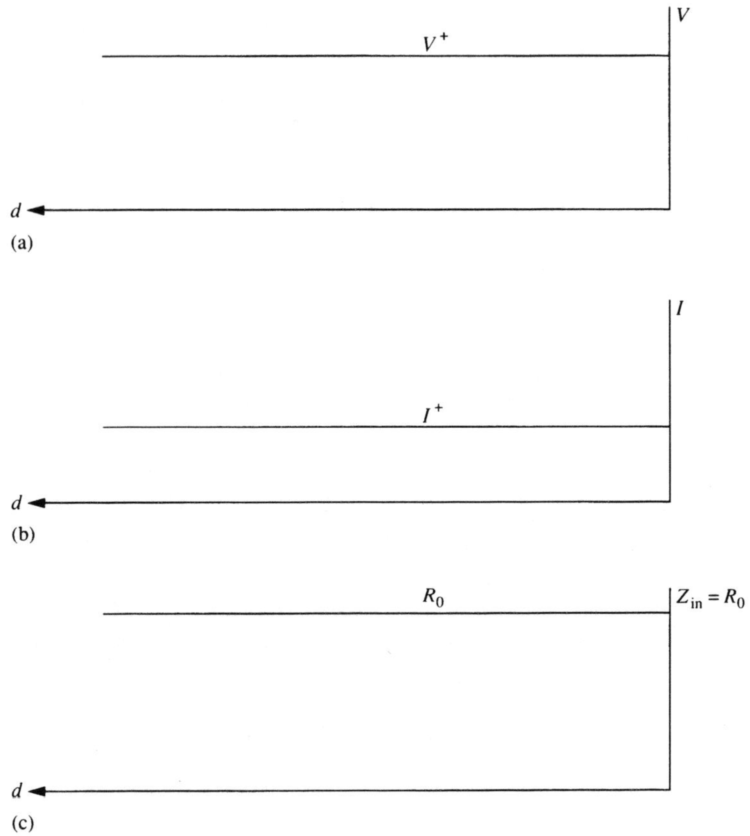

FIGURE 4–9 Forms of voltage and current magnitudes and input impedance for a line matched at load.

The input impedance of a line matched at the load is determined from (4–54) to be

$$Z_{\text{in}} = R_0 \tag{4–82}$$

Thus, *the input impedance to a lossless line terminated in its characteristic impedance is a constant value equal to the characteristic impedance at all points.* This concept is illustrated in Fig. 4–9(c).

The analysis of a lossless line terminated in a matched load is particularly straightforward as illustrated in Fig. 4–10. An equivalent circuit is drawn at the generator end with the line replaced by a resistance R_0. The input phasor current $\bar{I}_1$ is determined from the circuit model. The input power P_1 must equal the load power P_2 and is given by

$$P_2 = P_1 = I_1^2 R_0 \tag{4–83}$$

where I_1 is the rms current magnitude.

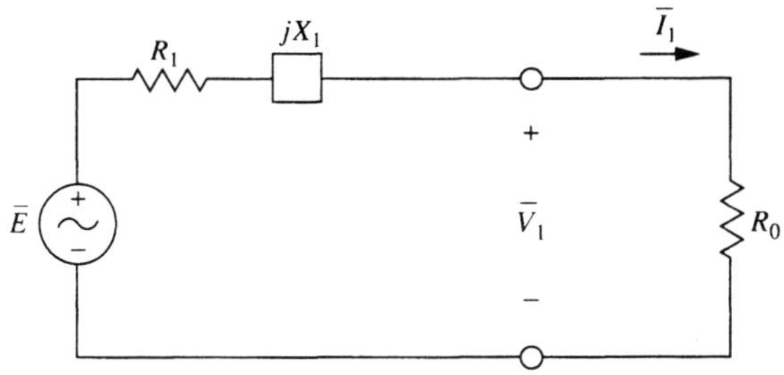

FIGURE 4–10 Equivalent circuit at generator end for line matched at load.

Consider now a generator having an rms open-circuit voltage magnitude E and an internal resistive impedance of R_1 as shown in Fig. 4–11. Assuming that R_1 is fixed, the maximum value of P_2 that can be delivered to a load resistance of value R_2 occurs when the value of R_2 is adjusted to equal R_1, i.e., when $R_2 = R_1$. It can be readily shown that this power, which is known as the **maximum available power** P_{av}, is given by

$$P_{av} = \frac{E^2}{4R_1} \tag{4–84}$$

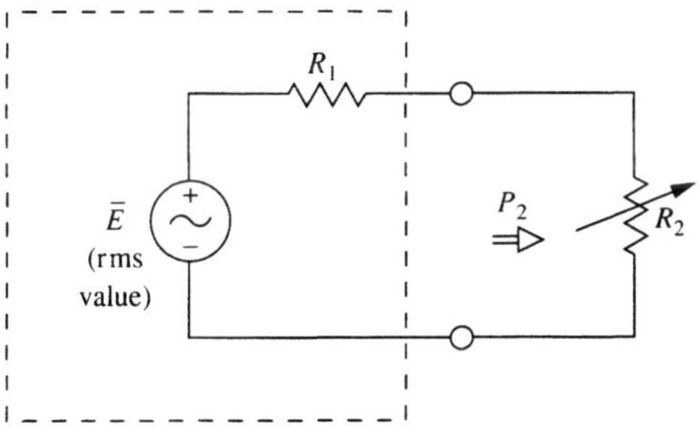

FIGURE 4–11 Determining maximum available power by adjusting load.

For a source having an internal resistance, it is desired that the characteristic impedance of the line be selected so that $R_0 = R_1$, and it is desired that the terminating impedance be $R_2 = R_0 = R_1$. In other words, *a system matched at both ends would not only eliminate reflections, but it would also result in the maximum available power from the generator being delivered to the load.*

Many transmission systems utilize common impedance levels to ensure compatibility between source, line, and load. For example, 50-Ω systems are quite common in RF equipment. Both input and output impedances of many units have resistive values of 50 Ω, and this allows 50-Ω cable to be interconnected between units without introducing mismatches.

EXAMPLE 4–9

Consider the system of Fig. 4–12 in which the 50-Ω resistance on the left is assumed to be the internal resistance of the generator. Determine (a) the VSWR, (b) the input impedance to the line, (c) the maximum available power of the generator, and (d) the power delivered to the load.

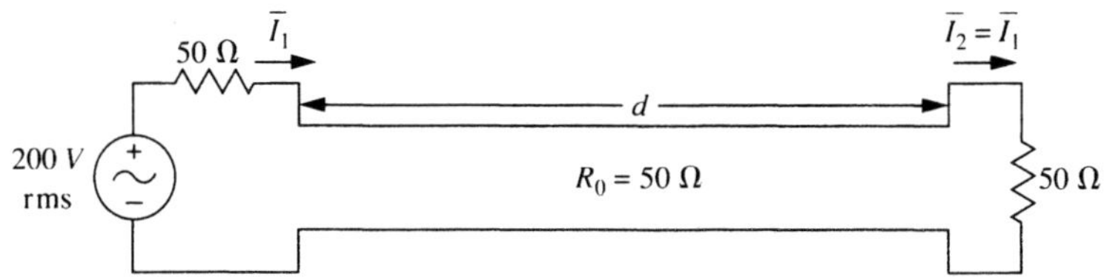

FIGURE 4–12 Circuit of Example 4–9.

Solution

(a) Since the line is matched at the load, the VSWR is

$$S = 1 \tag{4–85}$$

(b) The input impedance for any *reasonable* length is

$$Z_{in} = 50\ \Omega \tag{4–86}$$

(c) The maximum available power from the generator is

$$P_{av} = \frac{E^2}{4R_1} = \frac{(200)^2}{4 \times 50} = 200\ \text{W} \tag{4–87}$$

(d) To determine the power delivered to the load, we first determine I_1:

$$\bar{I}_1 = \frac{200}{50 + 50} = 2\ \text{A} \tag{4–88}$$

The power delivered to the load is

$$P_2 = P_1 = I_1^2 R_{in} = (2)^2 \times 50 = 200\ \text{W} \tag{4–89}$$

It is immediately obvious that $P_2 = P_{av}$. The reason is that the line is matched both at the load and at the generator. Thus, when a lossless line is matched at both ends, the power delivered to the load is the maximum available power from the generator.

4–6 ■ SHORT-CIRCUITED LINE

Two special cases of interest are (a) the short-circuited line and (b) the open-circuited line. The former will be discussed in this section, and the latter will be discussed in the next section.

A **short-circuited line** is one in which the load is a short circuit; i.e., $Z_2 = 0$. The reflection coefficient at the load is

$$\overline{\Gamma}_2 = \frac{Z_2 - R_0}{Z_2 + R_0} = \frac{0 - R_0}{0 + R_0} = -1 \tag{4–90}$$

The voltage and current on such a line when excited by a source can be determined from (4–33) and (4–34) by substituting (4–90) and $x = -d$ in these equations. The initial form can be expressed as

$$\overline{V} = \overline{V}^{+}(e^{j\beta d} - e^{-j\beta d}) \tag{4–91}$$

$$\overline{I} = \overline{I}^{+}(e^{j\beta d} + e^{-j\beta d}) \tag{4–92}$$

Let $V_{sc} = |\overline{V}|$ represent the magnitude of the voltage along the short-circuited line, and let $I_{sc} = |\overline{I}|$ represent the magnitude of the current along the line. It can be shown that these functions can be written as

$$V_{sc} = 2V^{+}|\sin \beta d| \tag{4–93}$$

$$I_{sc} = 2I^{+}|\cos \beta d| \tag{4–94}$$

where the vertical bars $|\,|$ indicate the magnitude of sin βd and cos βd.

The forms of V_{sc} and I_{sc} are shown in (a) and (b) of Fig. 4–13. An important reason to emphasize the magnitude variation is that most instruments would respond to the magnitude (or their corresponding rms values).

Some important points to note about the voltage and current on a short-circuited line are as follows:

1. The voltage magnitude at the load is zero, which is a necessary requirement of a short circuit.
2. The current magnitude at the load is at a maximum.
3. At $\lambda/4$ back from the load, the voltage is at a maximum, and the current is zero.
4. At $\lambda/2$ back from the load, the voltage is zero again, and the current is at a maximum.
5. The variations of voltage and current magnitude are periodic distance-varying functions having a distance of $\lambda/2$ between corresponding points.

Input Impedance

The input impedance of a short-circuited line is an important quantity and can be determined from (4–53) by substituting $\overline{\Gamma}_2 = -1$. The initial form of this function is

$$Z_{in}(d) = R_0 \frac{(1 - e^{-j2\beta d})}{(1 + e^{-j2\beta d})} \tag{4–95}$$

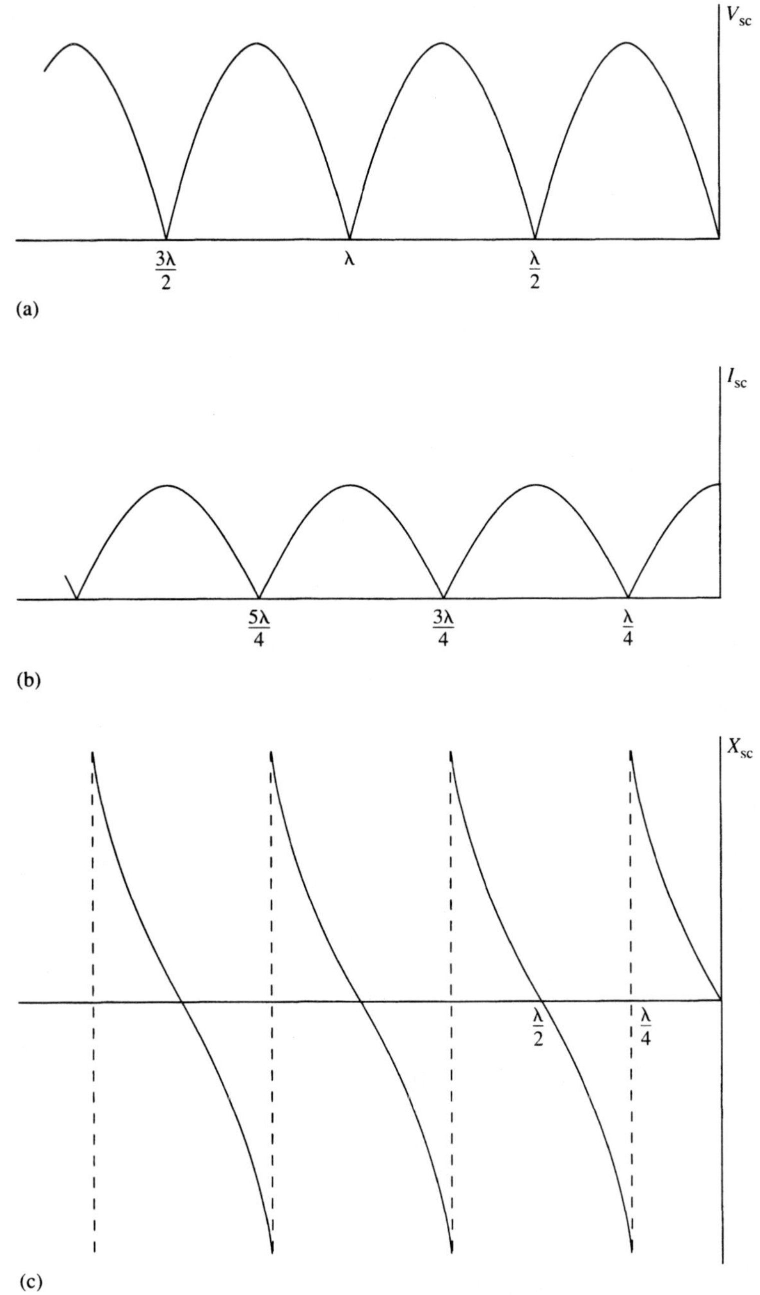

FIGURE 4–13 Voltage magnitude, current magnitude, and reactance for short-circuited line.

It will be left as an exercise for the reader to show that this function can be expressed as

$$Z_{\text{in}}(d) = jR_0 \tan(\beta d) \tag{4–96}$$

This impedance is purely imaginary for any combination of β and d, so it is convenient to write it in the form

$$Z_{sc} = jX_{sc} \tag{4–97}$$

in which case X_{sc} can be expressed as

$$X_{sc} = R_0 \tan(\beta d) = R_0 \tan\left(\frac{2\pi d}{\lambda}\right) \tag{4–98}$$

The variation of X_{sc} as a function of d is shown in Fig. 4–13(c). Several important points about this reactance can be deduced as follows:

1. In the first quarter-wavelength from the load, the reactance is *positive* (i.e., *inductive*) and assumes all positive values from zero to infinity.
2. At a length of $\lambda/4$ from the load, the line appears as an *open circuit.*
3. As the line length exceeds $\lambda/4$, the reactance becomes *negative* (i.e., *capacitive*) and assumes all possible negative values over the next quarter-wavelength.
4. At a distance $\lambda/2$ from the load, the line appears as a short circuit again.
5. The preceding pattern continues to repeat itself over each interval of $\lambda/2$.

There is a direct correlation between the voltage and current variation of (a) and (b) and the reactance variation of (c) in Fig. 4–13. At any point at which the voltage is zero and the current is maximum, the reactance is zero. Conversely, at any point at which the voltage is maximum and the current is zero, the reactance is infinite; i.e., the line acts like an open circuit.

At frequencies above a few hundred megahertz, lumped inductors and capacitors are difficult to use because of their small sizes and the presence of distributed effects. However, short-circuited lines (and open-circuited lines, as we will see in the next section) can be used to achieve the same effects. For example, if an inductive reactance is required, the length of a shorted line can be determined such that the required value of inductive reactance is achieved. Normally, a length in the first quarter-wavelength would be used, but one of the other lengths that produce a positive reactance could also be used. Conversely, if a capacitive reactance is required, a length between $\lambda/4$ and $\lambda/2$ can normally be used, but other lengths are possible.

Assume that a given value X_{sc} is to be realized. The value of d can be determined from (4–98) to yield the following result:

$$d = \frac{\lambda}{2\pi} \tan^{-1}\frac{X_{sc}}{R_0} \tag{4–99}$$

where the angle corresponding to $\tan^{-1}$ must be expressed in radians. One must be careful to choose the proper angle according to the sign of X_{sc}.

4–7 ■ OPEN-CIRCUITED LINE

The open-circuited line will be discussed in this section. As in the case of the short-circuited line, an open-circuited lossless line produces a purely reactive circuit for which no real power is dissipated.

An **open-circuited line** is one in which the line is left open at the load. It is represented in engineering analysis by indicating that $Z_2 = \infty$. The reflection coefficient at the load is determined from the basic definition by letting Z_2 increase without limit, which results in

$$\overline{\Gamma}_2 = 1 \tag{4–100}$$

The voltage and current on such a line when excited by a source can be determined from (4–33) and (4–34) by substituting $\overline{\Gamma}_2 = 1$ and $x = -d$. The initial forms can be expressed as

$$\overline{V} = \overline{V}^+(e^{j\beta d} + e^{-j\beta d}) \tag{4–101}$$

$$\overline{I} = \overline{I}^+(e^{j\beta d} - e^{-j\beta d}) \tag{4–102}$$

Let $V_{oc} = |\overline{V}|$ represent the magnitude of the voltage along the open-circuited line, and let $I_{oc} = |\overline{I}|$ represent the magnitude of the current along the line. It can be shown that these functions can be expressed as

$$V_{oc} = 2V^+|\cos \beta d| \tag{4–103}$$

$$I_{oc} = 2I^+|\sin \beta d| \tag{4–104}$$

The forms of V_{oc} and I_{oc} are shown in (a) and (b) of Fig. 4–14. Some important points to note about the voltage and current on an open-circuit line are as follows:

1. The current magnitude at the load is zero, which is a necessary requirement of an open circuit.
2. The voltage magnitude at the load is a maximum.
3. At $\lambda/4$ back from the load, the current is a maximum, and the voltage is zero.
4. At $\lambda/2$ back from the load, the current is zero again, and the voltage is a maximum.
5. The variations of voltage and current magnitudes are periodic, distance-varying functions having a distance of $\lambda/2$ between corresponding points.
6. The variation of current on an open-circuited line follows a distance variation similar to that of voltage on a short-circuited line, and vice versa.

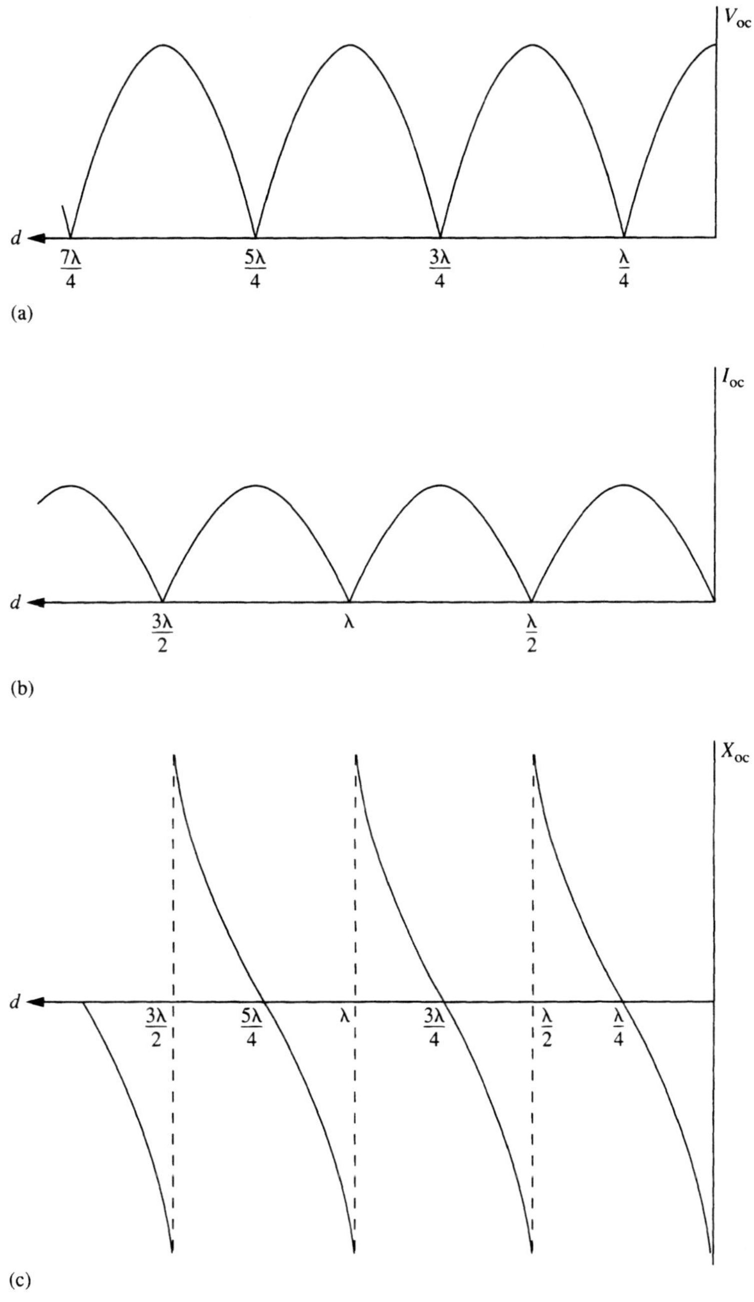

FIGURE 4–14 Voltage magnitude, current magnitude, and reactance for open-circuited line.

Input Impedance

The input impedance of an open-circuited line can be determined from (4–53) by substituting $\overline{\Gamma}_2 = 1$. The initial form of this function is

$$Z_{in}(d) = R_0 \frac{(1 + e^{-j2\beta d})}{(1 - e^{-j2\beta d})} \tag{4–105}$$

It will be left as an exercise for the reader to show that this function can be expressed as

$$Z_{oc} = -jR_0 \cot(\beta d) \tag{4–106}$$

which is purely imaginary. This impedance can be expressed as

$$Z_{oc} = jX_{oc} \tag{4–107}$$

where X_{oc} can be expressed as

$$X_{oc} = -R_0 \cot(\beta d) = -R_0 \cot\left(\frac{2\pi d}{\lambda}\right) \tag{4–108}$$

The variation of X_{oc} as a function of d is shown in Fig. 4–14(c). Several important points about the reactance can be deduced as follows:

1. In the first quarter-wavelength from the load, the reactance is *negative* (i.e., *capacitive*) and assumes all possible negative values.
2. At the length of $\lambda/4$ from the load, the line appears as a *short circuit.*
3. As the line length exceeds $\lambda/4$, the reactance becomes *positive* (i.e., *inductive*) and assumes all possible positive values over the next quarter-wavelength.
4. At a distance $\lambda/2$ from the load, the line appears as an open circuit again.
5. The preceding pattern continues to repeat itself over each interval of $\lambda/2$.
6. The variation of reactance on an open-circuit line follows the same type of pattern as that of a short-circuit line, but with a shift of $\lambda/4$ for corresponding lengths.

EXAMPLE 4–10

An assumed-lossless cable has a characteristic impedance $R_0 = 50\ \Omega$ and a velocity factor of 0.67. For a shorted line at a frequency of 100 MHz, determine the input reactance as the length is varied in 10-cm steps from 0 to 100 cm.

Solution

A velocity factor of 0.67 corresponds to a velocity of 2×10^8 m/s. The wavelength λ is

$$\lambda = \frac{\nu}{f} = \frac{2 \times 10^8}{10^8} = 2\text{ m} \tag{4–109}$$

The phase constant β is

$$\beta = \frac{2\pi}{\lambda} = \frac{2\pi}{2} = \pi \text{ rad/m} \tag{4–110}$$

For any distance d, the input reactance is determined from (4–98) to be

$$X_{sc} = R_0 \tan(\beta d) = 50 \tan(\pi d) \tag{4–111}$$

where d must be in meters.

The values of d from 0 to 100 cm (0 to 1 m) can then be determined from Eq. (4–111) as follows:

d (cm)	0	10	20	30	40	50	60	70	80	90	100
X_{sc} (Ω)	0	16.25	36.33	68.82	153.88	∞	−153.88	−68.82	−36.33	−16.25	0

The pattern of values corresponds to that of Fig. 4–13(c). The reactance for the first quarter-wavelength ($0 < d < 50$ cm) is positive (inductive) and increases with d. For the next quarter-wavelength (50 cm $< d <$ 100 cm), the reactance is negative (capacitive), and its magnitude decreases. At a quarter-wavelength (d = 50 cm), the line looks like an open circuit, and at a half-wavelength (d = 100 cm), it looks like a short circuit.

EXAMPLE 4–11

For the line of Example 4–10, determine the equivalent value of inductance corresponding to a length of 30 cm.

Solution

The value of inductance L_{sc} must satisfy

$$\omega L_{sc} = X_{sc} \tag{4–112}$$

or

$$L_{sc} = \frac{X_{sc}}{\omega} = \frac{68.82}{2\pi \times (100 \times 10^6)} = 109.5 \text{ nH} \tag{4–113}$$

EXAMPLE 4–12

For the line of Example 4–10, determine the equivalent value of capacitance corresponding to a length of 70 cm.

Solution

The value of capacitance C_{sc} must satisfy

$$\frac{1}{\omega C_{sc}} = |X_{sc}| \tag{4–114}$$

or

$$C_{sc} = \frac{1}{\omega|X_{sc}|} = \frac{1}{2\pi \times (100 \times 10^6) \times 68.82} = 23.13\ \text{pF} \qquad \textbf{(4–115)}$$

EXAMPLE 4–13

For the coaxial cable of Example 4–10 at a frequency of 100 MHz, determine the shortest length of shorted cable that produces an inductive reactance of 200 Ω.

Solution

We require that

$$X_{sc} = 50\tan(\pi d) = 200\ \Omega \qquad \textbf{(4–116)}$$

Rearranging yields

$$\tan(\pi d) = \frac{200}{50} = 4 \qquad \textbf{(4–117)}$$

Therefore,

$$\pi d = \tan^{-1}4 = 1.326 \qquad \textbf{(4–118)}$$

or

$$d = \frac{1.326}{\pi} = 0.4220\ \text{m} = 42.20\ \text{cm} \qquad \textbf{(4–119)}$$

EXAMPLE 4–14

For the coaxial cable of Example 4–10 at a frequency of 100 MHz, determine the shortest length of cable that produces a reactance of −200 Ω (i.e., a capacitive reactance).

Solution

The solution begins in the same manner as in Example 4–13; i.e., we require that

$$X_{sc} = 50\tan(\pi d) = -200 \qquad \textbf{(4–120)}$$

Rearranging yields

$$\tan(\pi d) = \frac{-200}{50} = -4 \qquad \textbf{(4–121)}$$

or

$$\pi d = \tan^{-1}(-4) \qquad \textbf{(4–122)}$$

A problem now arises with most calculators because the arctangent function is usually programmed to give the principal value of the angle, which will be in the fourth quadrant when the argument is negative. The author's calculator gave an angle of −1.326 rad, which is simply the negative of the angle in Example 4–13. This would correspond to a negative length, which is impossible.

The preceding dilemma is resolved by adding an integer multiple of π rad to the angle determined from the calculator to make the net angle positive. This is a result of the property that the tangent function has a period of π rad.

The smallest integer times π that would make the angle positive in this case is π, so the required angle is $\pi - 1.326 = 1.816$. We continue as follows:

$$\pi d = 1.816 \quad \text{or} \quad d = 0.5780 \text{ m} = 57.80 \text{ cm} \tag{4–123}$$

4–8 ■ HALF-WAVE AND QUARTER-WAVE LINES

Two special cases are of special interest for lossless lines terminated in arbitrary loads: (a) the **half-wave line** and (b) the **quarter-wave line.** The properties of such lines can be exploited in a number of special applications.

Half-Wave Line

Consider the situation depicted in Fig. 4–15 in which a load Z_2 is connected to a lossless line of length $\lambda/2$ with characteristic impedance R_0. The input impedance can be expressed as

$$Z_{in}\left(\frac{\lambda}{2}\right) = R_0 \frac{[1 + \overline{\Gamma}_2 e^{-j(2\beta\lambda/2)}]}{[1 - \overline{\Gamma}_2 e^{-j(2\beta\lambda/2)}]} \tag{4–124}$$

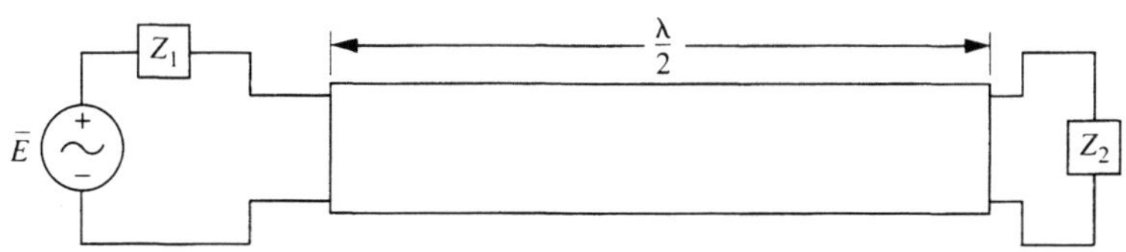

FIGURE 4–15 Half-wavelength line.

Since $\beta = 2\pi/\lambda$, the argument of the exponential is $-j\beta\lambda = -j(2\pi/\lambda) \times (\lambda) = -j2\pi$. Further, $e^{-j2\pi} = 1\underline{/-2\pi} = 1$. Substituting this value in (4–124) results in

$$Z_{in}\left(\frac{\lambda}{2}\right) = R_0 \frac{(1 + \overline{\Gamma}_2)}{(1 - \overline{\Gamma}_2)} \tag{4–125}$$

where

$$\overline{\Gamma}_2 = \frac{Z_2 - R_0}{Z_2 + R_0} \tag{4–126}$$

Substituting (4–126) in (4–125) and reducing leads to the following simple result:

$$Z_{in}\left(\frac{\lambda}{2}\right) = Z_2 \tag{4–127}$$

It can be readily shown that the same results hold for *any integer multiple of* $\lambda/2$. Stated in words, *the input impedance of any line having a length equal to an integer number of half-wavelengths is equal to the load impedance.* The process acts as if the line were not even present.

This property may be used to advantage in cases where the load impedance may be closely matched to the generator impedance, but in which no transmission line having a comparable characteristic impedance is available. In such a case, a line having a different characteristic impedance may be used, provided that the length is an integer number of half-wavelengths at the operating frequency.

Quarter-Wave Line

Next consider the system of Fig. 4–16 in which the load is connected through a lossless line of length $\lambda/4$. The input impedance is

$$Z_{in}\left(\frac{\lambda}{4}\right) = R_0\frac{[1 + \overline{\Gamma}_2 e^{-j(2\beta\lambda/4)}]}{[1 - \overline{\Gamma}_2 e^{-j(2\beta\lambda/4)}]} \tag{4–128}$$

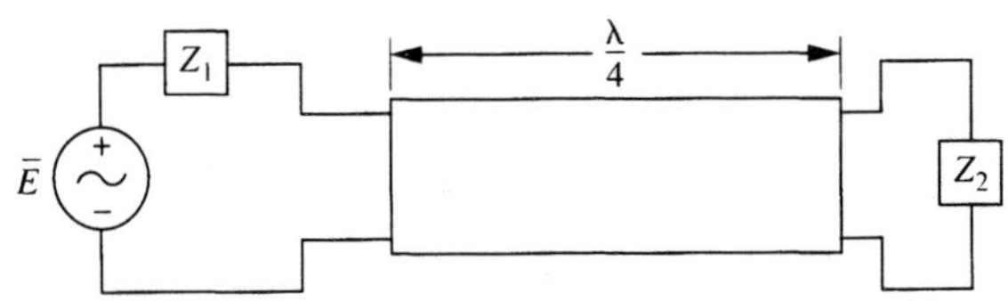

FIGURE 4–16 Quarter-wavelength line.

The argument of the exponentials is $-j\beta\lambda/2 = -j(2\pi/\lambda) \times (\lambda/2) = -j\pi$. Further, $e^{-j\pi} = 1\underline{/-\pi} = -1$. Substituting this value in (4–128) results in

$$Z_{in}\left(\frac{\lambda}{4}\right) = R_0\frac{(1 - \overline{\Gamma}_2)}{(1 + \overline{\Gamma}_2)} \tag{4–129}$$

Substituting (4–126) in (4–129) and reducing leads to the following result:

$$Z_{in}\left(\frac{\lambda}{4}\right) = \frac{R_0^2}{Z_2} \tag{4–130}$$

It can be readily shown that the same result holds for *any odd integer multiple of* $\lambda/4$. (Even multiples of $\lambda/4$ would not apply since they would be integer multiples of $\lambda/2$.) Stated in words, *the input impedance of any line having a length equal to an odd integer number of quarter-wavelengths is inversely proportional to the load impedance.* This inversion property holds at $\lambda/4$, $3\lambda/4$, $5\lambda/4$, and so on.

The inversion property means that *a very low impedance is transformed to a very high impedance, and vice versa.* Further, the sign of the reactive is changed. Thus, *an inductance reactive transforms to a capacitive reactance, and vice versa.*

Quarter-Wave Transformer

An important element in microwave integrated circuits (MIC) is the quarter-wave transformer. This is a section of transmission line whose characteristic impedance R_0 is created to form a match between a source resistance R_1 and a load of resistance R_2. By requiring that the input impedance of the line be the same as that of the generator, the required value of R_0 is given by

$$R_0 = \sqrt{R_1 R_2} \tag{4–131}$$

EXAMPLE 4–15

For the circuit of Fig. 4–17, at a frequency of 150 MHz, determine (a) the input impedance and (b) the power delivered to the load.

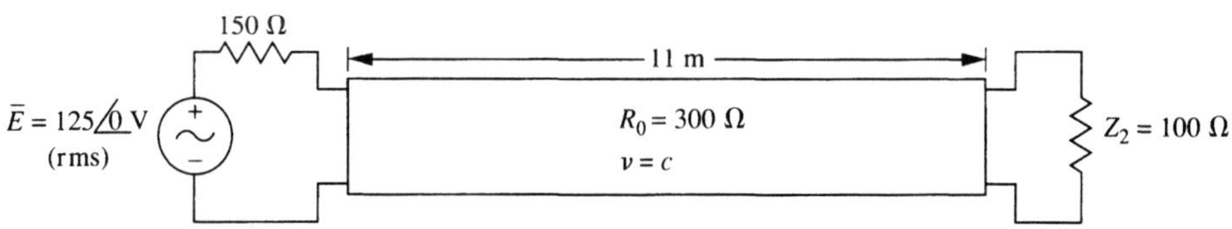

FIGURE 4–17 Circuit of Example 4–15.

Solution

(a) As a first step, the wavelength is determined to be

$$\lambda = \frac{c}{f} = \frac{3 \times 10^8}{150 \times 10^6} = 2\text{ m} \tag{4–132}$$

Further inspection reveals that the line length is an integer number of half-wavelengths. Specifically, $\lambda/2 = 1$ m, so the line has a length of 11 half-wavelengths. Thus, the input impedance is simply

$$Z_{\text{in}} = Z_2 = 100\ \Omega \tag{4–133}$$

(b) The rms current phasor $\bar{I}_1$ flowing from the generator is

$$\bar{I}_1 = \frac{125\angle 0}{150 + 100} = 0.5\angle 0 \text{ A} \tag{4–134}$$

The output power is the same as the input power and is

$$P_2 = P_1 = I_1^2 R_{in} = (0.5)^2 \times 100 = 25 \text{ W} \tag{4–135}$$

EXAMPLE 4–16

Assume in the circuit of Example 4–15 that the line length is shortened by 50 cm. Repeat the calculations of Example 4–15.

Solution

(a) A length of 50 cm corresponds to $\lambda/4$. The new length is 11 m − 0.5 m = 10.5 m which is an odd integer multiple of $\lambda/4$. Specifically, it has a length of 21 quarter-wavelengths. The input impedance is

$$Z_{in} = \frac{R_0^2}{Z_2} = \frac{(300)^2}{100} = 900\ \Omega \tag{4–136}$$

(b) The rms current phasor $\bar{I}_1$ is

$$\bar{I}_1 = \frac{125\angle 0}{150 + 900} = 0.1190 \text{ A} \tag{4–137}$$

The power is

$$P_2 = P_1 = I_1^2 R_{in} = (0.1190)^2 \times 900 = 12.76 \text{ W} \tag{4–138}$$

■ PROBLEMS

4–1. A 50-Ω lossless line is terminated in $Z_2 = R_2 = 30\ \Omega$. Determine the following:
(a) the reflection coefficient at the load
(b) the reflection coefficient at distances $\lambda/8$, $\lambda/4$, and $\lambda/2$ back from the load

4–2. A 75-Ω lossless line is terminated in $Z_2 = R_2 = 125\ \Omega$. Determine the following:
(a) the reflection coefficient at the load
(b) the reflection coefficient at distances $\lambda/8$, $\lambda/4$, and $\lambda/2$ back from the load

4–3. A 50-Ω lossless line is terminated in $Z_2 = 40 - j30\ \Omega$. Determine the following:
(a) the reflection coefficient at the load
(b) the reflection coefficient at distances $\lambda/8$ and $\lambda/4$ back from the load

4–4. A 75-Ω lossless line is terminated in $Z_2 = 100\angle 0.3375\pi\ \Omega$. Determine the following:
(a) the reflection coefficient at the load
(b) the reflection coefficient at distances $\lambda/8$ and $\lambda/4$ back from the load

4–5. The reflection coefficient of a load connected to a 50-Ω lossless line is known to be $\overline{\Gamma}_2 = 0.5\underline{/\pi/4}$. Determine Z_2.

4–6. The reflection coefficient of a load connected to a 300-Ω lossless line is known to be $\overline{\Gamma}_2 = 0.6\underline{/0.7\pi}$. Determine Z_2.

4–7. Determine the VSWR on the line of Problem 4–1.

4–8. Determine the VSWR on the line of Problem 4–2.

4–9. Determine the VSWR on the line of Problem 4–3.

4–10. Determine the VSWR on the line of Problem 4–4.

4–11. Determine the VSWR on the line of Problem 4–5.

4–12. Determine the VSWR on the line of Problem 4–6.

4–13. The VSWR on a 75-Ω lossless line is measured as $S = 3$. Determine the magnitude of the reflection coefficient.

4–14. The VSWR on a 50-Ω lossless line is measured as $S = 5$. Determine the magnitude of the reflection coefficient.

4–15. For the system of Problem 4–1, determine the input impedances for the following lengths:
(a) 0
(b) 0.02λ
(c) 0.125λ
(d) 0.25λ
(e) 0.5λ

4–16. For the system of Problem 4–2, determine the input impedances for the following lengths:
(a) 0
(b) 0.03λ
(c) 0.125λ
(d) 0.25λ
(e) 0.5λ

4–17. For the system of Problem 4–3, determine the input impedance for a length of 0.18λ.

4–18. For the system of Problem 4–9, determine the input impedance for a length of 0.42λ.

4–19. Consider the circuit of Fig. P4–19. Using the results of Problem 4–15, determine the power P_2 that would be delivered to the load R_2 for each of the lengths in Problem 4–15.

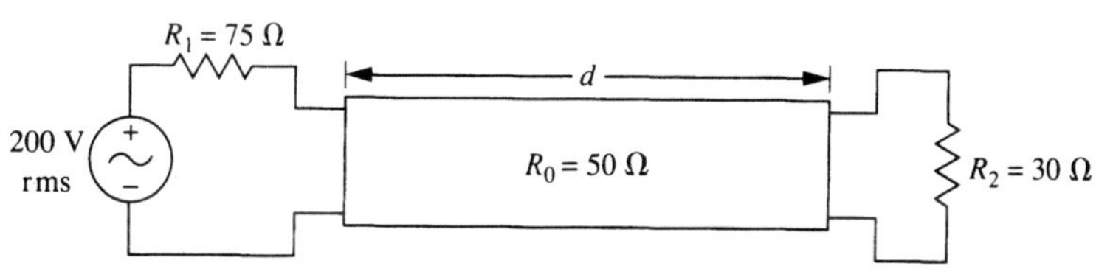

FIGURE P4–19

4–20. Consider the circuit of Fig. P4–20. Using the results of Problem 4–16, determine the power P_2 that would be delivered to the load R_2 for each of the lengths in Problem 4–16.

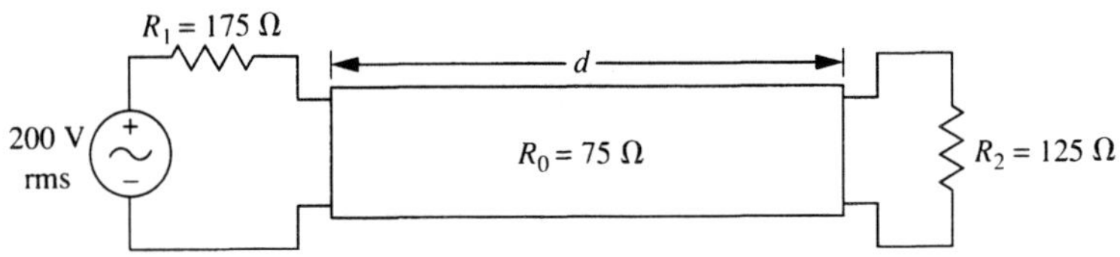

FIGURE P4–20

4–21. A lossless 75-Ω coaxial cable has a velocity factor of 0.67. For a shorted line at a frequency of 200 MHz, determine the input reactance for the following lengths:
(a) 20 cm
(b) 40 cm

4–22. A lossless 300-Ω parallel line has a velocity equal to that of free space. For a shorted line at a frequency of 150 MHz, determine the input reactance for the following lengths:
(a) 35 cm
(b) 80 cm

4–23. For the line of Problem 4–21, determine the equivalent value of inductance or capacitance corresponding to a length of 20 cm.

4–24. For the line of Problem 4–22, determine the equivalent value of inductance or capacitance corresponding to a length of 80 cm.

4–25. For the coaxial cable of Problem 4–21 at a frequency of 200 MHz, determine the shortest length of shorted cable that produces an inductive reactance of 30 Ω.

4–26. For the parallel line of Problem 4–22 at a frequency of 150 MHz, determine the shortest length of shorted line that produces an inductive reactance of 450 Ω.

4–27. For the coaxial cable of Problem 4–21 at a frequency of 200 MHz, determine the shortest length of shorted cable that produces a capacitive reactance of −100 Ω.

4–28. For the parallel line of Problem 4–22 at a frequency of 150 MHz, determine the shortest length of shorted line that produces a capacitive reactance of −75 Ω.

4–29. For the system of Problem 4–21, repeat the analysis based on an *open-circuited* line.

4–30. For the system of Problem 4–22, repeat the analysis based on an *open-circuited* line.

5

Smith Chart and Applications

OBJECTIVES

After completing this chapter, the reader should be able to:

1. Explain the organization of a **Smith chart,** including the identification of the **real line, imaginary circle, circles of constant resistance, circles of constant reactance,** and **operating circle.**
2. Plot an impedance or an admittance on a Smith chart and convert from one to the other.
3. Determine the VSWR and the angle of the reflection coefficient on a Smith chart.
4. Explain how to rotate either in the direction of the generator or in the direction of the load.
5. Use a Smith chart to determine the input impedance for an arbitrary length of line with an arbitrary load impedance.
6. Use a Smith chart to determine the load impedance for an arbitrary length of line if the input impedance is known.
7. Determine the positions of voltage, current, impedance, or admittance maxima and minima using a Smith chart on either an impedance or an admittance basis.
8. Determine the input impedance or admittance for either a shorted line or an open line using a Smith chart.
9. Design a **quarter-wavelength matching transformer** using a Smith chart.
10. Design a **single-stub matching system** using a Smith chart.

5–1 ■ SMITH CHART

In the preceding several chapters, numerous steady-state ac calculations for transmission lines were made. Many of these calculations are unwieldy in nature and do not always provide insight on the optimum approach when a system is being designed or tuned for maximum efficiency.

A major contribution to the evolution of transmission line theory was made by P. H. Smith in the form of a transmission line chart.* This chart, which has become known all over the world as the **Smith chart,** is an important tool for visualizing conditions on a transmission line. Even with all the modern computer technology of today, the Smith chart remains a principal tool for working with the analysis and design of transmission systems. In fact, some high-frequency computer programs provide a Smith chart display on the screen. Some instruments also provide output data directly on a Smith chart. Anyone working with RF and microwave systems should learn to visualize system operation in terms of the Smith chart orientation.

As is true with most graphical techniques, accuracy is limited to probable errors of the order of 1% to 5% or so on most portions of the chart and even larger errors on some portions of the chart. However, this is usually not a serious limitation for two reasons:

1. Most RF and microwave parameters normally cannot be measured to much greater accuracy anyway, so the Smith chart values are often as accurate as any of the data employed on the chart.
2. When one desires to obtain more accurate values, the Smith chart may be supplemented by the analytical techniques developed in preceding chapters.

However, the use of the Smith chart may serve as an important guide in knowing at which points to perform computations, and so on. In other words, the *visualization* provided by the Smith chart may provide insight that can never be seen solely from the mathematical techniques.

Although the Smith chart can be adapted on a limited basis to lines with losses, by far the most widely employed applications of the chart are in RF systems in which the lines are assumed to be lossless. The treatment in this text will be limited to this very common situation. The actual derivation of the Smith chart is beyond the scope of this text but is provided in advanced texts on transmission line theory. However, one can learn to use the chart without understanding its origin, and that will be the approach in the text.

A reduced-size reproduction of a typical Smith chart is shown in Fig. 5–1. There are several variations on the market, but this particular one, produced by Analog Instruments Company, will serve as the basis for this text. Do not be intimidated by all the numbers and curves appearing on the chart, because the necessary understanding will be developed gradually. In fact, unless one

*P. H. Smith, "Transmission-line Calculator," *Electronics* 12, January 1939, pp. 29–31.

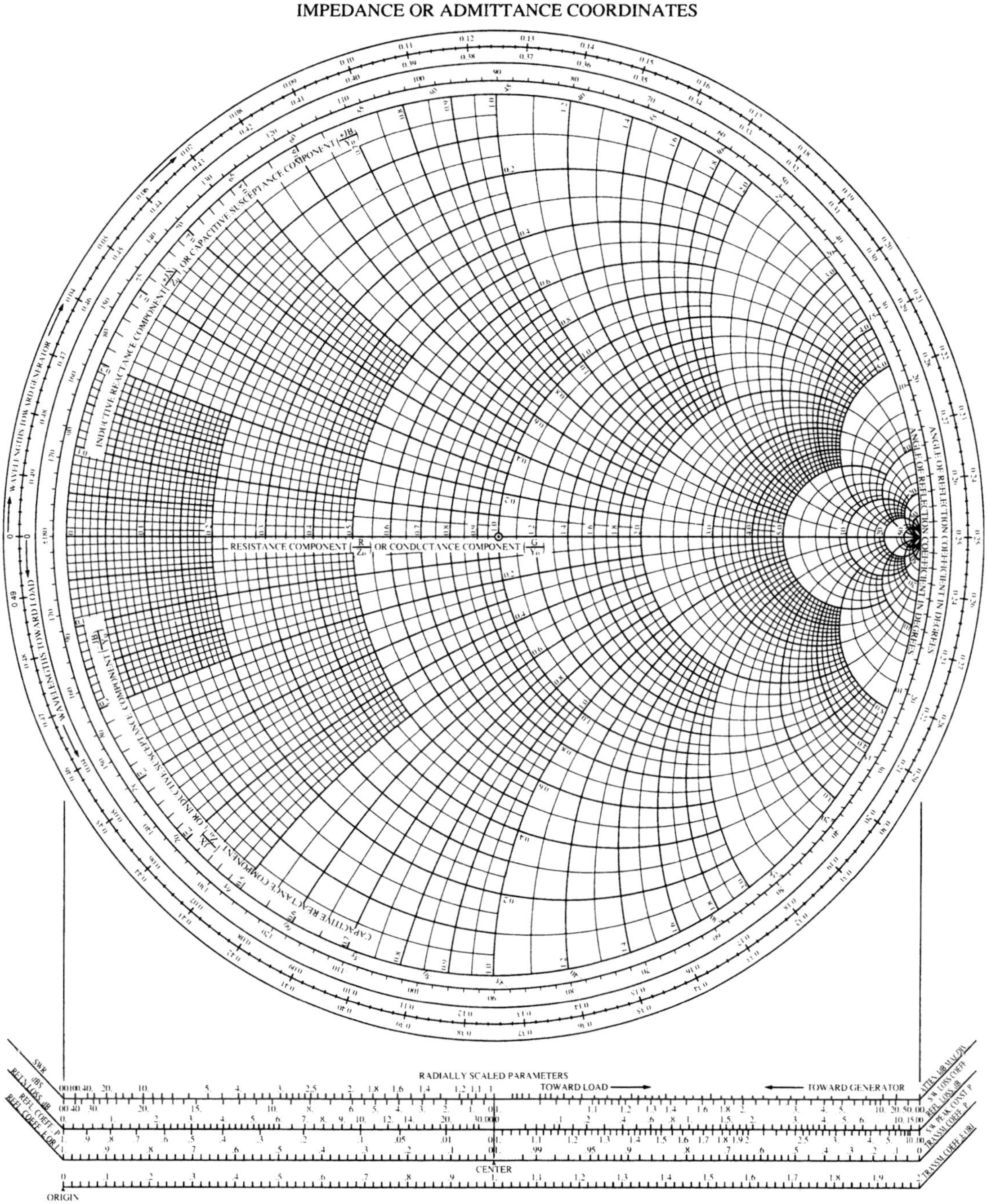

FIGURE 5–1 Smith chart.

specializes in RF and microwave systems, there are portions of the main plot and/or the charts underneath that may never be used.

5–2 ■ PLOTTING IMPEDANCE AND ADMITTANCE

It is the authors' experiences that most errors made by students on the Smith chart are in reading values correctly as the values appear. This mistake is understandable since the chart is a "busy" one, having a combination of several types of curves and data values.

First, let us be certain that we are clear on some of the definitions and signs for steady-state impedance and admittance. Let $Z = R + jX$ represent a complex impedance, and let $Y = 1/Z = G + jB$ represent the corresponding admittance. The complex *impedance* may be visualized as a *series* combination of the resistance R and the reactance X. The complex *admittance* may be visualized in terms of a *parallel* combination of the conductance G and the susceptance B. The following pattern for X and B should be noted and committed to memory (if not already known):

X	B	**Nature**
+	−	Inductive
−	+	Capacitive

The reversal of signs occurs automatically when impedance is converted to admittance, and vice versa. For example, if $Z = 3 + j4\ \Omega$, $Y = 1/Z = 0.12 - j0.16$ S (after some manipulation).

The Smith chart, stripped of all curves but one, is shown in Fig. 5–2. The horizontal line extending across the chart will be referred to as the **real line.** Any impedance or admittance appearing on this line will have only a real part; i.e., the imaginary part is zero. The point at the center will be appropriately referred to as the **center** of the chart. Later, when a compass is used on the chart, the pivot point for the compass will be placed at the center point.

The Smith chart, stripped again of all curves but one, is shown in Fig. 5–3. This curve, which is a circle surrounding the entire Smith chart, will be referred to as the **imaginary circle.** Do not confuse this curve with the several other scale circles having larger radii. The imaginary circle is an actual part of the Smith chart, while the others simply provide rotational scales and other auxiliary information.

An impedance or admittance appearing on this circle will have only an imaginary part; i.e., the real part is zero. Thus, all impedances or admittances on this circle will have only reactive or susceptive values.

Another partially stripped version of the Smith chart is shown in Fig. 5–4. The curves shown here represent **circles of constant real part.**

A different stripped version of the Smith chart is shown in Fig. 5–5. These curves represent **circles of constant imaginary part.** (It is not obvious that these are circles since their centers are off the chart.) The curves above the

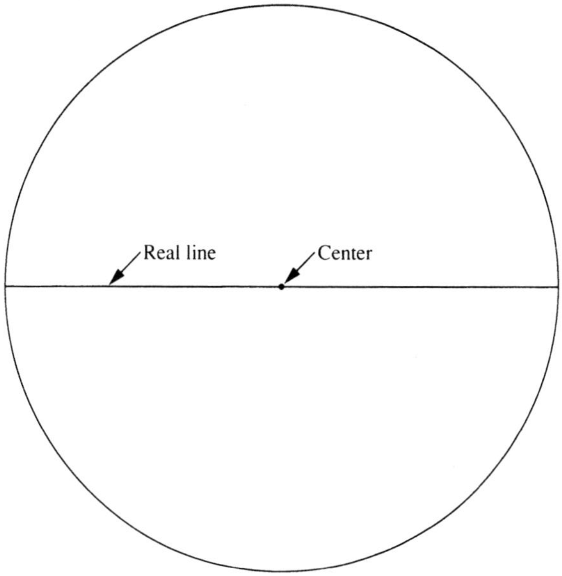

FIGURE 5–2 Real line and center of Smith chart.

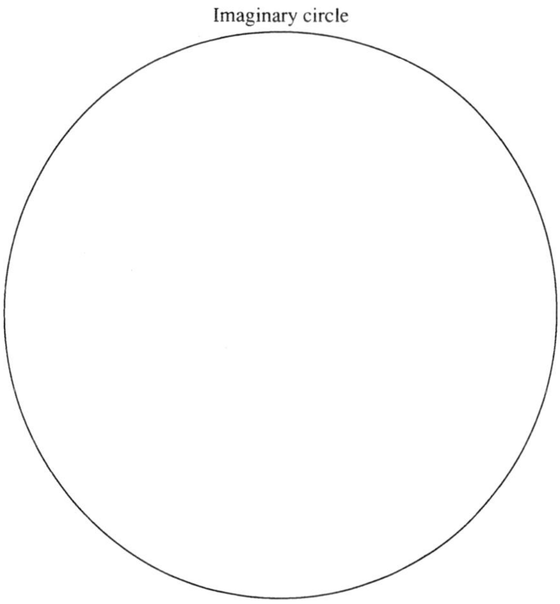

FIGURE 5–3 Imaginary circle of Smith chart.

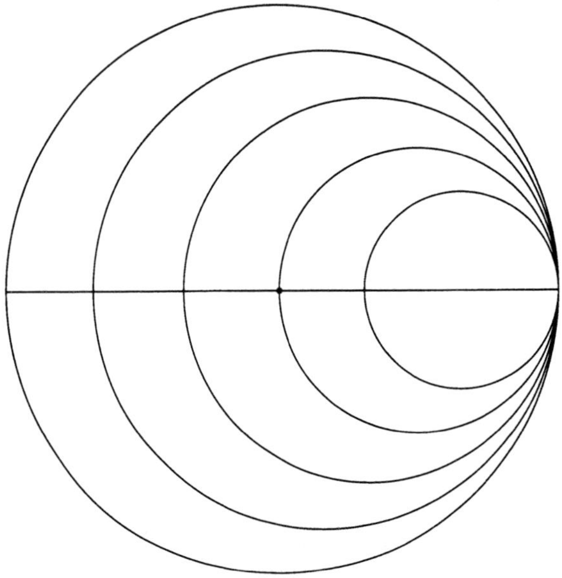

FIGURE 5–4 Circles of constant real part on Smith chart.

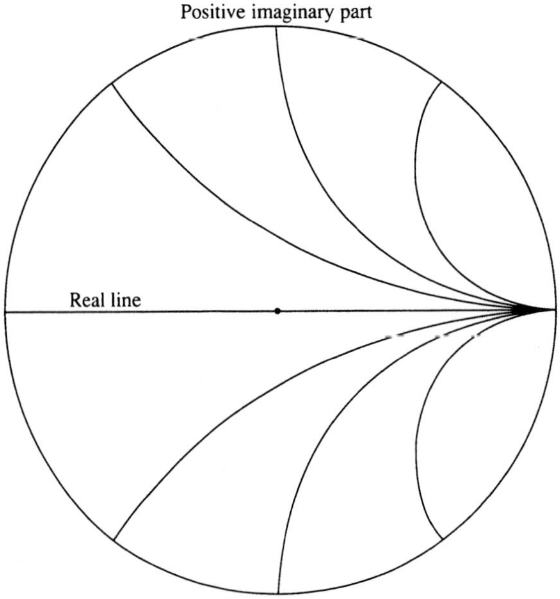

FIGURE 5–5 Circles of constant imaginary part on Smith chart.

real line represent *positive* imaginary parts, and the curves below the real line represent *negative* imaginary parts.

Finally, consider the stripped version shown in Fig. 5–6. The circle shown has been added to the chart by the use of a compass with the pivot point located at the center. The circle will be denoted as an **operating circle** or **VSWR circle.** Every point on this curve corresponds to a fixed value of VSWR and a fixed magnitude (but not angle) of the reflection coefficient.

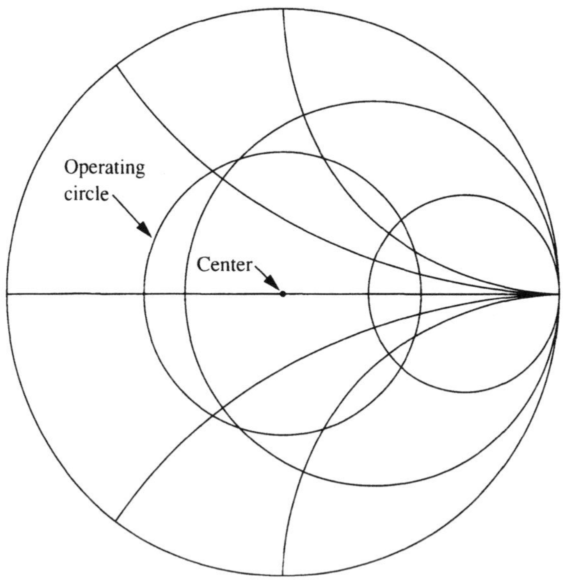

FIGURE 5–6 Operating circle added to Smith chart.

The Smith chart may be used to plot *either impedance or admittance.* However, major errors have been made by people momentarily forgetting which one is being plotted. Therefore, at each step in a problem, get in the habit of pausing to review whether the chart represents impedance or admittance. Some design procedures call for a switch from impedance to admittance (or vice versa) in the middle of a problem, so it is important to keep the perspective clear.

As a continual way of enforcing this process, we will frequently use the terms **impedance basis** and **admittance basis** to remind ourselves of the convention employed.

Impedances or admittances are plotted or read as *normalized* values. Assume that R_0 is the **characteristic impedance** of the lossless line under consideration. Let G_0 represent the corresponding **characteristic admittance.** Since both the impedance and the admittance are real, they are related by

$$G_0 = \frac{1}{R_0} \tag{5–1}$$

Assume that a given impedance Z is to be represented on the chart. The quantity actually plotted is the **normalized impedance** z, given by

$$z = \frac{Z}{R_0} = \frac{R + jX}{R_0} = r + jx \tag{5–2}$$

where $r = R/R_0$ is the normalized resistance and $x = X/R_0$ is the normalized reactance.

Assume next that a given admittance Y is to be represented on the chart. The quantity actually plotted is the **normalized admittance** y, given by

$$y = \frac{Y}{G_0} = \frac{G + jB}{G_0} = g + jb \tag{5–3}$$

where $g = G/G_0$ is the normalized conductance and $b = B/G_0$ is the normalized susceptance.

To convert a normalized impedance back to an actual impedance, multiply the normalized impedance by R_0. Likewise, to convert a normalized admittance back to an actual admittance, multiply the normalized admittance by G_0 (which is equivalent to dividing by R_0). It should be noted that the normalized quantities have no dimensions and, therefore, can be stated simply as dimensionless numbers.

It is left as an exercise for the reader to show that

$$y = \frac{1}{z} \tag{5–4}$$

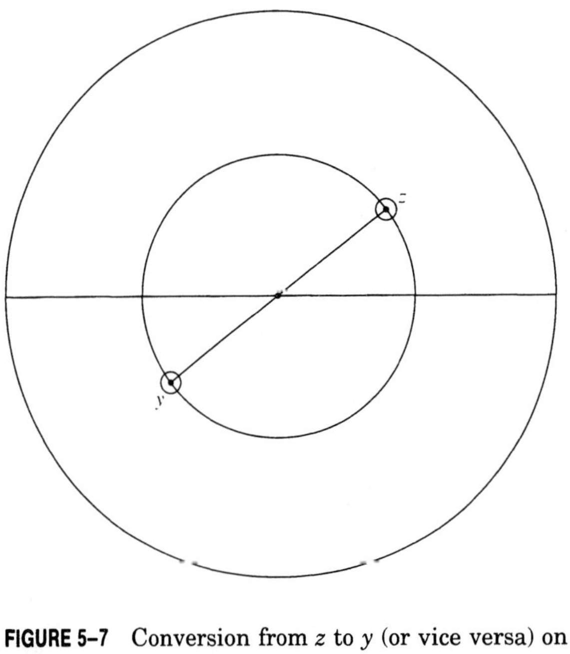

FIGURE 5–7 Conversion from z to y (or vice versa) on Smith chart.

Stated in words, *the reciprocal of the normalized impedance is the normalized admittance.*

The preceding operation can be easily accomplished on the Smith chart. Refer to Fig. 5–7 and assume that a given normalized impedance z has been plotted. Its reciprocal y is a point having exactly the same radius from the center and rotated by one-half of a full rotation. Thus, an operating circle may be temporarily constructed by connecting between the center and z and rotating around 180° on the chart. The value at the new location will be y. In effect, this shifts operation from an impedance basis to an admittance basis.

The preceding may be summarized as follows: *At any point in a problem solution, to switch from an impedance basis to an admittance basis, rotate any point 180° on the chart at a constant radius, and the value will then be defined on an admittance basis.* Conversion from an admittance basis to an impedance basis follows the same procedure.

EXAMPLE 5–1

A 50-Ω lossless line is terminated in $Z_2 = 40 + j70\ \Omega$. (a) Plot the normalized impedance z_2 on a Smith chart. (b) Using the Smith chart, determine the corresponding normalized admittance y_2. (c) Using the result of (b), determine the actual admittance Y_2.

Solution

Refer to Fig. 5–8 in the analysis that follows.

(a) The normalized load impedance is

$$z_2 = \frac{Z_2}{R_0} = \frac{40 + j70}{50} = 0.8 + j1.4 \tag{5–5}$$

To plot the point, locate the circle corresponding to a real part 0.8, and move along the circle until it intersects the imaginary part 1.4. This results in point $P1$ shown on Fig. 5–8. Irrespective of whatever additional analysis is to be done, it is recommended that the operating circle be established at this time as follows: Place the point of the compass at the center, set the radius out to point $P1$, and rotate a full circle around the center.

(b) The normalized admittance y_2 is a point on the operating circle 180° around the circle from $P1$, and this operation is best achieved with a straightedge. The resulting point is $P2$ on Fig. 5–8. The normalized admittance is estimated as

$$y_2 = 0.32 - j0.53 \tag{5–6}$$

(c) Before determining the actual admittance, we note that

$$G_0 = \frac{1}{R_0} = \frac{1}{50} = 0.02\ \text{S} \tag{5–7}$$

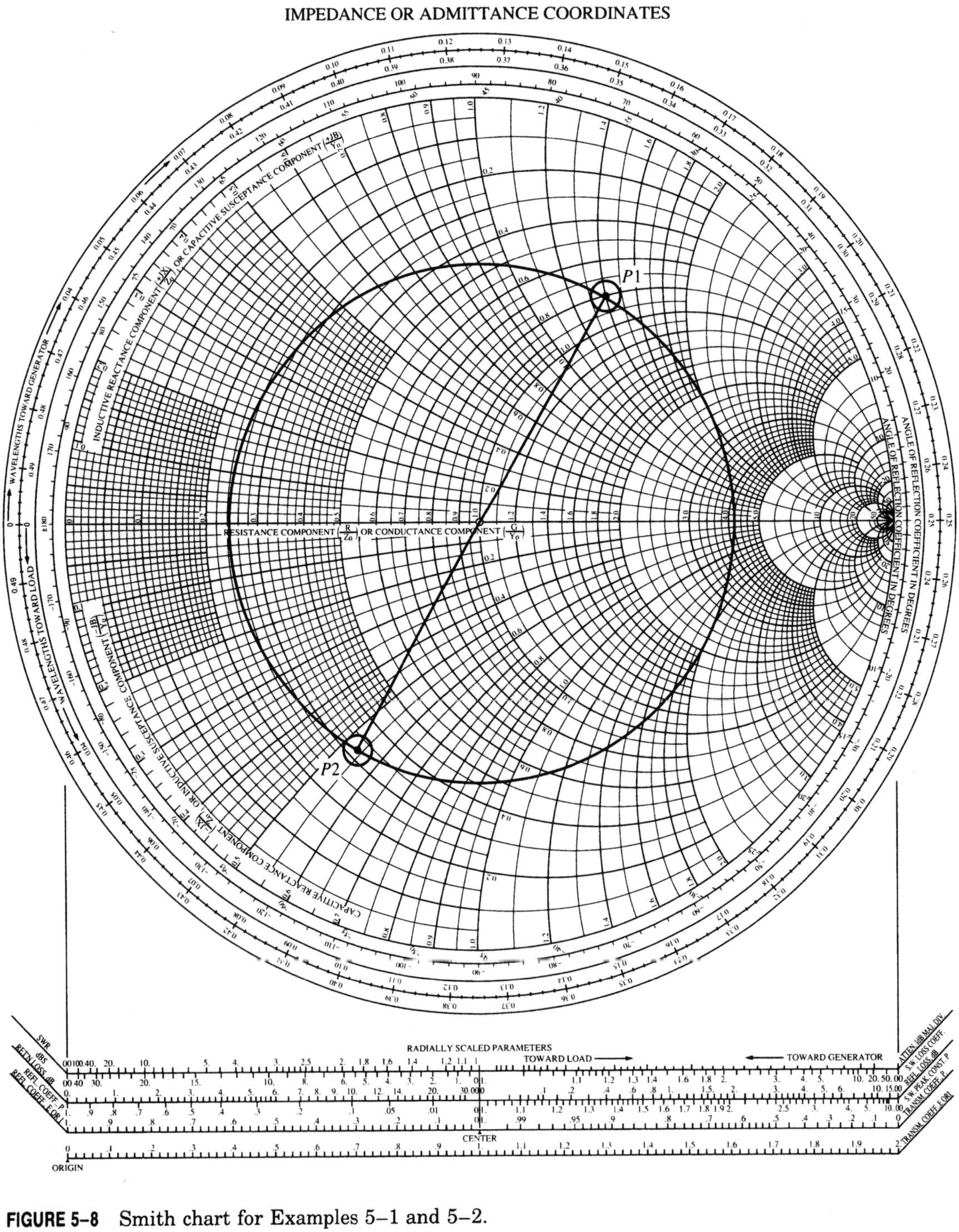

FIGURE 5–8 Smith chart for Examples 5–1 and 5–2.

The actual admittance is then

$$Y_2 = G_0 y_2 = 0.02(0.32 - j0.53) \quad \textbf{(5–8a)}$$
$$= (6.4 \times 10^{-3}) - j(10.6 \times 10^{-3})\ \mathrm{S} \quad \textbf{(5–8b)}$$

The reader may wish to verify analytically that a more exact value of the admittance is $Y_2 = (6.154 \times 10^{-3}) - j(10.77 \times 10^{-3})$ S. The error in the real part is about 3.9%, while the error in the imaginary part is about 1.5%.

5-3 ■ VSWR AND REFLECTION COEFFICIENT

The VSWR and the reflection coefficient may be readily determined from the Smith chart after the operating circle is constructed. Refer to Fig. 5–9 in the discussion that follows.

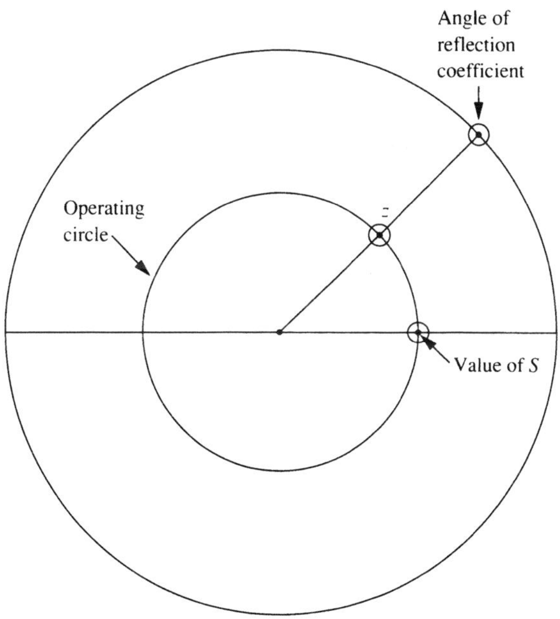

FIGURE 5–9 Determination of VSWR and angle of reflection coefficient.

Assume that a normalized load impedance z_2 is plotted on the chart and that the operating circle has been constructed as shown. One of the circles encompassing the entire Smith chart is labeled "Angle of reflection coefficient" (in degrees). A straight line drawn between the center, the impedance z_2, and this large circle intersects at the value of the reflection coefficient angle. *This particular value of angle is correct only when an impedance basis is employed.* Thus, if admittance is being plotted, it is necessary to switch back to impedance temporarily (by rotating 180° on the operating circle) to read the reflection coefficient angle. One can then convert back to admittance if desired. The remainder of the operations discussed here apply on either an impedance or an admittance basis.

The VSWR is determined by noting the point at which the operating circle intersects the real line on the half where the real part is greater than unity, as shown in Fig. 5–9. The value of this real part is the value of the VSWR.

The final parameter of interest at this time is the magnitude of the reflection coefficient. Most Smith charts have a scale at the bottom with a label such as "Reflection coefficient." By transferring the radius of the operating circle to this scale with the compass, one can read the magnitude from the scale. However, this particular parameter Γ_2 may be readily calculated from the VSWR value by

$$\Gamma_2 = \frac{S-1}{S+1} \tag{5–9}$$

It is questionable whether one can perform this latter operation faster with the chart than with a calculator. The same could be said about certain other values that may be determined from the scales on the bottom. Remember that at the time the Smith chart was developed, computational aids were very limited, and the major computational device was the slide rule.

EXAMPLE 5–2

For the system of Example 5–1, use the Smith chart (along with possible simple calculations) to determine the VSWR and the reflection coefficient.

Solution

Refer to Fig. 5–8 in Example 5–1. The operating circle intersects the real line on the right at a value of about 4.2. Thus,

$$S = 4.2 \tag{5–10}$$

The corresponding magnitude of the reflection coefficient is determined as

$$\Gamma_2 = \frac{S-1}{S+1} = \frac{4.2-1}{4.2+1} = 0.615 \tag{5–11}$$

The reader who wishes to check the use of the "Reflection coefficient" scale can transfer the radius to that scale and estimate a value of $\Gamma_2 \approx 0.61$.

The impedance was plotted as point $P1$. When a straight line is drawn from the center through $P1$, it intersects the reflection coefficient circle at an angle of 60°. Thus, the reflection coefficient can be expressed as

$$\Gamma_2 = 0.615\underline{/60^\circ} = 0.615\underline{/\pi/3} \tag{5–12}$$

where the last form has the angle in radians.

5–4 ■ ROTATION ON THE SMITH CHART

Most operations on the Smith chart require rotation on an operating circle by a certain distance. The distance is measured on the chart in wavelengths. Thus, any specified distance in length units should first be converted to the corresponding number of wavelengths.

A complete rotation of 360° on the chart corresponds to one half-wavelength. Therefore, if a given rotation is greater than 0.5λ, it is easier to simply subtract the largest integer number of half-wavelengths and rotate only by this shorter distance. For example, suppose that we desire to rotate a distance of 3.2λ. There are 6 integer numbers of half-wavelengths contained in 3.2λ, so we subtract $6 \times (0.5\lambda) = 3\lambda$ from 3.2λ, and we need rotate only 0.2λ. As a second example, suppose that we desire to rotate 3.7λ. There are 7 integer numbers of half-wavelengths contained in 3.7λ, so we subtract $7 \times (0.5\lambda) = 3.5\lambda$ from 3.7λ. The answer in this case is the same as in the preceding example, i.e., 0.2λ.

Rotation on the Smith chart can be in either of two directions: *clockwise* or *counterclockwise*. Clockwise rotation corresponds to rotation *toward the generator*. Counterclockwise rotation corresponds to the rotation *toward the load*. These concepts are illustrated in Fig. 5–10.

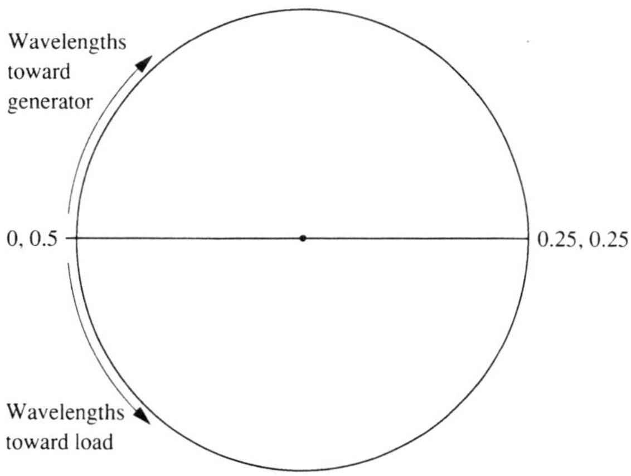

FIGURE 5–10 Scales indicating rotational distance toward generator and toward load.

Errors are frequently made by reading the wrong scale or by misinterpreting the results. We will look at each of the scales individually.

First, consider the "Wavelengths toward generator" circle shown in Fig. 5–10. The value in wavelengths is read clockwise starting from 0 on the left-hand end of the real line, increasing to 0.25 on the right-hand end of the real line, and increasing back to 0.5 on the left again.

Suppose that we desire to rotate a distance R from an initial point I to reach a final point F. (All values are measured in wavelengths.) The coordinate value associated with F is initially determined simply as $F = I + R$. If this resulting value is less than 0.5λ, simply use a straightedge connecting between center and the desired coordinate. For example, suppose that the initial coordinate is 0.1λ and that we desire to rotate by 0.3λ. The final coordinate is $0.1\lambda + 0.3\lambda = 0.4\lambda$.

If the value of F is greater than 0.5λ, simply subtract 0.5λ or the smallest integer number of half-wavelengths that provides a value on the scale. For example, suppose that the initial coordinate is 0.4λ and that we desire to rotate 0.3λ. The final coordinate appears to be $0.4\lambda + 0.3\lambda = 0.7\lambda$, which is greater than 0.5λ. We thus subtract 0.5λ and obtain the coordinate of the final position as $0.7\lambda - 0.5\lambda = 0.2\lambda$.

Next, consider the "Wavelengths toward load" circle shown in Fig. 5–10. The value in wavelengths is read counterclockwise starting from 0 on the left-hand end of the real line, increasing to 0.25 on the right-hand end of the real line, and increasing back to 0.5 on the left again. Use of this scale follows the same procedure discussed for the other rotational scale except, of course, that rotation is always counterclockwise.

5–5 ■ DETERMINING INPUT IMPEDANCE FROM LOAD IMPEDANCE

A basic operation with the Smith chart is to determine the input impedance of a specified length of transmission line from a knowledge of the load impedance. The operation to be described applies equally well to determining input admittance from the load admittance, but to simplify the discussion that follows, an impedance basis will be assumed.

Assume that the load impedance $Z_2 = R_2 + jX_2$ and the characteristic impedance R_0 are given. The procedure follows:

1. Calculate the normalized load impedance as

$$z_2 = \frac{R_2 + jX_2}{R_0} = r_2 + jx_2 \tag{5–13}$$

   and plot it on the Smith chart.
2. Using the compass, construct the operating circle.
3. Using a straightedge and the concepts discussed in the preceding section, rotate *clockwise* (i.e., toward the generator) by the length of the line.
4. Read the value of the normalized input impedance $z_{in} = r_{in} + jx_{in}$ at the point where the straightedge intersects the operating circle when the final coordinate is reached.
5. The actual input impedance Z_{in} may be calculated as

$$Z_{in} = R_0 \times z_{in} = R_0 r_{in} + jR_0 x_{in} \tag{5–14}$$

6. Any other desired parameters may be determined as previously discussed if desired (e.g., VSWR, Γ, and so on).

EXAMPLE 5–3

A 50-Ω lossless transmission line is terminated in a real load impedance $Z_2 = R_2 = 10\ \Omega$. Using the Smith chart, determine the input impedance for each of the following line lengths: (a) 0.1λ, (b) 0.25λ, (c) 0.375λ, (d) 0.5λ, and (e) 7.6λ.

Solution

Refer to Fig. 5–11 in the steps that follow. First, the normalized impedance z_2 is calculated as

$$z_2 = \frac{Z_2}{R_0} = \frac{10}{50} = 0.2 \tag{5–15}$$

Since the imaginary part is zero, this impedance is plotted on the real line and is represented as point P on Fig. 5–11. The operating circle is then constructed as shown. To determine the input impedance for any arbitrary length, we rotate clockwise on the operating circle and read the corresponding normalized input impedance. The actual input impedance is then determined by multiplying the normalized value by 50.

In this problem, the rotational coordinate has the simplest possible form because the load point P corresponds to a distance of 0 on the "Wavelengths toward generator" scale.

(a) 0.1λ

A clockwise rotation of 0.1λ results in point $P1$ on Fig. 5–11. The normalized impedance is estimated as

$$z_{in} = 0.3 + j0.68 \tag{5–16}$$

The actual input impedance is

$$Z_{in} = R_0 z_{in} = 50 \times (0.3 \times j0.68) = 15 + j34\ \Omega \tag{5–17}$$

(b) 0.25λ

A clockwise rotation of 0.25λ results in point $P2$, and the normalized input impedance is a real value of

$$z_{in} = 5 \tag{5–18}$$

The actual input impedance is

$$Z_{in} = 50 \times 5 = 250\ \Omega \tag{5–19}$$

This case could have been most easily solved with the formula for a quarter-wavelength line, but it is interesting to see the transformation take place on the Smith chart.

There is another point worth noting in the preceding computation: Rotation of an impedance by λ/4 corresponds on the Smith chart to the same

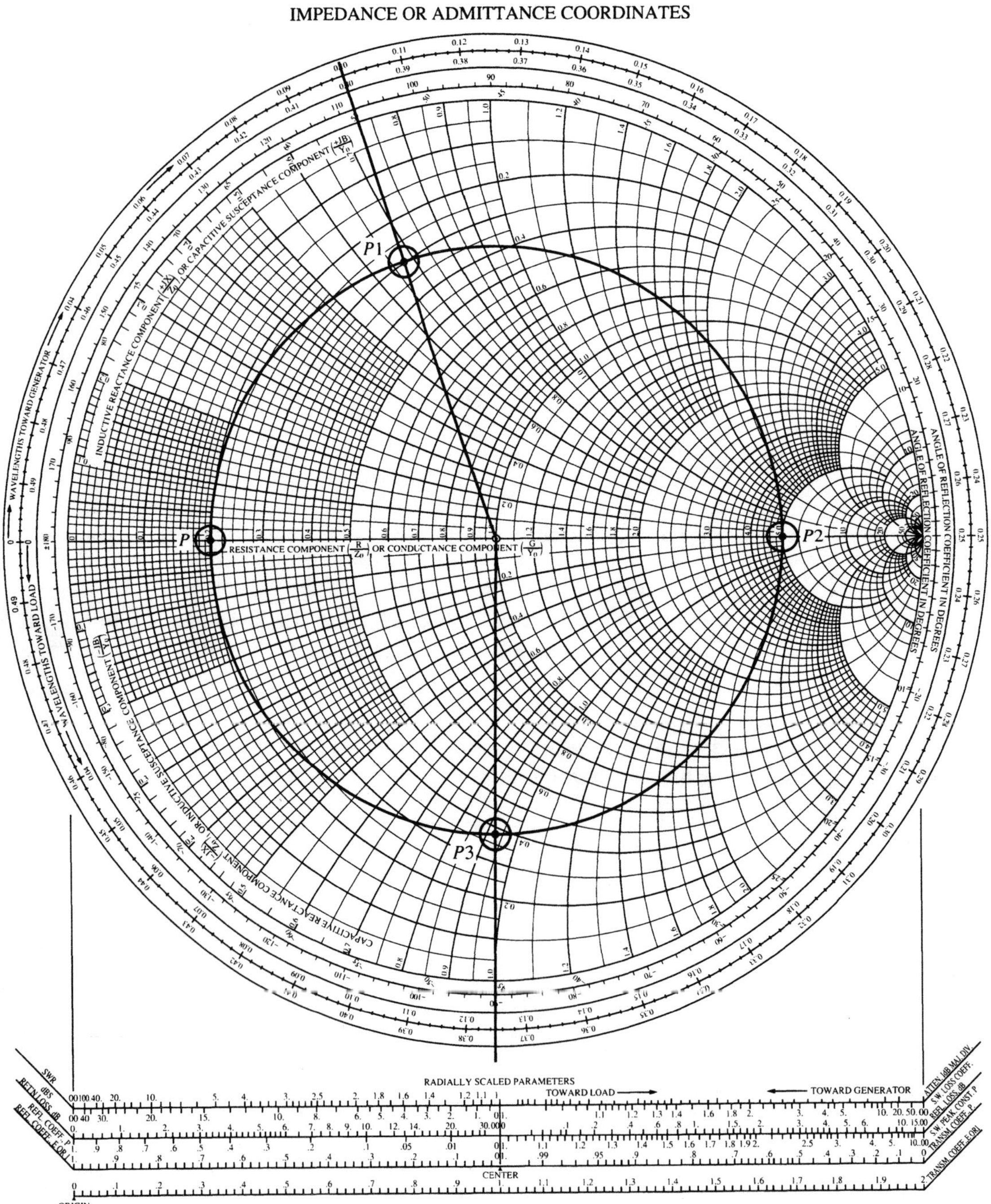

FIGURE 5–11 Smith chart of Example 5–3.

physical process as switching from impedance to admittance. However, the results are interpreted differently. In this example, rotation of a normalized impedance of 0.2 by $\lambda/4$ results in a normalized *impedance* of 5. If, instead, this rotation had been performed solely for converting impedance to admittance, the *impedance* value of 0.2 would then have been read as an *admittance* of 5, which is obviously correct. It is often helpful to visualize the similarity of the $\lambda/4$ transformation to the process of switching from impedance to admittance (or vice versa).

(c) 0.375λ

A clockwise rotation of 0.375λ results in point $P3$ on Fig. 5–11. The normalized input impedance is estimated as

$$z_{in} = 0.38 - j0.93 \quad \textbf{(5–20)}$$

The actual input impedance is approximately

$$Z_{in} = 50 \times (0.38 - j0.93) = 19 - j46.5\ \Omega \quad \textbf{(5–21)}$$

(d) 0.5λ

We certainly do not need a Smith chart to predict the result in this case, but to ensure that all readers understand the concept, rotation carries us back around to point P. Thus, the input impedance is

$$Z_{in} = 10\ \Omega \quad \textbf{(5–22)}$$

(e) 7.6λ

Since this length is greater than 0.5λ, we can first subtract the largest integer number of half-wavelengths. Specifically, there are 15 integer multiples of $\lambda/2$, that is, 7.5λ, contained in 7.6λ. Thus, we rotate only $7.6\lambda - 7.5\lambda = 0.1\lambda$. This value is the one considered in part (a), so the impedance is simply

$$Z_{in} = 15 + j34\ \Omega \quad \textbf{(5–23)}$$

Incidentally, it is not necessary to subtract out the value 7.5λ prior to rotation. Instead, one could rotate "around and around" for 15 complete rotations plus 0.1λ and still arrive at the same point!

EXAMPLE 5–4

A lossless 75-Ω line is terminated in a load impedance $Z_2 = 105 - j150\ \Omega$. (a) Using the Smith chart, determine the input impedance for a line of length 0.35λ. (b) Determine the VSWR and the reflection coefficient at the load.

Solution

Refer to Fig. 5–12 for the calculations that follow.

(a) The normalized load impedance is

$$z_2 = \frac{105 - j150}{75} = 1.4 - j2 \quad \textbf{(5–24)}$$

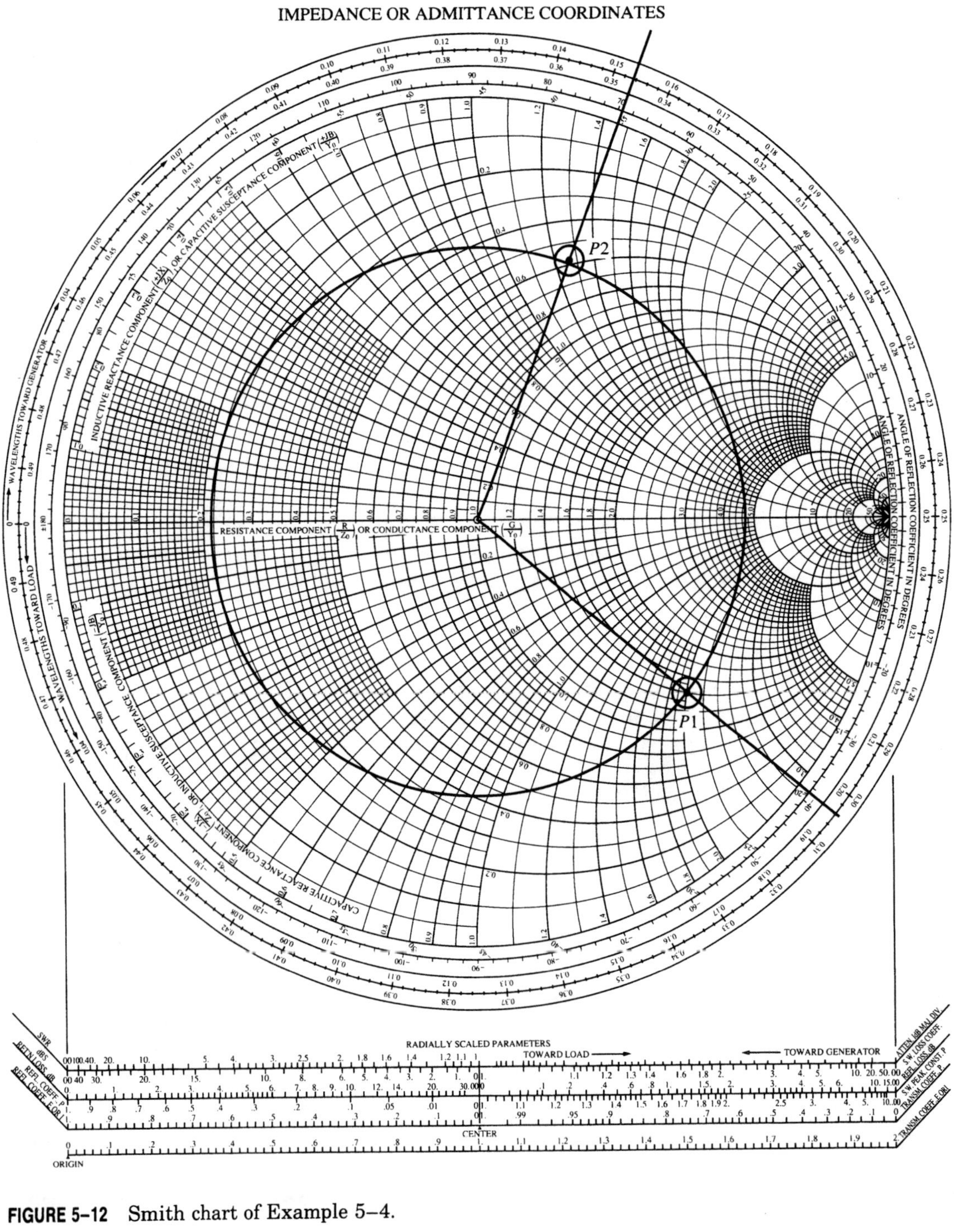

FIGURE 5–12 Smith chart of Example 5–4.

This value is shown in $P1$ on Fig. 5–12, and the operating circle is then constructed.

The initial coordinate on the "Wavelengths toward generator" scale is about 0.304λ. We must rotate clockwise by 0.35λ, so this would call for a final coordinate of 0.304λ + 0.35λ = 0.654λ. However, this value exceeds 0.5λ, so we subtract 0.5λ to obtain a coordinate of 0.654λ − 0.5λ = 0.154λ. The corresponding point is $P2$ on Fig. 5–12. The value of this normalized impedance is read as

$$z_{in} = 0.58 + j1.28 \tag{5–25}$$

The corresponding actual impedance is

$$Z_{in} = 75 \times (0.58 + j1.28) = 43.5 + j96\ \Omega \tag{5–26}$$

(b) The VSWR is read from the intersection of the operating circle with the right half of the real line and is approximately

$$S = 4.7 \tag{5–27}$$

The magnitude of the reflection coefficient is determined as

$$\Gamma_2 = \frac{S-1}{S+1} = \frac{4.7-1}{4.7+1} = 0.649 \tag{5–28}$$

The angle of the reflection coefficient at the load is about −38.5°. Thus, the reflection coefficient at the load is about

$$\overline{\Gamma}_2 = 0.649\underline{/-38.5^\circ} = 0.649\underline{/-0.672} \tag{5–29}$$

where the latter angle is −0.672 rad.

EXAMPLE 5–5

The antenna in an RF system can be modeled by the circuit shown in Fig. 5–13. The antenna is fed by a lossless 50-Ω line. Determine the input impedance for a line of length 0.15λ.

Solution

Be careful here! It would be too easy to quickly express the impedance (incorrectly) as 40 − j25 Ω, but that value would be correct *only if this were a series connection.* One could first calculate the parallel equivalent impedance and

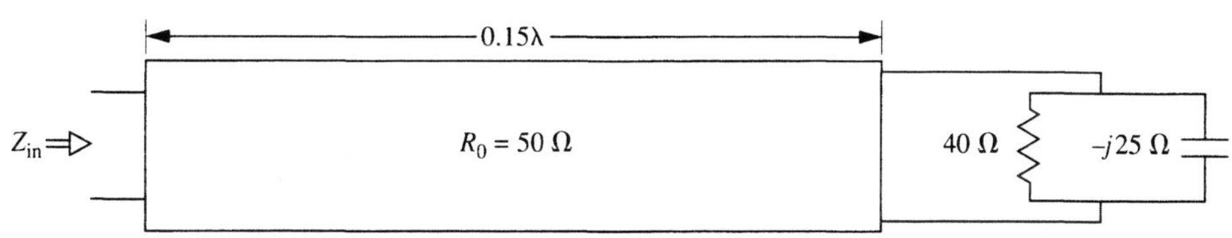

FIGURE 5–13 Circuit model of Example 5–5.

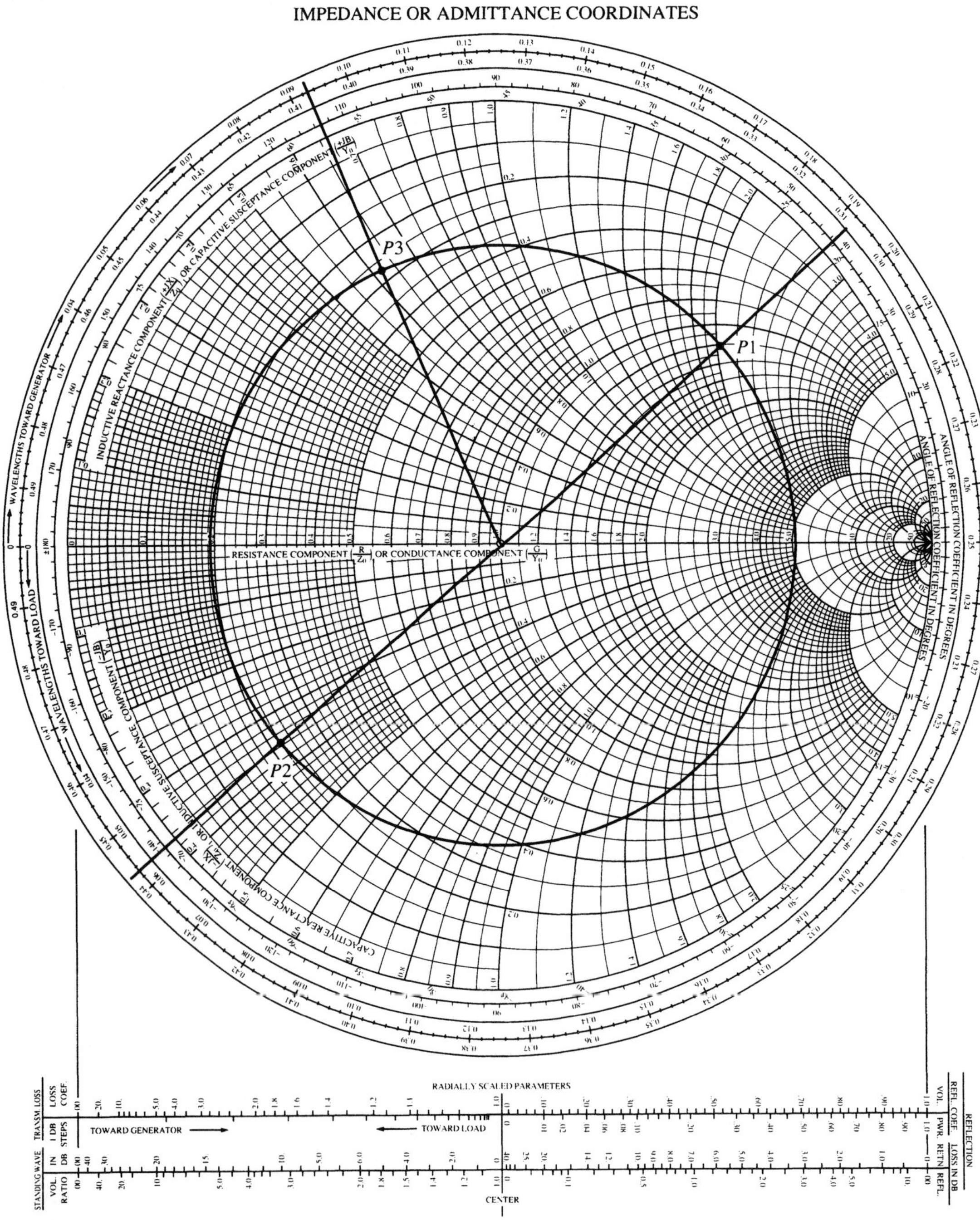

FIGURE 5–14 Smith chart of Example 5–5.

plot its normalized value. However, a simpler approach on the Smith chart is to first express the load impedance as a load admittance and then convert that value to impedance. For this purpose, the individual impedances are reciprocated and added. The admittance corresponding to 40 Ω is 1/40 = 0.025 S, and the admittance corresponding to $-j25$ Ω is $1/(-j25) = j0.04$ S. Thus, the load admittance is

$$Y_2 = 0.025 + j0.04 \text{ S} \tag{5–30}$$

The characteristic admittance is $G_0 = 1/R_0 = 1/50 = 0.02$ S. The normalized load admittance is

$$y_2 = \frac{0.025 + j0.04}{0.02} = 1.25 + j2 \tag{5–31}$$

This value is shown as point $P1$ on Fig. 5–14.

Since the desired answer is to be an impedance, two rotations are necessary: (1) conversion of y_2 to a corresponding z_2 and (2) transformation of the load through a length of 0.15λ. The two required rotations may be performed in either order, but we will arbitrarily choose the order just mentioned. Thus, we first rotate 180° on the chart to convert y_2 to z_2, and this results in point $P2$. If desired, we can read the normalized value of z_2 at that point, and it is about $z_2 = 0.22 - j0.36$.

The initial coordinate on an impedance basis is about 0.443λ. Rotation by 0.15λ toward the generator would result in a final coordinate of 0.443λ + 0.15λ = 0.593λ. However, this value exceeds 0.5λ, so we subtract 0.5λ to obtain a final coordinate 0.593λ − 0.5λ = 0.093λ. This is shown as $P3$, and the normalized input impedance is estimated as

$$z_{in} = 0.28 + j0.63 \tag{5–32}$$

The actual input impedance is

$$Z_{in} = 50(0.28 + j0.63) = 14 + j31.5\ \Omega \tag{5–33}$$

5-6 ■ DETERMINING LOAD IMPEDANCE FROM INPUT IMPEDANCE

The inverse operation of that considered in the preceding section is the process of determining the load impedance from a knowledge of the input impedance. This process is often necessary when the actual load impedance is inaccessible, e.g., an antenna mounted on a tower. As in the preceding section, the operation applies equally well for both impedance and admittance, but it will be explained on an impedance basis.

Assume that the input impedance $Z_{in} = R_{in} + jX_{in}$ and the characteristic impedance R_0 are given. (In most practical applications, the input impedance can be measured with an RF bridge.) The procedure follows:

1. Calculate the normalized input impedance

$$z_{in} = \frac{R_{in} + jX_{in}}{R_0} = r_{in} + jx_{in} \tag{5–34}$$

2. Using a compass, construct the operating circle.
3. Using a straightedge and the concepts discussed in Section 5–4, rotate *counterclockwise* (i.e., toward load) by the length of the line.
4. Read the value of the normalized load impedance $z_2 = r_2 + jx_2$ at the point where the straightedge intersects the operating circle.
5. The actual load impedance Z_2 may be calculated as

$$Z_2 = R_0 \times z_2 = R_0 r_2 + jR_0 x_2 \tag{5–35}$$

6. Any other desired parameters may be determined as previously discussed if desired (e.g., VSWR, reflection coefficient, and so on).

EXAMPLE 5–6

An inaccessible RF load is connected through a 300-Ω assumed-lossless line to a measurement point. The line has a length of 6.9λ. The input impedance is measured as $Z_{in} = 450 - j150\ \Omega$. Determine (a) the value of the load impedance, (b) the VSWR, and (c) the reflection coefficient at the load.

Solution

Refer to Fig. 5–15 in the work that follows.

(a) The normalized value of the input impedance is

$$z_{in} = \frac{Z_{in}}{R_0} = \frac{450 - j150}{300} = 1.5 - j0.5 \tag{5–36}$$

This value is shown as point $P1$ in Fig. 5–15, and the operating circle is constructed as shown.

The required rotation exceeds 0.5λ, and the largest integer multiple of 0.5λ is 6.5λ. Thus, we need only rotate 6.9λ − 6.5λ = 0.4λ. Rotation in this case is counterclockwise, i.e., on the "Wavelengths toward load" scale. The initial coordinate corresponds to $P1$ and is about 0.203λ. The final coordinate is about 0.203λ + 0.4λ = 0.603λ. However, this value exceeds 0.5λ, so we subtract 0.5λ to obtain 0.603λ − 0.5λ = 0.103λ. Thus, the final coordinate is 0.103λ on the "Wavelengths toward load" scale, and this results in point $P2$. The normalized load impedance is read as

$$z_2 = 0.74 - j0.46 \tag{5–37}$$

The actual load impedance is

$$Z_2 = 300 \times (0.74 - j0.46) = 222 - j138\ \Omega \tag{5–38}$$

IMPEDANCE OR ADMITTANCE COORDINATES

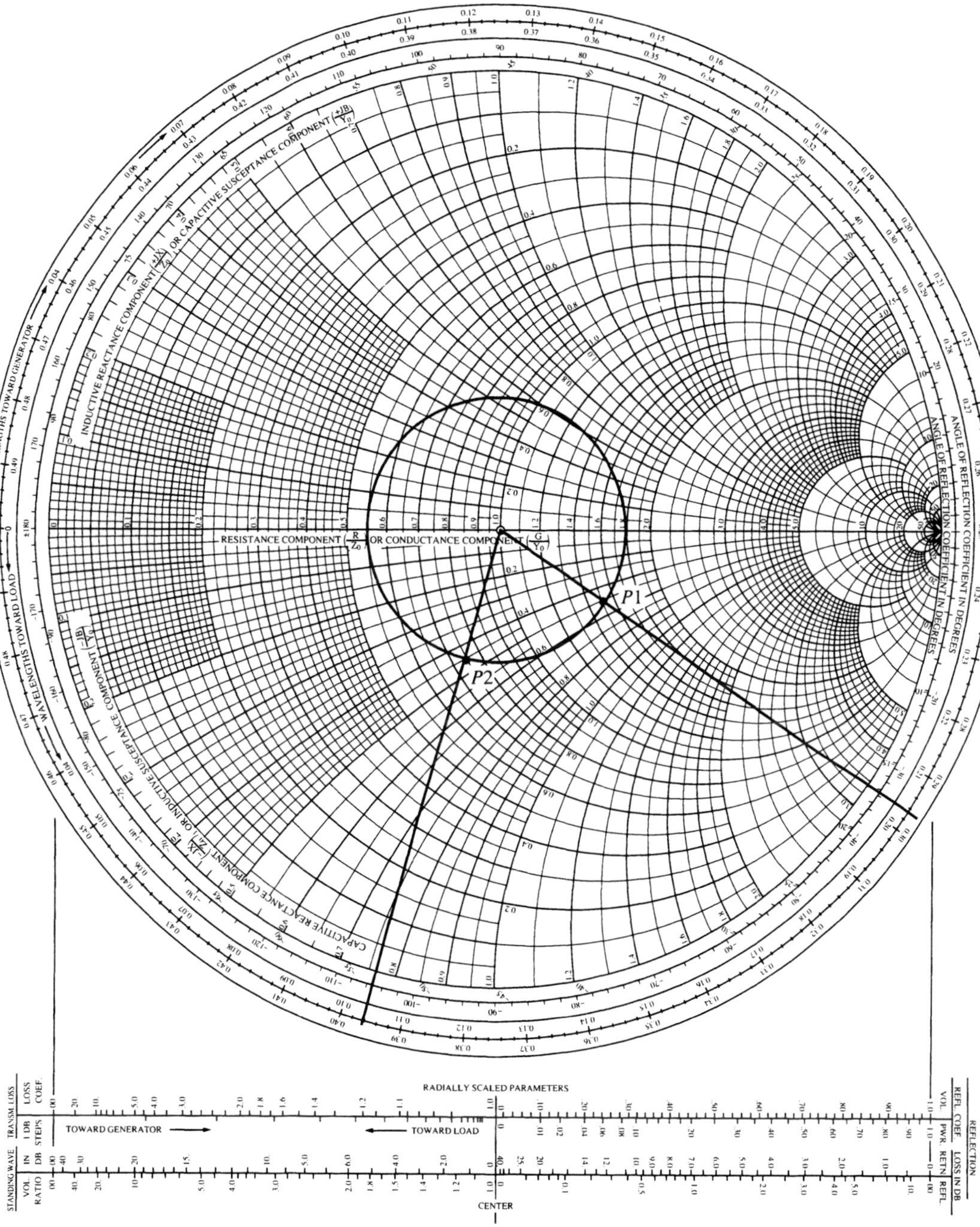

FIGURE 5–15 Smith chart of Example 5–6.

(b) The VSWR is estimated as

$$S = 1.78 \tag{5–39}$$

(c) The magnitude of the reflection coefficient Γ_2 is calculated as

$$\Gamma_2 = \frac{1.78 - 1}{1.78 + 1} = 0.281 \tag{5–40}$$

The angle of the reflection coefficient at the load is estimated as $-105.5° = -1.84$ rad. The complex load reflection coefficient is then

$$\bar{\Gamma}_2 = 0.281\underline{/-105.5°} = 0.281\underline{/-1.84} \tag{5–41}$$

5–7 ■ VOLTAGE AND CURRENT MAXIMA AND MINIMA

When a transmission line connected to an RF load is excited by a sinusoidal source, a standing wave of voltage and current will appear on the line unless the line is matched at the load. As noted in Chapter 4, the distance between successive maxima or minima is 0.5λ. These points may be located easily with the help of the Smith chart, as will be demonstrated next.

First, let us establish certain properties that will assist in the process:

1. Both maxima and minima of voltage and current occur at points at which the imaginary part of the impedance or admittance is zero; i.e., they occur at intersections of the operating circle with the real line on the Smith chart.
2. A voltage *minimum* occurs at the same point as a current *maximum,* and this corresponds to a point of *minimum resistance* or *maximum conductance.*
3. A voltage *maximum* occurs at the same point as a current *minimum,* and this corresponds to a point of *maximum resistance* or *minimum conductance.*

The preceding points may be displayed graphically on the real lines of two separate Smith charts as shown in Fig. 5–16. The line of (a) indicates the locations of various maxima and minima when the chart is defined on an *impedance basis,* and (b) indicates the corresponding properties when the chart is defined on an *admittance basis.* Instead of memorization, it is suggested that you think of the properties previously listed and reason out the conditions in a logical way.

EXAMPLE 5–7

Suppose that we desire to determine the value of a certain unknown load impedance connected to an assumed-lossless 50-Ω line. The VSWR is measured as $S = 3$. It is determined that a voltage minimum is located at a distance of 3.15λ from the load. Determine the value of the load impedance.

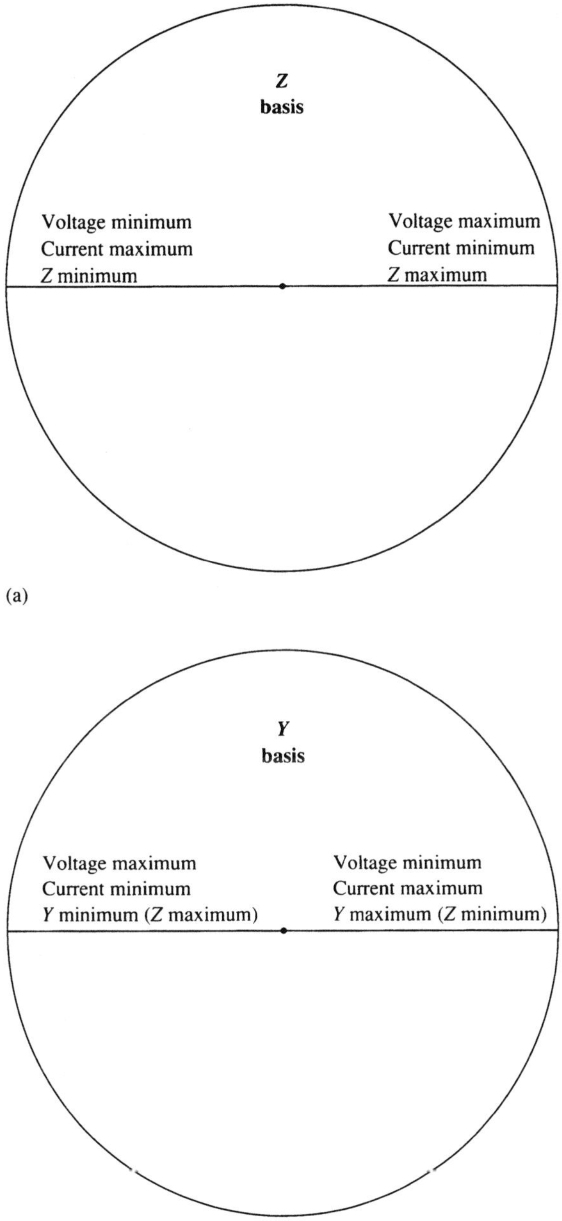

FIGURE 5–16 Locations of maximum and minimum values on Smith chart.

IMPEDANCE OR ADMITTANCE COORDINATES

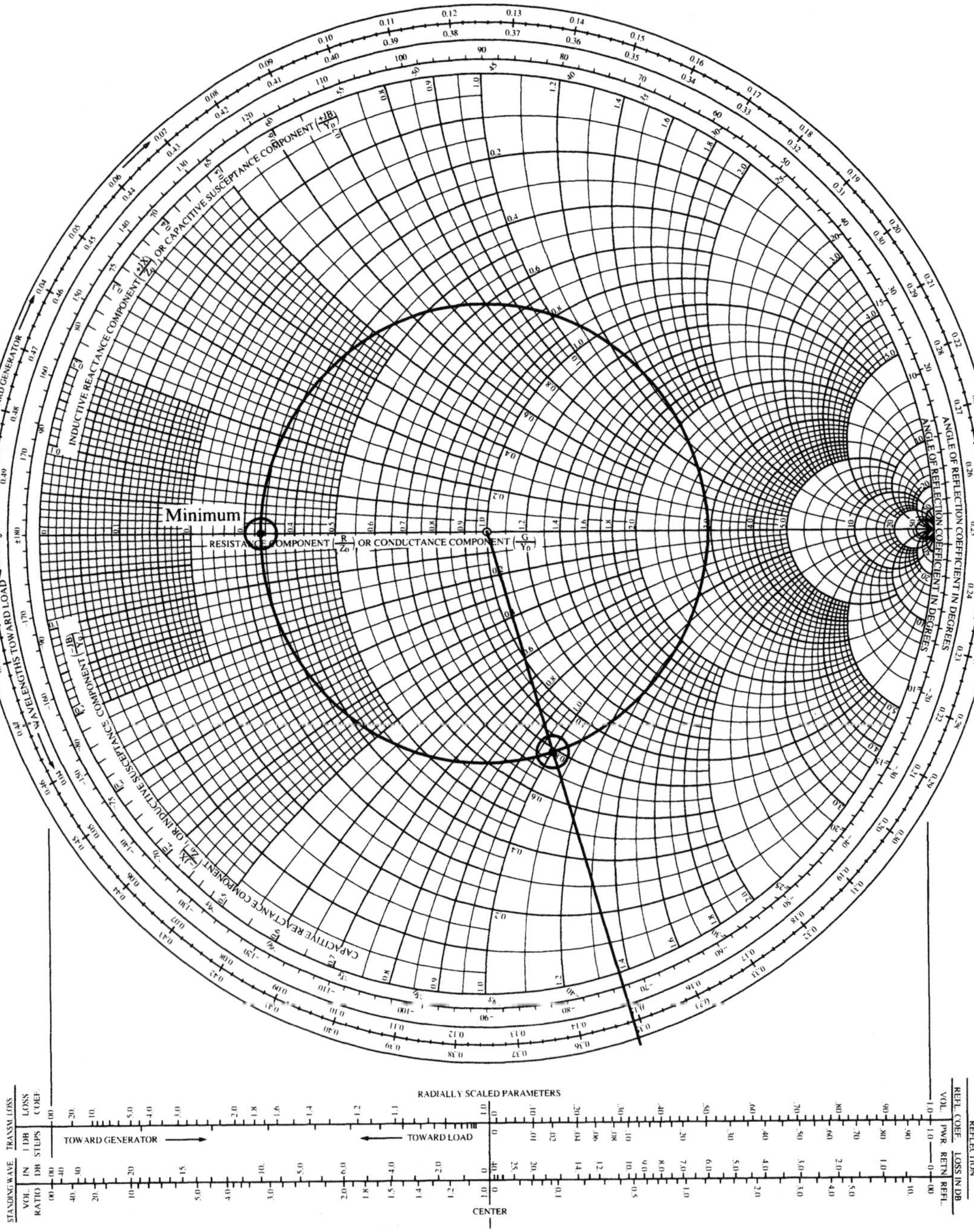

FIGURE 5–17 Smith chart of Example 5–7.

Solution

Refer to Fig. 5–17 in the analysis that follows. Since the result is desired as an impedance, the Smith chart will be defined on an *impedance basis*. Thus, voltage minima, which correspond to current maxima and resistance minima, will be on the left-hand side of the real line, as indicated.

In order to reach the load from a voltage minimum, we should rotate back toward the load by the required amount. Since 3.15λ exceeds 0.5λ, we subtract 3λ and rotate $3.15\lambda - 3\lambda = 0.15\lambda$ as shown. The normalized impedance z_2 is estimated as

$$z_2 = 0.78 - j1 \tag{5–42}$$

The actual impedance Z_2 is

$$Z_2 = 50(0.78 - j1) = 39 - j50\ \Omega \tag{5–43}$$

5–8 ■ SHORTED AND OPEN LINES ON SMITH CHART

The property was developed in Chapter 4 that the input impedance or admittance of a lossless line terminated by either a short circuit or an open circuit is reactive at all points on the line; i.e., the real part is zero. It was noted in Section 5–2 that the larger inner circle corresponds to a real part of zero. Hence, the impedance or admittance of all short or open lines will be located on this circle as illustrated in Fig. 5–18.

As in the preceding section, it is necessary to first establish bearings on whether an impedance or an admittance basis is used and whether a short or an open line is being considered. It is helpful to think of a short circuit as a point of zero impedance or infinite admittance, and an open circuit as a point of infinite impedance or zero admittance. This leads to the scheme shown in (a) of Fig. 5–18 for an impedance basis and that of (b) for an admittance basis.

EXAMPLE 5–8

An assumed-lossless 50-Ω line is terminated in a short circuit. Determine the input impedance for a line of length 2.9λ.

Solution

Refer to the Smith chart of Fig. 5–19 in the work that follows. The chart will be defined on an impedance basis since the desired result is to be expressed as an impedance. A shorted line corresponds to the point $P1$ on the left-hand end of the real line. The operating circle is the large inner circle. Since the length is 2.9λ, we subtract 2.5λ and rotate $(2.9 - 2.5)\lambda = 0.4\lambda$ in the direction of the generator. The resulting normalized input impedance is found at point $P2$ and is estimated as

$$z_{in} = -j0.72 \tag{5–44}$$

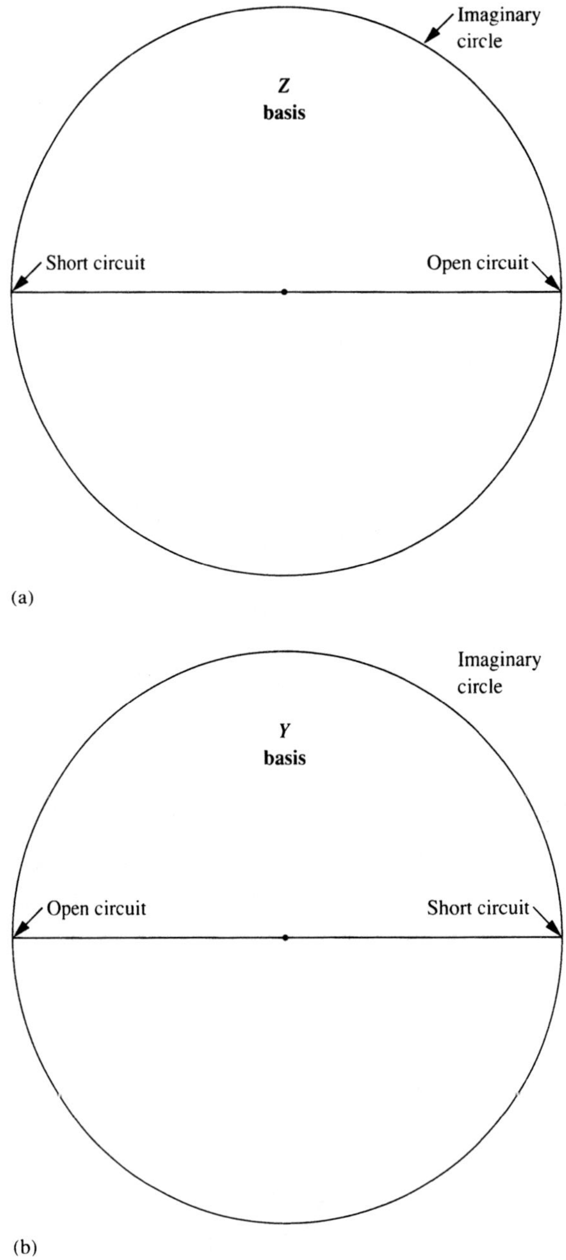

FIGURE 5–18 Locations of short and open circuits for Z and Y bases on Smith chart.

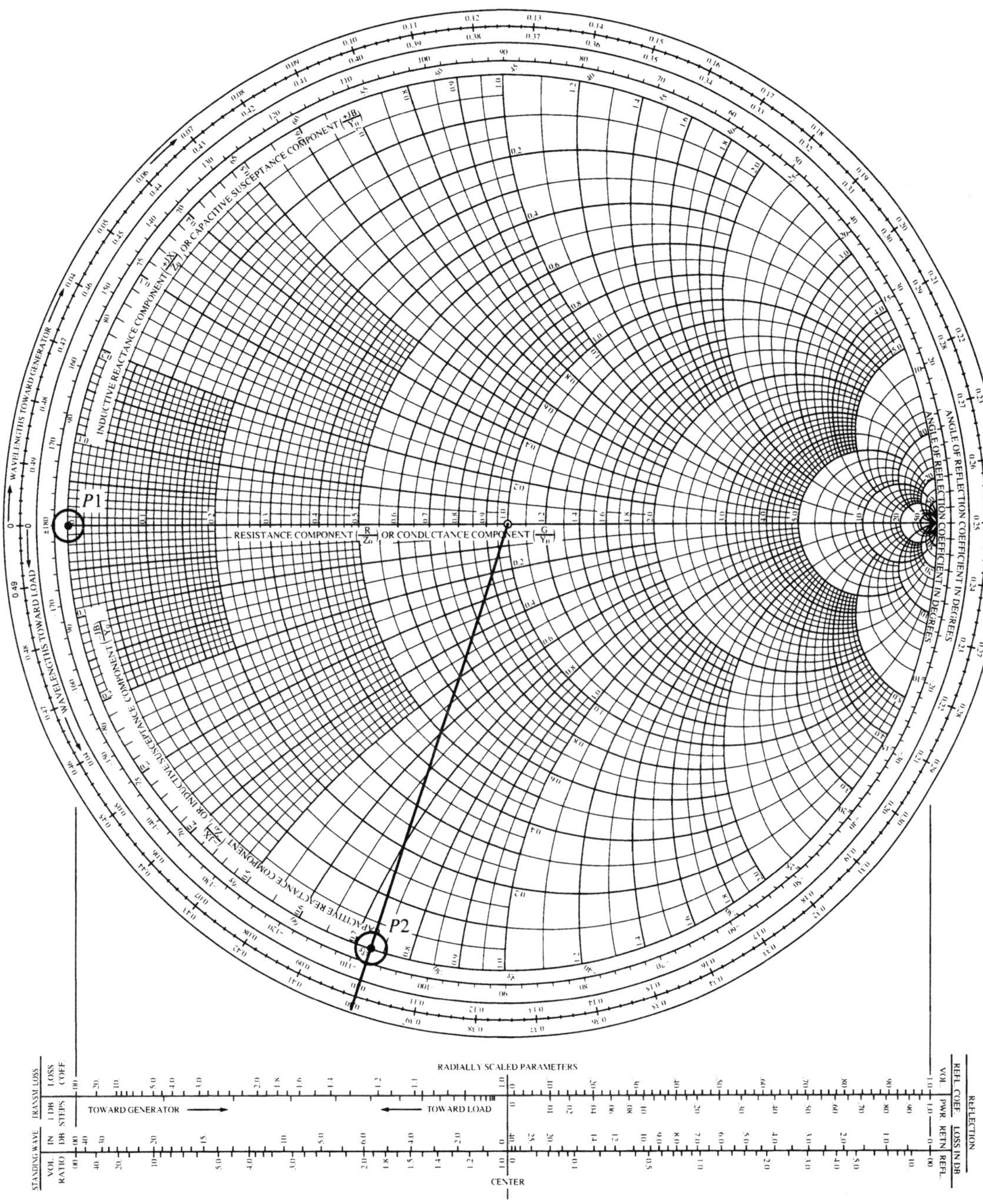

FIGURE 5–19 Smith chart of Example 5–8.

The actual input impedance is

$$Z_{\text{in}} = 50(-j0.72) = -j36\ \Omega \tag{5–45}$$

5–9 ■ QUARTER-WAVE TRANSFORMER

A common technique for impedance matching in microwave-integrated circuits (MIC) is to use a quarter-wave transformer. Such use is practical because microstrip transmission line sections may be fabricated to produce an arbitrary value of characteristic impedance (within reasonable limits).

The concept is illustrated in Fig. 5–20. Assume that the input is a generator with resistive internal impedance R_1 and that the load is an impedance $Z_2 = R_2 + jX_2$.

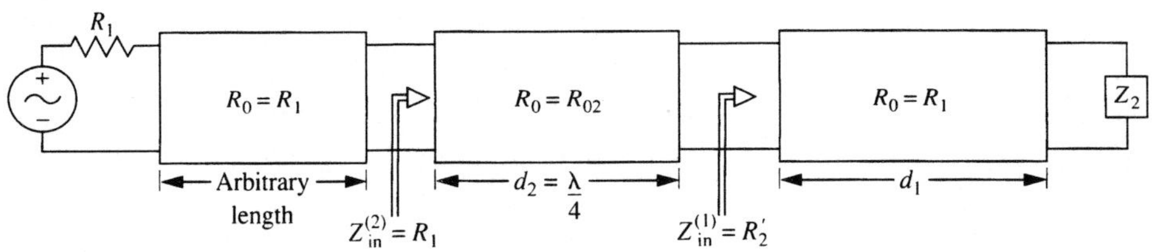

FIGURE 5–20 Block diagram illustrating design approach for quarter-wave transformer.

Maximum power transfer cannot be achieved directly because of (a) the presence of a reactive component at the load and (b) different impedance levels between generator and load.

Starting at the load, the first task is to cancel out the reactive component of the load impedance. This can be achieved by determining a length d_1 for the first section at which the input impedance is real. We will assume that the characteristic impedance of this section is selected as a value R_1, which corresponds to the internal resistance of the generator. (However, the concept will work with other values.)

Assume that at a distance d_1 from the load, the input impedance is real and has the value

$$Z_{\text{in}}^{(1)} = R_2' \tag{5–46}$$

where R_2' will, in general, be different from R_2. A section of line of length $d_2 = \lambda/4$ is then fabricated to yield a characteristic impedance R_{02} that satisfies

$$R_{02} = \sqrt{R_1 R_2'} \tag{5–47}$$

The input impedance of this section will then be given as

$$Z_{\text{in}}^{(2)} = R_1 \tag{5–48}$$

Any additional coupling between the generator and this point may be achieved with a length of line with characteristic impedance $R_0 = R_1$, and the system is matched.

EXAMPLE 5-9

Refer to the circuit of Fig. 5–21. A 10-GHz MIC amplifier is to be driven by a generator whose output impedance is 50 Ω resistive. The input impedance to the amplifier is $Z_2 = 20 - j60\ \Omega$. To ensure maximum power transfer, a quarter-wave matching system is to be designed based on the procedure of this section. The effective dielectric constant of the microstrip transmission line used in the MIC is 2.25. Determine (a) the shortest length for d_1, (b) R_{02}, and (c) d_2.

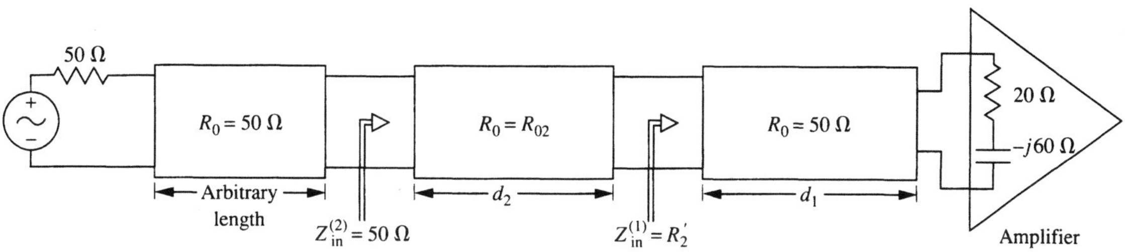

FIGURE 5–21 Circuit of Example 5–9.

Solution

As a few preliminary calculations, the velocity of propagation for the MIC is

$$v = \frac{3 \times 10^8}{\sqrt{2.25}} = 2 \times 10^8 \text{ m/s} \tag{5–49}$$

The wavelength λ is

$$\lambda = \frac{v}{f} = \frac{2 \times 10^8}{10 \times 10^9} = 0.02 \text{ m} = 2 \text{ cm} \tag{5–50}$$

The generator internal impedance $R_1 = 50\ \Omega$ will be used as the basis for matching. The normalized input impedance to the amplifier is

$$z_2 = \frac{Z_2}{50} = \frac{20 - j60}{50} = 0.4 - j1.2 \tag{5–51}$$

Refer to Fig. 5–22 in the steps that follow.

(a) The value of z_2 is plotted as point $P1$ on Fig. 5–22, and the operating circle is constructed. We then rotate in the direction of the generator until a point is reached at which the reactance is zero. This corresponds to the points at which the operating circle intersects the real line. The solution corresponding to the shortest length is point $P2$ on Fig. 5–22. The distance

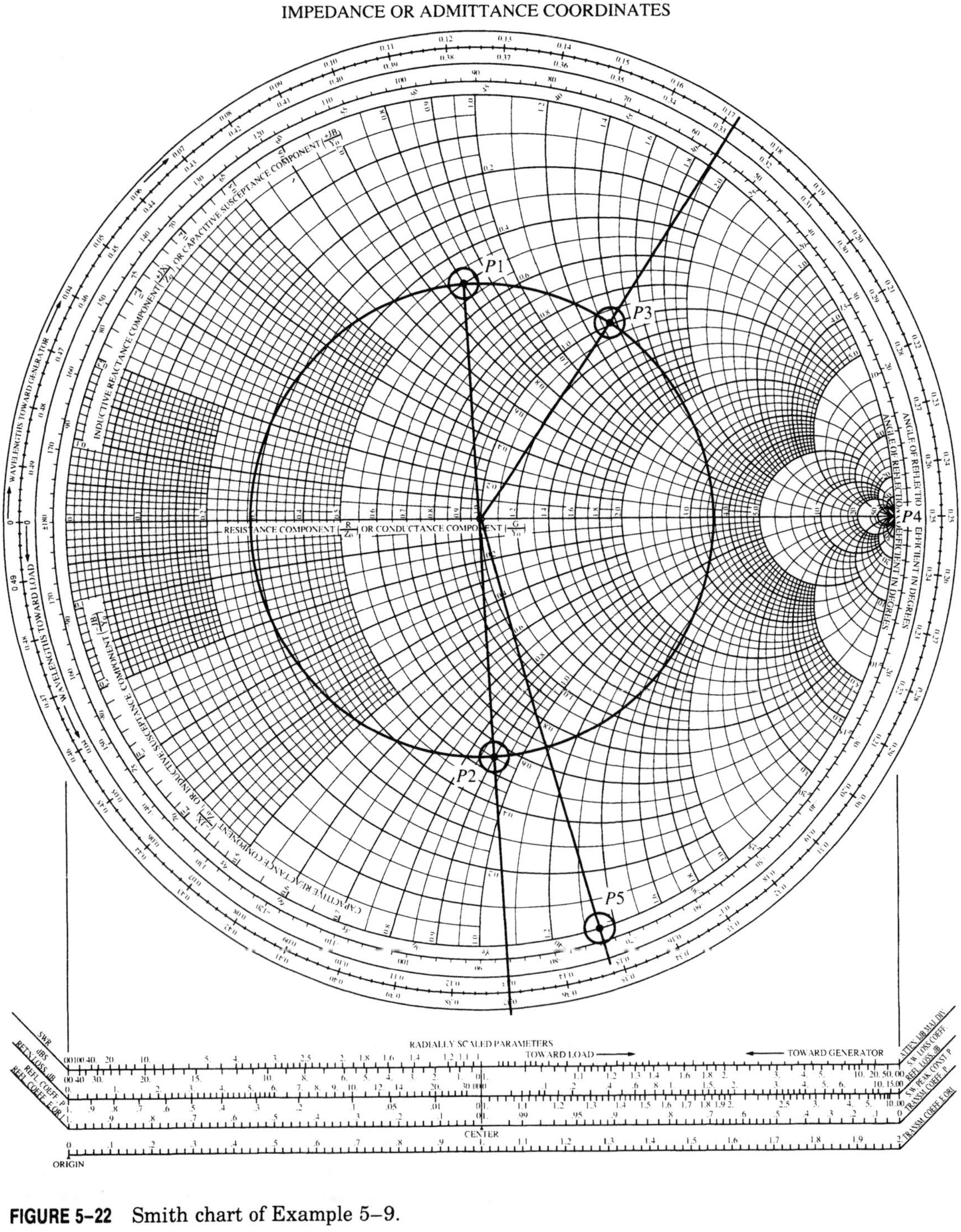

FIGURE 5–22 Smith chart of Example 5–9.

in wavelengths is determined from the coordinate on the "Wavelengths toward generator" scale as 0.5λ − 0.355λ = 0.145 wavelength. The value of d_1 is then

$$d_1 = 0.145\lambda = 0.145 \times 2\text{ cm} = 0.29\text{ cm} \tag{5–52}$$

(b) The normalized value of input impedance at point $P2$ is

$$z_{\text{in}}^{(2)} = r_2' = 0.16 \tag{5–53}$$

The actual input impedance is

$$Z_{\text{in}}^{(2)} = R_2' = 50 \times 0.16 = 8\ \Omega \tag{5–54}$$

The required value for R_{02} is then

$$R_{02} = \sqrt{R_1 R_2'} = \sqrt{50 \times 8} = 20\ \Omega \tag{5–55}$$

(c) The length d_2 is

$$d_2 = \frac{\lambda}{4} = \frac{2\text{ cm}}{4} = 0.5\text{ cm} \tag{5–56}$$

5-10 ■ SINGLE-STUB MATCHING

A classical technique for matching a load to the characteristic impedance involves the use of a shunt short-circuited or open-circuited stub located at an optimum point on the line. Theoretically, a stub could be connected in either series or parallel, but in order to maintain proper balance on the line, a shunt stub is almost always employed. In view of that fact, an admittance basis is used in the discussion that follows.

The concept is illustrated in Fig. 5–23. Assume that all lines shown have a common characteristic admittance $G_0 = 1/R_0$. Starting from the load, a distance d_1 is found such that the real part of the admittance looking to the right is G_0. In general, however, there will be a susceptive component of admittance at the same point. Either a short-circuited or an open-circuited stub of length d_2 is selected such that its susceptance is equal in magnitude and opposite in sign. This susceptance will cancel that offered by the input to the line of length d_1, and the net susceptance of the parallel junction will be zero. In effect then, the line to the left of the junction will be a matched line since it will be terminated in an admittance G_0 or an impedance R_0.

The design is best performed on a normalized basis with a Smith chart and can be summarized as follows:

1. Establish the Smith chart on an admittance basis by plotting the normalized load admittance:

$$y_2 = \frac{Y_2}{G_0} = \frac{G_2 + jB_2}{G_0} = g_2 + jb_2 \tag{5–57}$$

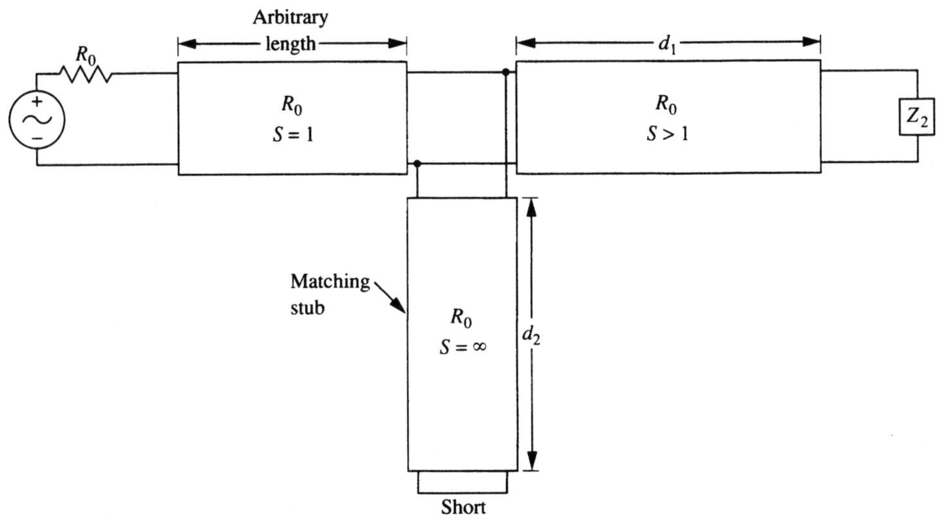

FIGURE 5–23 Block diagram illustrating design approach for single-stub matching.

Then, draw the operating circle. (*Note:* If the load impedance Z_2 is given first, its normalized value $z_2 = Z_2/R_0$ can be plotted on an impedance basis, and it can be converted to an admittance by rotating 180° on the chart.)

2. Starting at y_2, rotate on the operating circle until the circle corresponding to a real part of unity, i.e., $g = 1$, is reached. The distance required to reach this point as determined from the coordinate scale is d_1. The normalized input admittance y_{in} at this point will thus be of the form

$$y_{in} = 1 + jb \tag{5–58}$$

where b could be either positive or negative.

3. Determine the length d_2 of a shorted or an open stub whose normalized admittance is

$$y_{stub} = -jb \tag{5–59}$$

4. When the stub is connected, the net shunt admittance y_2' is given by

$$y_2' = 1 + jb - jb = 1 \tag{5–60}$$

The line is now matched at the junction point, and the VSWR to the left is $S = 1$.

There are many possible solutions to the single-stub problem. Within the first half-wavelength from the load, there are two separate intersections with the $g = 1$ circle: one corresponding to positive susceptance and one corresponding to negative susceptance. Further, these conditions continue to repeat at intervals of $\lambda/2$. In practice, the location closest to the load is usually selected whenever practicable. The reason is that there are always some losses, which have been neglected in the analysis. Losses increase with an increase in VSWR, so it is prudent to keep the section of the line with high VSWR as short as feasible.

EXAMPLE 5–10

Refer to the circuit of Fig. 5–24. At a frequency of 100 MHz, an antenna has an input impedance of $Z_2 = 25 + j40\ \Omega$. A short-circuited stub is to be located at the shortest possible distance from the antenna. The velocity of propagation for the cable is 2×10^8 m/s. Determine the lengths d_1 and d_2.

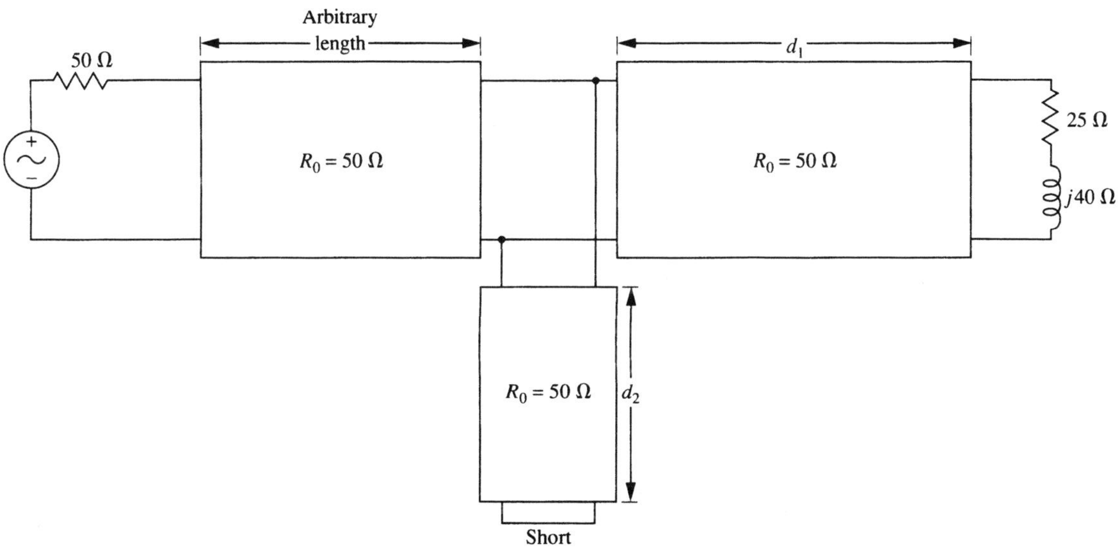

FIGURE 5–24 Circuit of Example 5–10.

Solution

The wavelength on the cable is

$$\lambda = \frac{2 \times 10^8}{100 \times 10^6} = 2\ \text{m} \tag{5–61}$$

The load is given as an impedance, so it is necessary to first establish the equivalent load admittance. Although this could be done analytically, we will choose to perform this operation on the Smith chart. The normalized load impedance is

$$z_2 = \frac{Z_2}{R_0} = \frac{25 + j40}{50} = 0.5 + j0.8 \tag{5–62}$$

This impedance is plotted as $P1$ on Fig. 5–25, and an operating circle is constructed.

The first required operation is to convert from an impedance basis to an admittance basis by rotating 180° on the chart. We could pause and determine the equivalent admittance if desired, but this is not really necessary. This admittance is indicated as $P2$ on Fig. 5–25.

The next step is to rotate from $P2$ in the direction of the generator until the $g = 1$ circle is intersected. This corresponds to point $P3$. The amount of rotation is the sum of the rotation from $P2$ to the end of the scale on the left-hand side of the real line, plus the additional rotation from that point to $P3$. The first portion of the rotation can be determined by subtracting the initial coordinate of $P2$ (0.37 wavelength) from 0.5 wavelength on the "Wavelengths toward generator" scale to yield $0.5\lambda - 0.37\lambda = 0.13$ wavelength. If one is careful, however, this distance could also be determined from momentarily using the "Wavelengths toward load" scale since it begins on the real line, and the value of 0.13λ could be determined directly.

The second component is the distance from 0 to point $P3$, which is readily determined as 0.172 wavelength. The net rotation is thus $0.13\lambda + 0.172\lambda =$ 0.302 wavelength. The distance d_1 is then

$$d_1 = 0.302\lambda = 0.302 \times 2 = 0.604 \text{ m} \tag{5–63}$$

At point $P3$, the normalized admittance is about

$$y_{\text{in}} = 1 + j1.35 \tag{5–64}$$

The required value of admittance for the stub is

$$y_{\text{in}} = -j1.35 \tag{5–65}$$

The stub must, therefore, provide a susceptance of -1.35 on a normalized basis.

The length of the stub is determined using the imaginary circle. A short circuit corresponds to infinite susceptance, so the starting point is at the right-hand of the real line at point $P4$ on Fig. 5–25. We rotate clockwise (toward the generator) to reach a value of -1.35 on the imaginary circle, which is point $P5$. This distance is determined as $0.352\lambda - 0.25\lambda = 0.102$ wavelength. The value of d_2 is thus

$$d_2 = 0.102\lambda = 0.102 \times 2 = 0.204 \text{ m} \tag{5–66}$$

Incidentally, the value of VSWR on the line to the right of the stub is about 3.5, and the VSWR on the stub is, of course, infinite. To the left of the stub, however, the VSWR is unity, and the line is matched.

IMPEDANCE OR ADMITTANCE COORDINATES

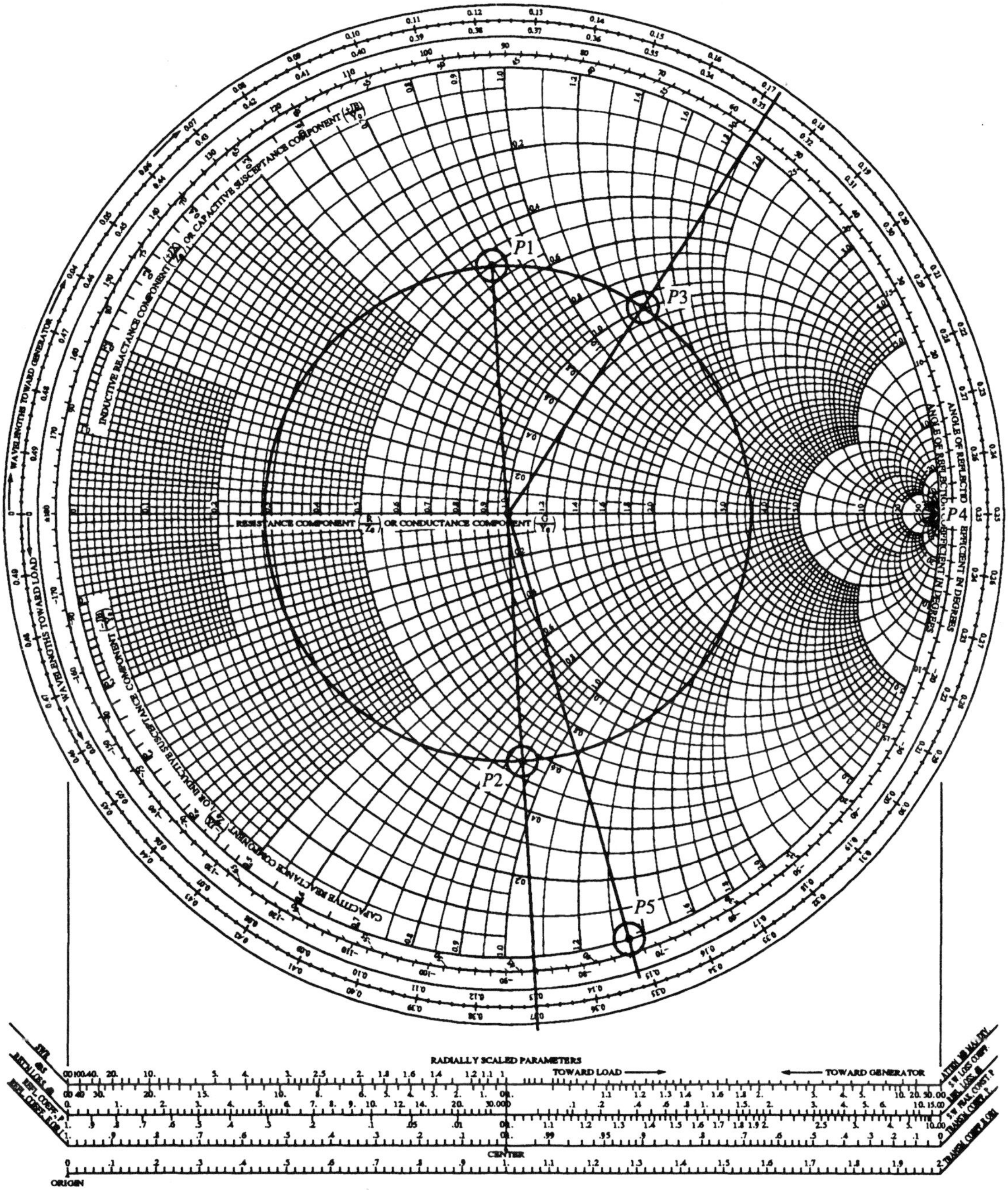

FIGURE 5–25 Smith chart of Example 5–10.

■ PROBLEMS

5–1. A 50-Ω lossless line is terminated in $Z_2 = 15 - j20\ \Omega$.
- **(a)** Plot the normalized impedance on a Smith chart.
- **(b)** Using the Smith chart, determine the corresponding normalized admittance y_2.
- **(c)** Using the result of (b), determine the actual admittance Y_2.

5–2. A 50-Ω lossless line is terminated in $Y_2 = 0.01 + j0.02$ S.
- **(a)** Plot the normalized admittance on a Smith chart.
- **(b)** Using the Smith chart, determine the corresponding normalized impedance z_2.
- **(c)** Using the result of (b), determine the actual impedance Z_2.

5–3. For the system of Problem 5–1, determine the VSWR and the reflection coefficient.

5–4. For the system of Problem 5–2, determine the VSWR and the reflection coefficient.

5–5. A 50-Ω lossless line is terminated in a real load impedance $Z_2 = R_2 = 150\ \Omega$. Using a Smith chart, determine the input impedance for each of the following line lengths:
- **(a)** 0.15λ
- **(b)** 0.25λ
- **(c)** 0.35λ
- **(d)** 0.45λ

5–6. A 50-Ω lossless line is terminated in a real load admittance $Y_2 = 0.01$ S. Using a Smith chart, determine the input admittance for the following line lengths:
- **(a)** 0.125λ
- **(b)** 0.25λ
- **(c)** 0.375λ
- **(d)** 0.50λ

5–7. A 50-Ω lossless line is terminated in a load impedance $Z_2 = 75 + j100\ \Omega$.
- **(a)** Using a Smith chart, determine the input impedance for a line of length 0.2λ.
- **(b)** Determine the VSWR and the reflection coefficient at the load.

5–8. A 300-Ω lossless line is terminated in a load impedance $Z_2 = 150 - j240\ \Omega$.
- **(a)** Using a Smith chart, determine the input impedance for a line of length 0.3λ.
- **(b)** Determine the VSWR and the reflection coefficient at the load.

6

Electric and Magnetic Fields

OBJECTIVES

After completing this chapter, the reader should be able to:

1. Discuss the limitations of lumped and distributed circuit theory and the significance of electromagnetic field theory.
2. Define **electric field intensity** and **magnetic field intensity.**
3. Define **electric flux density** and **magnetic flux density.**
4. Define **conduction current density** and **conductivity.**
5. State and explain the microscopic form of Ohm's law.
6. State the four equations known as **Maxwell's equations.**
7. Explain the significance of Maxwell's equations.
8. Explain the concept of a **plane wave** propagating in a lossless medium.
9. Define **intrinsic impedance.**
10. Define **power density** and the **Poynting vector.**
11. Discuss the analogies between plane wave propagation and a wave propagating on a transmission line.
12. Explain the factors that distinguish a material as being a **dielectric,** a **conductor,** or a combination of both.
13. Discuss the phenomenon of **skin effect.**
14. Define the **skin depth** or **depth of penetration.**
15. Analyze the behavior of a plane wave normal to the boundary between two dielectric materials in terms of **incident power, reflected power,** and **transmitted power.**

6–1 ■ GENERAL DISCUSSION

Most readers should now be familiar with lumped-parameter circuits (from previous texts or courses) and with distributed-parameter circuits (from the first five chapters of this text). The fact is, however, that both lumped-parameter and distributed-parameter circuits are special cases that arise from a more general framework of analysis based on **electric** and **magnetic fields.** This is not to suggest that circuit theory is inadequate as a technical science. Indeed, the majority of engineering applications utilize the vast body of circuit theory, both lumped and distributed, to implement a vast array of meaningful systems. However, the discriminating engineer or technologist should strive to understand those situations in which basic circuit theory is limited and in which the use of electric and magnetic field theory is mandated.

Early in the text, it was observed that lumped circuit theory was inadequate in the frequency range in which the dimensions of the circuit were of the same order of magnitude as the wavelength. At that point, our attention was directed toward distributed-parameter networks, i.e., transmission lines. At still higher frequencies, however, even two-conductor transmission lines become difficult to use due to losses that increase markedly with frequency. The transition point varies with the level of acceptable losses and with the length and quality of the lines, but as a rough estimate, use of two-wire transmission lines is limited to the frequency range below about 1 GHz.

The frequency of 1 GHz marks the rough boundary for the **microwave** range. At a frequency of 1 GHz, the free-space wavelength is 30 cm (about 11.8 inches).

In the microwave region, conventional two-wire transmission lines are virtually useless. Instead, electrical energy is propagated via **waves.** Such waves may be confined to the limits of hollow pipes called **waveguides,** or they may be propagated through **free space** as **electromagnetic waves.** At even higher frequencies, energy may be propagated through **optical fibers.** These concepts will be studied throughout the remainder of the text.

Understand that free-space wave propagation is utilized over a much wider frequency range than at microwave frequencies. Most of the commercial radio and television transmission is achieved through free-space wave propagation, but at the lower frequencies, much larger antenna structures are required. It is even possible to employ free-space wave propagation at frequencies in the upper audio range, but the sizes of the antennas require lengths of several miles! The point is that wave concepts can, theoretically, be employed at *all* frequencies, but they become the primary mode of transmission in the microwave region and above.

6–2 ■ STATIC FIELDS

The simplest situation for introducing the concepts of electric and magnetic fields is through the **static field.** *A static field is one in which none of the field* *quantities are varying with time.* Any voltages that exist between two points are constant, and any currents flowing are steady (i.e., dc).

Electric Field

Consider the case of two conducting surfaces separated by a dielectric or an insulating region as shown in Fig. 6–1. Assume that a constant voltage exists between the two surfaces, resulting from a deficiency of electrons on the upper surface and a surplus of electrons on the lower surface. Thus, the voltage between the surface is positive at the upper surface with respect to the lower surface. The effect shown here is basically that of a charged capacitor.

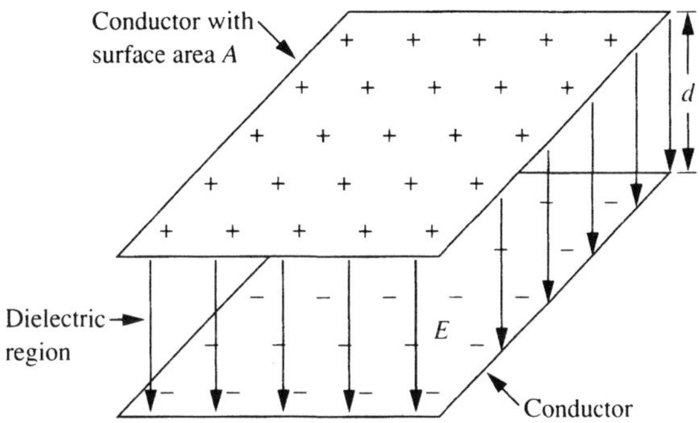

FIGURE 6–1 Electric field lines between charged surfaces.

Because of the potential difference between the two surfaces, there exists a force field of attraction between them. In the SI system, force is measured in **newtons (N).** The force per unit charge is defined as the **electric field intensity,** denoted by E. The basic units for E are **newtons/coulomb (N/C).** However, newtons/coulomb are dimensionally equivalent to **volts/meters (V/m),** and the latter units are much more commonly used in electrical theory. Thus, our practice will be to express electric field intensity in *volts/meters* (V/m) throughout the remainder of the book.

Electric field lines may be visualized in the static case as originating on the positively charged surface and terminating on the negatively charged surface as shown in Fig. 6–1.

There are advanced numerical procedures for determining the field intensity at different points for arbitrary surface configurations. We will limit our consideration here to the simple case of two parallel plates separated by a constant distance d and with area A on each surface. If there exits a voltage V between the two plates, and if the charge is uniformly distributed, the electric field intensity E is simply

$$E = \frac{V}{d} \tag{6–1}$$

Magnetic Field

Consider the case of a conductor carrying a constant value of current I as shown in Fig. 6–2. The flow of current produces a circular force field surrounding the wire as shown. Unlike the electric field, this field does not end or terminate on any type of charge. Furthermore, it is in a direction perpendicular to the direction of current. The field under consideration is called the **magnetic field,** and its strength is referred to as **magnetic field intensity.** The symbol for magnetic field intensity is H, and the units are **amperes/meter (A/m).**

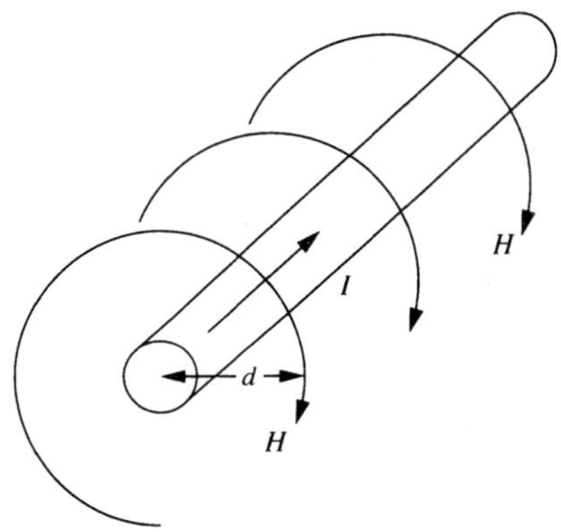

FIGURE 6–2 Magnetic field lines surrounding a current-carrying conductor.

The direction of the magnetic field may be determined from one version of the **right-hand rule.** If the thumb of the right hand is pointed in the direction of the positive current flow, the other four fingers, when curled, will point in the direction of the magnetic field lines. Thus, the magnetic field lines in this case form circles in which the magnitude is constant at a given radius.

For the situation depicted in Fig. 6–2, the magnetic field intensity at a distance d from the center of the conductor may be determined from the formula

$$H = \frac{I}{2\pi d} \tag{6–2}$$

To summarize the major points of this section, there are two basic field quantities: **electric field intensity** E (measured in volts/meters) and **magnetic field intensity** H (measured in amperes/meter). In the static field case, electric field lines originate and terminate on charges at rest. Magnetic field lines, however, have no point sources but, instead, arise from the flow of current and are always perpendicular (normal) to the direction of current flow.

EXAMPLE 6–1

The voltage between two parallel plates separated by a distance of 2 cm is 100 V. Determine the electric field intensity.

Solution

Using basic units, the separation is $d = 2 \times 10^{-2}$ m. The electric field strength is

$$E = \frac{V}{d} = \frac{100\text{ V}}{2 \times 10^{-2}\text{ m}} = 5000\text{ V/m} \tag{6–3}$$

EXAMPLE 6–2

A direct current of 80 A is flowing in a conductor. Determine the magnetic field intensity at a distance of 40 cm from the conductor.

Solution

Using basic units, the distance is $d = 0.4$ m. The magnetic field strength is

$$H = \frac{I}{2\pi d} = \frac{80}{2\pi \times 0.4} = 31.83\text{ A/m} \tag{6–4}$$

EXAMPLE 6–3

When the capacitance of two parallel plates of area A separated by a distance d is derived, the following formula is obtained:

$$C = \frac{\epsilon A}{d} \tag{6–5}$$

where ϵ is the permittivity of the dielectric medium. To better grasp the size of this quantity, determine the value of the capacitance of a parallel-plate capacitor whose plates measure 5 cm by 6 cm separated by a distance of 2 mm with polyethylene as a dielectric ($\epsilon_r = 2.25$).

Solution

As always, we should revert to basic units when using basic formulas. The area should be expressed in m^2, and this value is

$$A = 0.05 \times 0.06 = 3 \times 10^{-3}\text{ m}^2 \tag{6–6}$$

The value of ϵ is

$$\epsilon = \epsilon_r\epsilon_0 = 2.25 \times 8.842 \times 10^{-12} = 19.89 \times 10^{-12}\text{ F/m} \tag{6–7}$$

The value of d is 2×10^{-3} m, and when all quantities are substituted in (6–5), the result is

$$C = \frac{\epsilon A}{d} = \frac{(19.89 \times 10^{-12}) \times (3 \times 10^{-3})}{(2 \times 10^{-3})} \tag{6–8a}$$

$$= 29.84 \times 10^{-12}\text{ F} = 29.84\text{ pF} \tag{6–8b}$$

This value is in the range of small fixed capacitors used in high-frequency circuits. (However, most stock capacitors in this range are constructed differently and are usually much smaller in physical size.)

6–3 ■ ELECTRIC AND MAGNETIC FLUX DENSITIES

In the previous section, the electric field intensity E and the magnetic field intensity H were introduced as the fundamental field quantities. In order to deal with all aspects of field relationships, however, it is necessary to introduce two accompanying field quantities: the **electric flux density** and the **magnetic flux density.**

Electric Flux Density

The electric flux density is denoted as D, and its units are **coulombs/meters^2 (C/m^2).** It is a measure of the electric field in terms of the equivalent charge per unit area. The electric flux density in a dielectric medium is directly proportional to the electric field intensity through the equation

$$D = \epsilon E \tag{6–9}$$

where ϵ is the permittivity in farads/meter (F/m) as introduced in Chapter 1.

The relationship of (6–9) can also serve as the basis for the definition of permittivity for a dielectric medium if D and E are separately defined. The value of ϵ can then be expressed as

$$\epsilon = \frac{D}{E} \tag{6–10}$$

Stated in words, *the permittivity of a dielectric material may be defined as the electric flux density per unit of electric field intensity within the material.*

Assume that the electric field intensity is perpendicular **(normal)** to an area A and is constant over that area as illustrated in Fig. 6–3. The total amount of electric flux Ψ in coulombs is given by

$$\Psi = DA = \epsilon EA \tag{6–11}$$

If the electric field intensity varies over the surface area, integration is required to determine the electric flux.

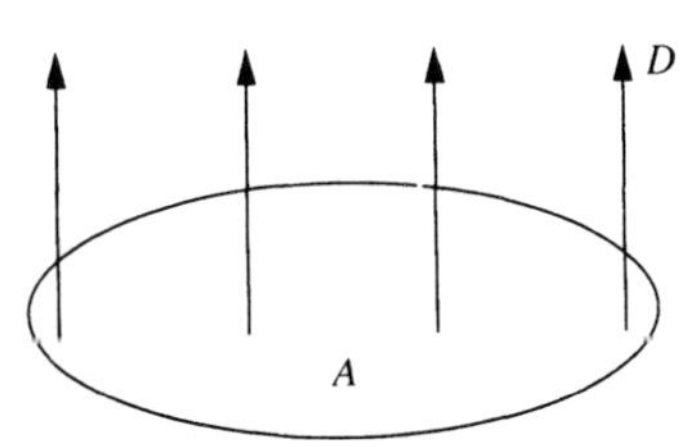

FIGURE 6–3 Electric flux density normal to an area A as a basis for determining total electric flux.

Magnetic Flux Density

The magnetic flux density is denoted as B, and its units have traditionally been **webers/meter² (Wb/m²)**.* Magnetic flux is a measure of the magnetic field in terms of the equivalent flux per unit area. The magnetic flux density is directly proportional to the magnetic field intensity through the equation

$$B = \mu H \tag{6–12}$$

where μ is the permeability in henries/meter (H/m) as introduced in Chapter 1.

The relationship of (6–12) can also serve as the basis for the definition of permeability if B and H are separately defined. The value of μ can then be expressed as

$$\mu = \frac{B}{H} \tag{6–13}$$

Stated in words, *the permeability of material may be defined as the magnetic flux density per unit of magnetic field intensity within the material.*

Assume that the magnetic field intensity is normal to an area A and is constant over that area as illustrated in Fig. 6–4. The total amount of magnetic flux ϕ in webers is given by

$$\phi = BA = \mu HA \tag{6–14}$$

If the magnetic field intensity varies over the surface area, integration is required to determine the total magnetic flux.

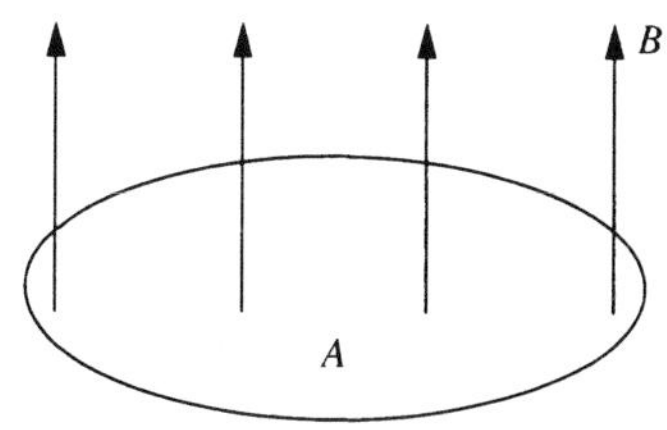

FIGURE 6–4 Magnetic flux density normal to an area A as a basis for determining total magnetic flux.

EXAMPLE 6–4

For the parallel plates of Example 6–1, determine the electric flux density if the dielectric is air.

*More recently, the tesla (T) has been adopted as the unit of magnetic flux density in which 1 T = 1 Wb/m². In order to retain the parallel relationships between electric and magnetic fields, we have chosen to use Wb/m² in this text.

Solution

For air, we assume that $\epsilon = \epsilon_0 = (1/36\pi) \times 10^{-9} = 8.842 \times 10^{-12}$ F/m. The electric flux density is then

$$D = \epsilon E = 8.842 \times 10^{-12} \times 5000 = 44.21 \text{ nC/m}^2 \tag{6–15}$$

EXAMPLE 6–5

For the current-carrying conductor of Example 6–2, determine the magnetic flux density at a distance of 40 cm from the conductor if the medium is air.

Solution

For air, we assume that $\mu = \mu_0 = 4\pi \times 10^{-7} = 1.2566 \times 10^{-6}$ H/m. The magnetic flux density is then

$$B = \mu H = 1.257 \times 10^{-6} \times 31.83 = 40\ \mu\text{Wb/m}^2 \tag{6–16}$$

EXAMPLE 6–6

The electric flux density normal to an area of 4 m^2 has a value of 2 μC/m^2. Determine the value of the electric flux across the area.

Solution

The electric flux is

$$\Psi = DA = 2 \times 10^{-6} \times 4 = 8 \times 10^{-6} \text{ C} = 8\ \mu\text{C} \tag{6–17}$$

EXAMPLE 6–7

The magnetic flux density normal to an area of 3 m^2 has a value of 2 nWb/m^2 (or 2 nT). Determine the value of the magnetic flux across the area.

Solution

The magnetic flux ϕ is

$$\phi = BA = 2 \times 10^{-9} \times 3 = 6 \times 10^{-9} \text{ Wb} = 6 \text{ nWb} \tag{6–18}$$

6–4 ■ CONDUCTION CURRENT DENSITY

The concept of current flow within a conductor is very basic to the study of electricity and should be very familiar to the reader. For the study of field theory, however, the concept of **conduction current density** is required.

Refer to the conductor of Fig. 6–5, and assume that the current I is uniformly distributed over the cross-sectional area A. The conduction current density is

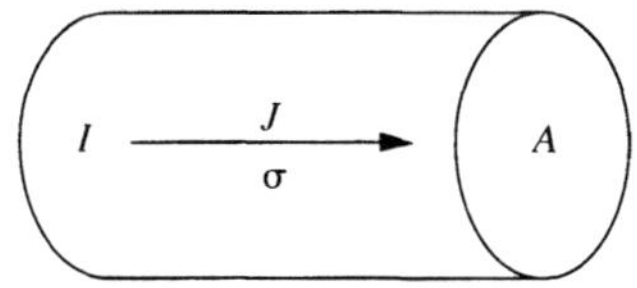

FIGURE 6–5 Current density within a conductor.

denoted as J and is measured in **amperes/meter² (A/m²).** For a current uniformly distributed over the area, the current density is given by

$$J = \frac{I}{A} \tag{6–19}$$

Conversely, if J is known, the total current I is given by

$$I = JA \tag{6–20}$$

If J varies over the area, integration is required to determine the total current flow.

Ohm's Law

The **microscopic form** of Ohm's law for a conductor can be stated as follows:

$$J = \sigma E \tag{6–21}$$

where J is the current density in amperes/meter² (A/m²) and E is the electric field intensity in volts/meter (V/m) in the direction of the current flow. The quantity σ is defined as the **conductivity** for the conducting medium and is a measure of the relative ease with which current can flow. The units of conductivity are siemens/meter (S/m). The conductivity is the reciprocal of the **resistivity factor,** which the reader may recognize from basic electricity, and in which the basic units are ohms · meters (Ω · m). Thus,

$$\sigma = \frac{1}{\text{resistivity factor}} \tag{6–22}$$

From this point on, all references to either resistivity or conductivity will utilize σ only as is customary in field theory.

It should be noted from (6–21) that the units of volts/meter for E times the units of siemens/meter for σ yield volts · siemens/meter² = amperes/meter² as required. When both sides of (6–21) are integrated over the appropriate area and lengths using vector calculus, more familiar forms of Ohm's law, $I = GV = V/R$, are obtained. The quantity G is the total conductance in siemens, and $R = 1/G$ is the total resistance in ohms, as we already know.

EXAMPLE 6–8

A current of 5 A is uniformly distributed over a circular conductor with radius $r = 1.2$ mm. Determine the current density.

Solution

The cross-sectional area is

$$A = \pi r^2 = \pi \times (1.2 \times 10^{-3})^2 = 4.524 \times 10^{-6}\ \mathrm{m}^2 \tag{6–23}$$

The current density is then

$$J = \frac{I}{A} = \frac{5}{4.524 \times 10^{-6}} = 1.105 \times 10^6\ \mathrm{A/m^2} \tag{6–24}$$

EXAMPLE 6–9

Assume that the conductivity for the conductor of Example 6–8 is 4×10^6 S/m. Determine the electric field intensity.

Solution

The microscopic form of Ohm's law as given by (6–21) is employed, and the value of E is determined as

$$E = \frac{J}{\sigma} = \frac{1.105 \times 10^6\ \mathrm{A/m^2}}{4 \times 10^6\ \mathrm{S/m}} = 0.2763\ \mathrm{V/m} \tag{6–25}$$

6–5 ■ SUMMARY OF FIELD QUANTITIES

Before we become too bewildered by all of these concepts, let us pause and summarize the quantities introduced thus far. Such a summary is provided in Table 6–1. Remember that E and H are being considered as the basic field intensity variables which, in some sense, correspond to the circuit variables of voltage and current, respectively. The quantities D and B are the corresponding flux density variables, which are proportional to E and H, respectively. The constants of proportionality are the permittivity and permeability, respectively. Finally, Ψ and ϕ are the total electric and magnetic flux quantities determined over a specific area.

The quantities J and σ are the current density and conductivity for a conductor. Together with the electric field intensity E, they form the basis for the microscopic form of Ohm's law.

The comments indicate that ϵ and D apply primarily to a dielectric and that σ and J apply primarily to a conductor. The term *primarily* should be explained.

In a perfect dielectric, there is no conduction, and σ and J have no significance. In this case, ϵ and D provide the basis for electric field behavior. Conversely, in a nearly perfect conductor, ϵ and D have no significance, and the quantities σ and J serve as a basis for electric field behavior.

TABLE 6–1 Summary of Basic Field Quantities.

Quantity	Symbol	Units	Comments
Electric field intensity	E	V/m	
Magnetic field intensity	H	A/m	
Permittivity	ϵ	F/m	Applies primarily to dielectric
Permeability	μ	H/m	
Electric flux (charge) density	D	C/m^2	Applies primarily to dielectric
Magnetic flux density	B	Wb/m^2	
Conductivity	σ	S/m	Applies primarily to conductor
Conduction current density	J	A/m^2	Applies primarily to conductor
Electric flux	Ψ	C	
Magnetic flux	ϕ	Wb	

The reality is that many materials are neither perfect dielectrics nor perfect conductors. In fact, the behavior of many materials is highly frequency dependent, as we will see in later sections. In other words, some materials may behave more like a dielectric over a certain frequency range and more like a conductor over a different frequency range. There may still be another frequency range in which the dielectric and conduction properties are both significant. Thus, it may be necessary to consider both the dielectric and conduction properties simultaneously in some cases.

6–6 ■ TIME-VARYING FIELDS

Assume now that the electric and magnetic fields are varying with time in some fashion. An electric field may then arise as a result of two different phenomena: (a) the presence of charge in the region as discussed earlier for static fields and (b) the presence of a time-varying magnetic field.

To reiterate the second case in a different manner, *a time-varying magnetic field creates an electric field, which is also time-varying*. This phenomenon is one statement of **Faraday's law,** and it is the basis of rotating generators, motors, and many other electrical devices.

The reader may have first studied Faraday's law from the concept that changing magnetic flux causes a voltage to be induced in a conductor. This earlier interpretation is the most useful one when dealing with lumped circuits. However, the present situation is more appropriate for dealing with electric and magnetic fields on a point-by-point basis as required in field theory.

When the fields are varying with time, a magnetic field may also arise as a result of two different phenomena: (a) the presence of current flowing in a conductor as discussed earlier in the chapter and (b) the presence of a time-varying electric field.

The first phenomenon is a result of Ampère's law. However, the second phenomenon was not known until the English scientist James Clerk Maxwell postulated its existence as a way of resolving certain inconsistencies in earlier laws. Without the existence of this phenomenon, electromagnetic wave propagation as we know it today would never have been possible.

6–7 ■ MAXWELL'S EQUATIONS

As mentioned in the last section, Maxwell resolved an inconsistency in the earlier laws of electricity and magnetism. He also unified the laws into a compact set of four mathematical expressions that describe electricity at its most fundamental level. These equations have become known worldwide as **Maxwell's equations,** and they have served as the basis for many new developments and concepts in electricity.

The results that follow could easily be interpreted as a "snow job" by the authors. It is not our intent to attempt to develop proficiency in the application of these powerful results, since we have devoted many facets of our careers to such endeavors, and neither of us claims to be an expert in the area. Many graduate courses in engineering and physics are devoted to detailed treatments of the basic laws of electricity utilizing Maxwell's equation as a basis, so any treatment in this short space is necessarily limited.

Our intent here is to provide the technologist or technician with a brief overview of the equations and some of their broad implications. If our goals are accomplished, the reader will gain some appreciation of the wonder surrounding the mathematical generalization of the various electrical laws into a broad perspective. He or she should then be able to at least communicate with others about the implications of these laws.

Do not be intimidated if you have difficulty comprehending the full significance of the equations. Frankly, the majority of graduate electrical engineers have only a casual knowledge of the concepts. Most persons having proficiency in the application of Maxwell's equations are those who have pursued intensive studies at the graduate level and who perform lifelong research in the area.

The compact forms of the equations to be presented here involve operations from an area of mathematics known as **vector analysis.** The treatment here will be limited to an overview and a qualitative discussion.

Thus far, E and H (along with D, B, and J) have been discussed in simplified terms. In the most general case as formulated in the equations that follow, these variables are considered as **vectors.** Thus, each quantity will have both a **magnitude** and a **direction.** The direction is an orientation in three-dimensional space. The vector nature is indicated by placing an arrow ($\rightarrow$) above the symbol. For example, the electric and magnetic field intensities are denoted by $\vec{E}$ and $\vec{H}$. Do not confuse this notation with **phasor** notation, which is indicated by an overbar (–) as used earlier in the text. Actually, any of these vectors could be simultaneously a space vector and an ac phasor, but we will not need the phasor representation for a while.

The four equations known as **Maxwell's equations** can be stated as follows:

Equation 1 $$\nabla \times \vec{E} = \frac{-\partial \vec{B}}{\partial t} \tag{6–26}$$

Equation 2 $$\nabla \times \vec{H} = \frac{\partial \vec{D}}{\partial t} + \vec{J} \tag{6–27}$$

Equation 3 $$\nabla \cdot \vec{D} = \rho_v \tag{6–28}$$

Equation 4 $$\nabla \cdot \vec{B} = 0 \tag{6–29}$$

The quantity ∇ (pronounced "del") is a vector operation whose nature will be briefly described shortly. The only variable in the preceding equations that has not been previously defined is ρ_v, which is the **volume charge density** in **coulombs/meter^3 (C/m^3).**

The various equations will now be described briefly. In each case, the equations will be repeated as given earlier, and a second form will be given in which $\vec{D}$ and $\vec{B}$ have been replaced with $\epsilon\vec{E}$ and $\mu\vec{H}$, respectively. The first form is the most compact form, but the second contains fewer variables and is the one that would likely be used in the actual solution of the equations.

Equation 1

The first equation is expressed two ways as

$$\nabla \times \vec{E} = \frac{-\partial \vec{B}}{\partial t} \tag{6–30a}$$

or

$$\nabla \times \vec{E} = -\mu\frac{\partial \vec{H}}{\partial t} \tag{6–30b}$$

While ∇ is combined with $\times$, the result ($\nabla\times$) is referred to as the **curl** of the vector quantity that follows. For students who have not studied partial differentiation, the quantity $\partial\vec{B}/\partial t$ represents the derivative of $\vec{B}$ with respect to t, while all other variables are considered to be constants. This concept is used when a quantity is a function of more than one variable.

Equation 1 provides a microscopic form of **Faraday's law.** It states that a time-varying magnetic field induces an electric field. The magnitude and the direction of the electric field, along with its spatial variation, are determined from the curl operation.

Equation 2

The second equation is expressed two ways as

$$\nabla \times \vec{H} = \vec{J} + \frac{\partial \vec{D}}{\partial t} \tag{6–31a}$$

or

$$\nabla \times \vec{H} = \sigma\vec{E} + \epsilon\frac{\partial \vec{E}}{\partial t} \tag{6–31b}$$

The first term on the right-hand side of (6–31a) or (6–31b) is referred to as the **conduction current density.** This term, coupled with the left-hand side of either equation, constitutes a vector form of **Ampère's law.**

The second term on the right-hand side of (6–31a) or (6–31b) is referred to as the **displacement current density.** This term was added by Maxwell to resolve an earlier scientific inconsistency, and it led to many exciting future discoveries (e.g., radio and television).

A general summary of Equation 2 is that *a magnetic field can be created either by current flowing in a conductor or by a time-varying electric field.*

Equation 3

The third equation may be expressed two ways as

$$\nabla \cdot \vec{D} = \rho_v \tag{6–32a}$$

or

$$\nabla \cdot \vec{E} = \frac{\rho_v}{\epsilon} \tag{6–32b}$$

When ∇ is combined with a dot $(\cdot)$, the result $(\nabla\cdot)$ is referred to as the **divergence.** The divergence operation on a vector yields a **scalar,** which is a quantity having magnitude but not direction. Thus, the volume charge density ρ_v is a scalar quantity.

Equation 3 indicates that an electric field may begin or end on electric charge. It is a microscopic form of **Gauss's law for electric fields.**

Equation 4

The fourth equation is expressed two ways as

$$\nabla \cdot \vec{B} = 0 \tag{6–33a}$$

or

$$\nabla \cdot \vec{H} = 0 \tag{6–33b}$$

This equation implies that magnetic fields, unlike electric fields, have no point sources on which the field lines could begin or end. Hence, magnetic field lines are always continuous. Equation (6–33b) is a microscopic form of **Gauss's law for magnetic fields.**

6–8 ■ PLANE WAVE PROPAGATION IN LOSSLESS MEDIUM

We now turn our attention to a very important special case referred to as **plane wave propagation.** The radiation from an antenna at normal receiving distances assumes the form of a plane wave. Light waves and various other forms of radiation tend to be of the form of a plane wave.

Refer to Fig. 6–6(a) for the discussion that follows. A plane wave is characterized by an electric field in one direction, a magnetic field perpendicular to the electric field, and the direction of propagation perpendicular to both the electric and the magnetic fields. In the frame of reference chosen, the electric field has only an x-component, and it is denoted as E_x. The magnetic field has only a y-component, and it is denoted by H_y. The direction of propagation in this case is in the z-direction as shown.

In general, the direction of propagation for a plane wave may be determined by taking the right hand (with the thumb on the left) and rotating from $\vec{E}$ to $\vec{H}$. The direction of the thumb aligned perpendicular to the plane containing $\vec{E}$ and $\vec{H}$ will represent the direction of propagation. The simplified diagram of Fig. 6–6(b) is useful for this purpose.

Assume now that the medium is lossless, i.e., that it is a perfect dielectric. In this case, a mathematical solution of Maxwell's equations yields a linear algebraic relationship between E_x and H_y. The relationship can be expressed as

$$\frac{E_x}{H_y} = \eta \qquad \textbf{(6–34)}$$

The quantity η is called the **intrinsic impedance.** Since the units of E_x are volts/meter and the units of H_y are amperes/meter, there is a cancellation of meters in the ratio. Thus, the units of η are volts/amperes = ohms, which leads to its definition as an impedance. The intrinsic impedance is to a single plane wave what characteristic impedance is to a single wave of voltage and current on a transmission line.

For a lossless dielectric medium, the intrinsic impedance for a plane wave is a real number given by

$$\eta = \sqrt{\frac{\mu}{\epsilon}} \qquad \textbf{(6–35)}$$

where μ is the permeability and ϵ is the permittivity.

When free-space plane wave propagation is considered, $\mu = \mu_0$, $\epsilon = \epsilon_0$, and the intrinsic impedance is denoted as η_0. Substitution of the values in (6–35) yields

$$\eta_0 = 120\pi \approx 377\ \Omega \qquad \textbf{(6–36)}$$

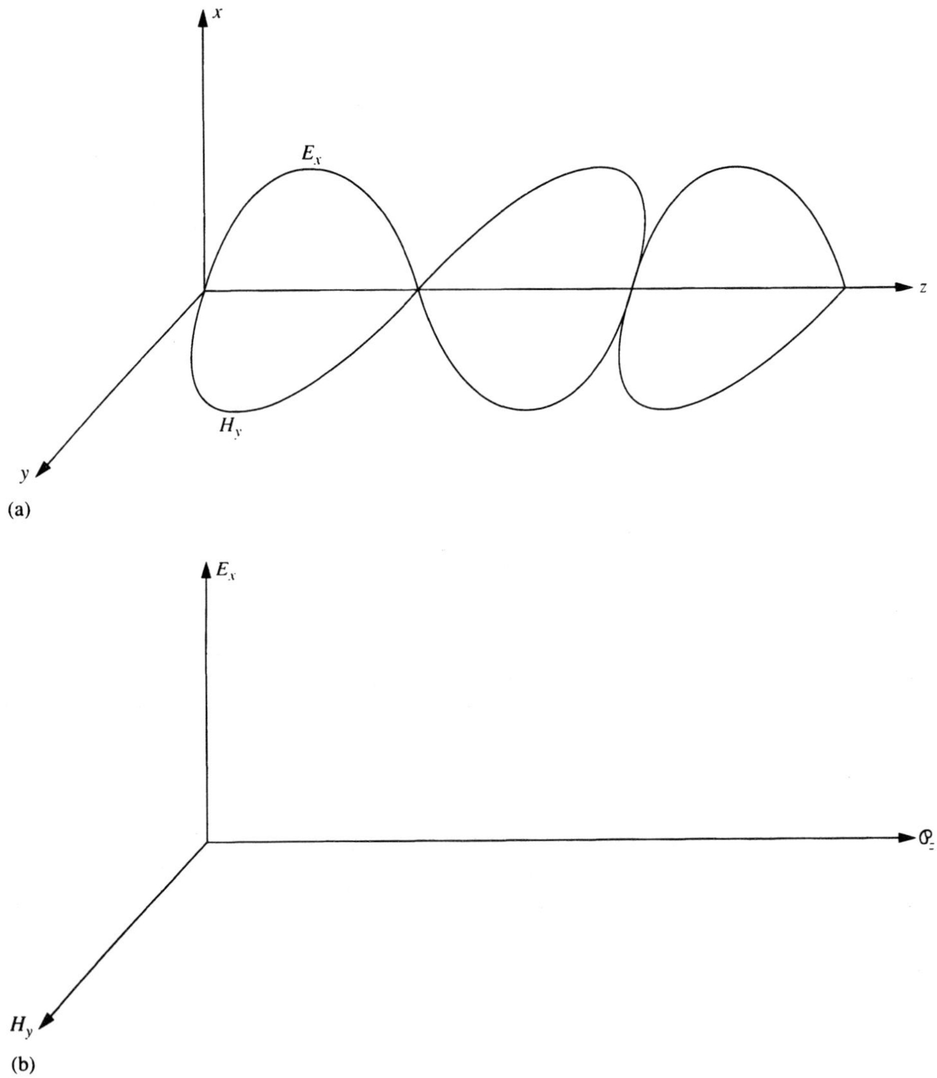

FIGURE 6–6 Plane wave propagating in z-direction with E_x and H_y components.

Unless indicated otherwise, this value will also be assumed as the intrinsic impedance of air. Assuming that $\mu = \mu_0$, the value of η for any dielectric can be expressed as

$$\eta = \frac{\eta_0}{\sqrt{\epsilon_r}} = \frac{377}{\sqrt{\epsilon_r}}\ \Omega \quad (6\text{–}37)$$

For a lossless dielectric medium, the velocity of propagation may be expressed as

$$\nu = \frac{1}{\sqrt{\mu\epsilon}} \tag{6–38}$$

When $\mu = \mu_0$ and $\epsilon = \epsilon_0$ are substituted in (6–38), the free-space propagation $c = 3 \times 10^8$ m/s is obtained as demonstrated in Chapter 1.

The plane wave is actually propagating a certain amount of energy in the form of electromagnetic radiation. This energy serves as the basis for the signal captured by a receiver antenna. (It also serves as the basis for the warmth felt by light rays from the sun, for the heating effect in a microwave oven, and many other phenomena, both natural and artificial.)

The propagated energy is best measured in terms of power, which is, of course, the rate of energy transmitted. For a plane wave, the concept of **power density** is used. The power density is the power per unit of area in **watts/meter² (W/m²)** as measured over an area normal to the direction of propagation of the wave. This power density will be denoted as $\vec{\mathcal{P}}$. In general, $\vec{\mathcal{P}}$ is a vector quantity having both a magnitude and a spatial direction and is called the **Poynting vector.** The Poynting vector is normal to the plane containing $\vec{E}$ and $\vec{H}$.

For the orientation of our assumed plane wave, the electric field E_x is in the positive x-direction, and the magnetic field H_y is in the positive y-direction. Thus, the Poynting vector will be in the positive z-direction as shown in Fig. 6–6. We will denote it as $\mathcal{P}_z$, and it is given by

$$\mathcal{P}_z = E_x H_y \tag{6–39}$$

Note the dimensions in (6–39). We have (volts/meter) × (amperes/meter) = volt·amperes/meter² = watts/meter² as required.

We see that (6–39) has the same form as $P = VI$ in circuit theory. In the same sense that the power in a resistance can also be expressed as V^2/R or I^2R, (6–39) can be expressed in terms of either E_x or H_y by using the definition of intrinsic impedance in (6–34), and two alternate expressions for the power density are obtained:

$$\mathcal{P}_z = \frac{E_x^2}{\eta} \tag{6–40}$$

and

$$\mathcal{P}_z = H_y^2 \eta \tag{6–41}$$

Assume that an arbitrary Poynting vector is normal to a surface area A as illustrated in Fig. 6–7, and assume that the value of $\mathcal{P}_z$ is constant over that area. The total power P is given by

$$P = \mathcal{P}_z A \tag{6–42}$$

If $\mathcal{P}_z$ varies over the area, integration is required to determine the total power.

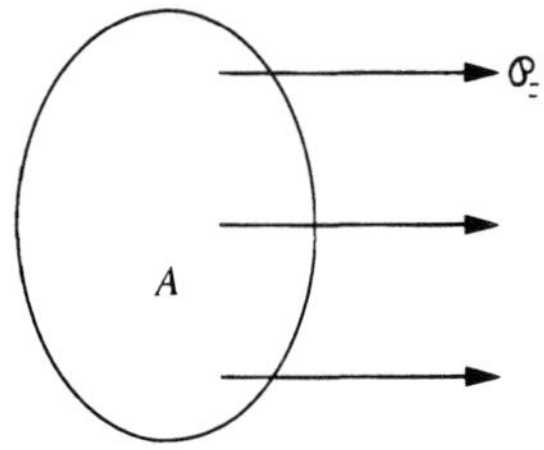

FIGURE 6–7 Power density normal to an area A as a basis for determining total power flow.

EXAMPLE 6–10

The rms magnitude of the electric field intensity of a plane wave in free space is $E_x = 2$ V/m. Assuming that $\overrightarrow{H}$ is in the positive y-direction, determine (a) H_y, (b) $\mathcal{P}_z$, (c) the total power transmitted through a rectangular surface of width 20 m × 30 m in the x-y plane over which the fields are constant.

Solution

(a) From (6–34) with $\eta = \eta_0$, the value of H_y is

$$H_y = \frac{E_x}{\eta_0} = \frac{2}{377} = 5.305 \times 10^{-3} \text{ A/m} = 5.305 \text{ mA/m} \quad \textbf{(6–43)}$$

(b) The power density could be determined from (6–39), (6–40), or (6–41). These alternatives give us the opportunity to make a point to the reader about good computational practices: When several equally simple formulas are available, it is usually wise to select the form that minimizes the use of quantities that have been calculated earlier and maximizes the use of the original data. This method minimizes the possibility of either a complete error or roundoff error due to other computations. The best choice in this case is (6–40), and the power density is

$$\mathcal{P}_z = \frac{E_x^2}{\eta} = \frac{E_x^2}{\eta_0} = \frac{(2)^2}{377} \quad \textbf{(6–44a)}$$

$$= 10.61 \times 10^{-3} \text{ W/m}^2 = 10.61 \text{ mW/m}^2 \quad \textbf{(6–44b)}$$

(c) The area A of the surface involved is

$$A = 20 \text{ m} \times 30 \text{ m} = 600 \text{ m}^2 \quad \textbf{(6–45)}$$

The power passing through this surface is

$$P = \mathcal{P}_z A = (10.61 \times 10^{-3}) \times 600 = 6.366 \text{ W} \quad \textbf{(6–46)}$$

EXAMPLE 6–11

In a lossless dielectric medium, the rms electric and magnetic field intensities are $E_x = 20$ V/m and $H_y = 80$ mA/m. Determine (a) the intrinsic impedance and (b) the power density. (c) Assuming that $\mu = \mu_0$, determine the dielectric constant.

Solution

(a) The intrinsic impedance is

$$\eta = \frac{E_x}{H_y} = \frac{20}{80 \times 10^{-3}} = 250\ \Omega \qquad \textbf{(6–47)}$$

(b) The power density can be determined by (6–39) as

$$\mathcal{P}_z = E_x H_y = (20) \times (80 \times 10^{-3}) = 1.6\ \text{W/m}^2 \qquad \textbf{(6–48)}$$

(c) The intrinsic impedance is related to that of free space by

$$\eta = \frac{377}{\sqrt{\epsilon_r}} \qquad \textbf{(6–49)}$$

Solving for ϵ_r, we have

$$\epsilon_r = \frac{(377)^2}{\eta^2} = \frac{(377)^2}{(250)^2} = 2.274 \qquad \textbf{(6–50)}$$

EXAMPLE 6–12

An **omnidirectional antenna** (also called an **isotropic radiator**) is one that radiates equal power density in all directions. The concept is very useful in dealing with antenna theory, as we will see later in the text. Consider the presence of such an antenna at the center of a fictitious sphere of radius r = 200 m as shown in Fig. 6–8. Assume that the power density is normal to the surface with a value of 50 μW/m^2 at the surface. Calculate the total power radiated out of the sphere.

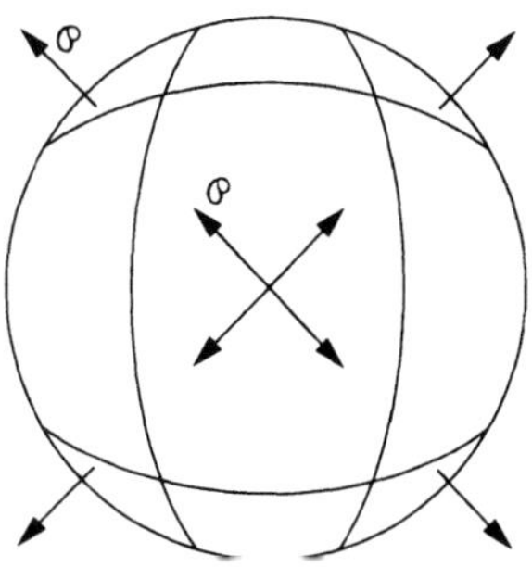

FIGURE 6–8 Sphere of Example 6–12.

Solution

All readers may not know how to calculate the surface area of a sphere, so we begin with the formula for that area:

$$A = 4\pi r^2 \qquad \textbf{(6–51)}$$

Since the power density is constant everywhere on the surface and is normal to the surface, the net power is

$$P = \mathcal{P}_z A = (50 \times 10^{-6}) \times 4\pi \times (200)^2 = 25.13 \text{ W} \tag{6–52}$$

If the medium within the sphere is lossless, this value must also be equal to the power delivered by the antenna.

6–9 ■ PLANE WAVE–TRANSMISSION LINE ANALOGIES

The mathematical forms of the field equations for a plane wave propagating in some arbitrary uniform medium are strikingly similar to the circuit equations for a wave propagating on a transmission line. Because of this similarity, a significant amount of analysis can be performed for a traveling wave based on the visualization of a transmission line analogy.

The analogy at this point will be limited to that of a single wave based on sinusoidal steady-state conditions. The electric field will be assumed in the positive x-direction, and the magnetic field will be assumed in the positive y-direction, as previously stated. Thus, the power flow will be in the positive z-direction. The medium will be assumed to be of such a large length with uniform properties that only a single wave need be considered.

Let $\overline{E}_x$ represent the phasor form of the electric field, and let $\overline{H}_y$ represent the corresponding phasor form of the magnetic field. Solutions of Maxwell's equations would yield expressions of the forms

$$\overline{E}_x = \overline{E}_x^+ e^{-\gamma z} \tag{6–53}$$

and

$$\overline{H}_y = \overline{H}_y^+ e^{-\gamma z} \tag{6–54}$$

From this point on, a transmission line analog can be used as the basis for analysis.

Refer to Table 6–2 in the discussion that follows. The analogies between transmission line and traveling wave quantities are shown. Note at the outset that $\overline{V}$ and $\overline{I}$ correspond, respectively, to $\overrightarrow{E}_x$ and $\overrightarrow{H}_y$ as previously noted.

There is no known field quantity that corresponds to the series resistance R in this particular analog. However, the conductivity σ corresponds to the shunt conductance G. The parameters L and C correspond, respectively, to μ and ϵ.

In the case of transmission lines in the sinusoidal steady state, the series impedance is $Z = R + j\omega L$. The corresponding field quantity is $0 + j\omega\mu = j\omega\mu$. (The "0" was shown to remind us that there is no direct analogy to R.) For transmission lines, the shunt admittance is $Y = G + j\omega C$. The corresponding field quantity is $\sigma + j\omega\epsilon$.

TABLE 6–2 Transmission Line–Plane Wave Analogies.

Transmission Line			*Traveling Wave*		
Quantity	**Symbol**	**Units**	**Quantity**	**Symbol**	**Units**
Voltage	$\overline{V}$	V	Electric field intensity	$\overline{E}$	V/m
Current	$\overline{I}$	A	Magnetic field intensity	$\overline{H}$	A/m
Series resistance	R	Ω/m	No known quantity		
Series inductance	L	H/m	Permeability	μ	H/m
Shunt conductance	G	S/m	Conductivity	σ	S/m
Shunt capacitance	C	F/m	Permittivity	ϵ	F/m
Series impedance	$Z = R + j\omega L$	Ω/m		$j\omega\mu$	Ω/m
Shunt admittance	$Y = G + j\omega C$	S/m		$\sigma + j\omega\epsilon$	S/m
Characteristic impedance	$Z_0 = \sqrt{\dfrac{R + j\omega L}{G + j\omega C}}$	Ω/m	Intrinsic impedance	$\eta = \sqrt{\dfrac{j\omega\mu}{\sigma + j\omega\epsilon}}$	S/m
Characteristic impedance (lossless line)	$Z_0 = R_0 = \sqrt{\dfrac{L}{C}}$	Ω	Intrinsic impedance (lossless medium)	$\eta = \sqrt{\dfrac{\mu}{\epsilon}}$	Ω
Propagation constant	$\gamma = \sqrt{(R + j\omega L)(G + j\omega C)}$	Np/m rad/m	Propagation constant	$\gamma = \sqrt{j\omega\mu(\sigma + j\omega\epsilon)}$	Np/m rad/m
Propagation constant (lossless line)	$\gamma = j\omega\sqrt{LC}$	rad/m	Propagation constant (lossless medium)	$\gamma = j\omega\sqrt{\mu\epsilon}$	rad/m
Velocity of propagation (lossless line)	$\nu = \dfrac{1}{\sqrt{LC}}$	m/s	Velocity of propagation (lossless medium)	$\nu = \dfrac{1}{\sqrt{\mu\epsilon}}$	m/s

The characteristic impedance for a transmission line with losses is $Z_0 = \sqrt{Z/Y}$. The corresponding field quantity is the **complex intrinsic impedance** η, which for the case of a lossy medium is

$$\eta = \sqrt{\frac{j\omega\mu}{\sigma + j\omega\epsilon}} \tag{6–55}$$

For the lossless case, $\sigma = 0$, and Equation (6–55) reduces to

$$\eta = \sqrt{\frac{j\omega\mu}{j\omega\epsilon}} = \sqrt{\frac{\mu}{\epsilon}} \tag{6–56}$$

which was given in an earlier section.

The propagation constant for a transmission line is

$$\gamma = \sqrt{ZY} = \sqrt{(R + j\omega L)(G + j\omega C)}$$

The corresponding field quantity for a lossy medium is

$$\gamma = \sqrt{j\omega\mu(\sigma + j\omega\epsilon)} \tag{6–57}$$

For the lossless case, $\sigma = 0$, and Equation (6–57) reduces to

$$\gamma = \sqrt{j\omega\mu(j\omega\epsilon)} = j\omega\sqrt{\mu\epsilon} = j\beta \tag{6–58}$$

which was given in an earlier section.

For this lossless case, it can be readily shown that the velocity of propagation is

$$v = \frac{\omega}{\beta} = \frac{\omega}{\omega\sqrt{\mu\epsilon}} = \frac{1}{\sqrt{\mu\epsilon}} \tag{6–59}$$

as stated earlier.

With the results provided in Table 6–2, along with the discussion in the text, many properties of plane waves propagating in lossy media may be predicted, as will be illustrated in the example that follows.

EXAMPLE 6–13

Sea water has typical values of dielectric constant and conductivity given by $\epsilon_r = 80$ and $\sigma = 4$ S/m. At a frequency of 100 MHz, determine the following quantities: (a) intrinsic impedance, (b) propagation constant, (c) attenuation constant, and (d) velocity of propagation.

Solution

We must first calculate $j\omega\mu$ and $\sigma + j\omega\epsilon$. The computations are as follows:

$$j\omega\mu = j[2\pi(100 \times 10^6)(4\pi \times 10^{-7})] \tag{6–60a}$$

$$= j789.6 = 789.6\underline{/1.5708} \tag{6–60b}$$

$$j\omega\epsilon = j[2\pi(100 \times 10^6)(80)(8.842 \times 10^{-12})] \tag{6–61a}$$

$$= j0.4444 = 0.4444\underline{/1.5708} \tag{6–61b}$$

$$\sigma + j\omega\epsilon = 4 + j0.4444 \tag{6–62a}$$

$$= 4.025\underline{/0.1106} \tag{6–62b}$$

(a) The complex intrinsic impedance is

$$\eta = \sqrt{\frac{j\omega\mu}{\sigma + j\omega\epsilon}} = \sqrt{\frac{789.6\underline{/1.5708}}{4.025\underline{/0.1106}}} \tag{6–63a}$$

$$= 14.01\underline{/0.7301}\ \Omega \tag{6–63b}$$

(b) The propagation constant is

$$\gamma = \sqrt{j\omega\mu(\sigma + j\omega\epsilon)} = \sqrt{(789.6\underline{/1.5708})(4.025\underline{/0.1106})} \tag{6–64a}$$

$$= \sqrt{3178\underline{/1.6814}} = 56.37\underline{/0.8407} = 37.60 + j42.00 \tag{6–64b}$$

(c) The attenuation constant in nepers/meter is

$$\alpha = 37.60 \text{ Np/m} \tag{6–65}$$

The corresponding value in decibels/meters is

$$\alpha_{dB} = 8.686\alpha = 8.686 \times 37.60 = 326.6 \text{ dB/m} \tag{6–66}$$

This is a very high attenuation constant, and it illustrates the difficulty of propagating high frequencies in sea water (more on this later).

(d) The velocity of propagation is

$$\nu = \frac{\omega}{\beta} = \frac{2\pi(100 \times 10^6)}{42.00} = 14.96 \times 10^6 \text{ m/s} \tag{6–67}$$

6–10 ■ DIELECTRIC OR CONDUCTOR?

In an earlier section, it was indicated that some materials may exhibit the properties of a dielectric over one frequency range, while in a different frequency range, the properties of a conductor may dominate. This phenomenon will be investigated in this section. First, we will consider the properties of a **perfect dielectric** and a **perfect conductor.**

Perfect Dielectric

A perfect dielectric is a material in which there is no conductivity, i.e., in which $\sigma = 0$. This situation is analogous to that of a lossless transmission line as discussed earlier in the text. In this case, the attenuation constant is $\alpha = 0$, and a plane wave encounters no attenuation and only phase shift. The intrinsic impedance is real with a value of $\eta = \sqrt{\mu/\epsilon}$. The properties of a number of low-loss materials such as polyethylene, mylar, and polystyrene can often be assumed to be perfect dielectrics in a number of applications.

Free space outside the atmosphere of planets behaves essentially like a perfect dielectric. Proof of this statement is demonstrated by the fact that we can see the light rays emitted by stars that are many light-years away.

Perfect Conductor

A perfect conductor is one in which there is no resistance at all, which would imply that $\sigma = \infty$ (or resistivity $= 0$). Neither an electric field nor a magnetic field can exist in a perfect conductor, so the current would all flow on the surface in this limiting case. Since the resistance would be zero, there would be no voltage drop along the length of such a conductor. There are fewer situations in which perfect conduction can be assumed as compared with the assumption of a perfect dielectric. However, superconducting materials cooled to a temperature close to absolute zero (−273°C) exhibit the properties of a perfect conductor.

General Materials

In the most general case, a material may exhibit properties of both a dielectric and a conductor. To investigate this possibility further, consider the expressions for intrinsic impedance and propagation constant as given by either (6–55) and (6–57) or from Table 6–2:

$$\eta = \sqrt{\frac{j\omega\mu}{\sigma + j\omega\epsilon}} \tag{6–68}$$

$$\gamma = \sqrt{j\omega\mu(\sigma + j\omega\epsilon)} \tag{6–69}$$

In general, both η and γ will be complex, and this implies both a dielectric property and a conduction property.

Now assume that $\omega\epsilon \gg \sigma$. In this case, the quantity $\sigma + j\omega\epsilon$ may be approximated as $\sigma + j\omega\epsilon \approx j\omega\epsilon$, and (6–68) and (6–69) reduce to

$$\eta \approx \sqrt{\frac{\mu}{\epsilon}} \tag{6–70}$$

and

$$\gamma \approx j\omega\sqrt{\mu\epsilon} = j\beta \tag{6–71}$$

This case has been previously encountered for the assumption of a lossless medium. Therefore, a material behaves very much like a dielectric when $\omega\epsilon \gg \sigma$, although there will still be some small losses.

Next assume that $\sigma \gg \omega\epsilon$. In this case, the quantity $\sigma + j\omega\epsilon$ may be approximated as $\sigma + j\omega\epsilon \approx \sigma$, and (6–68) and (6–69) reduce to

$$\eta \approx \sqrt{\frac{j\omega\mu}{\sigma}} \tag{6–72}$$

and

$$\gamma \approx \sqrt{j\omega\mu\sigma} \tag{6–73}$$

Both of the preceding equations involve taking the square root of j. For reasons of later convenience, the value of η will be retained in polar form, while that of γ will be expanded in rectangular form. The reader is invited to show that

$$\eta \approx \sqrt{\frac{\omega\mu}{\sigma}} \angle(\pi/4) \tag{6–74}$$

and

$$\gamma \approx \sqrt{\frac{\omega\mu\sigma}{2}} + j\sqrt{\frac{\omega\mu\sigma}{2}} = \alpha + j\beta \tag{6–75}$$

in which

$$\alpha = \sqrt{\pi f\mu\sigma} \tag{6–76}$$

and

$$\beta = \sqrt{\pi f \mu \sigma} \tag{6–77}$$

The preceding results imply the following phenomena:

1. The electric field E_x leads the magnetic field H_y by $\pi/4$ rad or 45°.
2. The intrinsic impedance increases with frequency.
3. Both the attenuation constant and the phase constant increase with frequency.

These properties are the properties of an imperfect conductor. Therefore, a material behaves like an imperfect conductor when $\sigma \gg \omega\epsilon$.

As a means of assessing the relative behavior of material, the ratio $\sigma/\omega\epsilon$ is used as a convenient measure. This quality is referred to in the material science area as the **loss tangent** and is denoted as $\tan\theta$; i.e.,

$$\tan\theta = \frac{\sigma}{\omega\epsilon} \tag{6–78}$$

The significance of using the tangent in the definition is the triangle shown in Fig. 6–9, which shows the mutually perpendicular nature of σ and $\omega\epsilon$. For $\tan\theta \ll 1$, the dominant behavior is that of a dielectric, while for $\tan\theta \gg 1$, the dominant behavior is that of a conductor. Although the range is somewhat arbitrary, a common assumption is based on a bound of 0.01 in one direction and 100 in the opposite direction. This particular criterion will be used here, and it can be stated as follows:

1. For $\tan\theta \leq 0.01$, the material will be assumed to be a *dielectric,* although there will be some losses unless $\tan\theta = 0$.
2. For $\theta \geq 100$, the material will be assumed to be a *conductor,* although there will be some resistance unless $\tan\theta = \infty$.
3. For $0.01 \leq \tan\theta \leq 100$, the material will be classified as a *quasi conductor,* and it exhibits the properties of both a dielectric and a conductor.

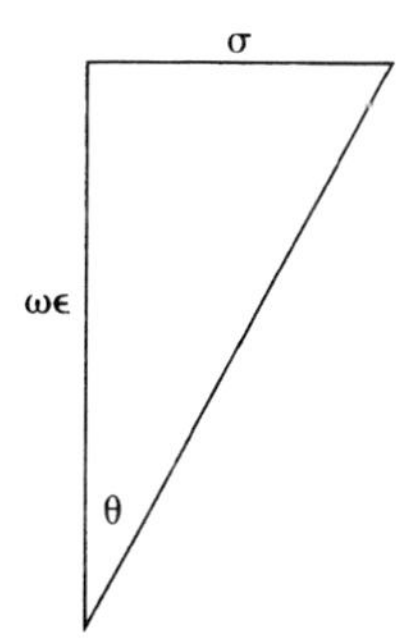

FIGURE 6–9 Triangle used in definition of loss tangent.

EXAMPLE 6–14

The conductivity and the permittivity of the ground near the surface of the earth vary considerably with various soil parameters. However, to illustrate some typical conditions and to provide "ball-park" estimates, assume that the parameters for a particular area are $\sigma = 0.008$ S/m and $\epsilon_r = 12$, which are typical values. Classify the behavior of this ground at the following frequencies: (a) 100 kHz, (b) 10 MHz, and (c) 10 GHz.

Solution

At any frequency f, the loss tangent $\tan\theta$ can be expressed as

$$\tan\theta = \frac{\sigma}{\omega\epsilon} = \frac{0.008}{2\pi f[(12)(8.842\times 10^{-12})]} = \frac{12\times 10^6}{f} \quad \textbf{(6–79)}$$

We can now substitute specific values of frequency in (6–79) to determine loss tangent values. The calculations are as follows:

(a) $f = 100$ kHz

$$\tan\theta = \frac{12\times 10^6}{10^5} = 120 \quad \textbf{(6–80)}$$

At 100 kHz, this ground behaves like a *conductor* (although it is certainly not a perfect conductor).

(b) $f = 10$ MHz

$$\tan\theta = \frac{12\times 10^6}{10^7} = 1.2 \quad \textbf{(6–81)}$$

At 10 MHz, this ground can be considered as a *quasi conductor.*

(c) $f = 10$ GHz

$$\tan\theta = \frac{12\times 10^6}{10^{10}} = 0.0012 \quad \textbf{(6–82)}$$

At 10 GHz, this ground can be considered as a *dielectric.*

6–11 ■ SKIN EFFECT

The work of the preceding section serves as the basis for the phenomenon of **skin effect,** which is an important property of conducting materials. Consider an imperfect conductor characterized by the inequality $\sigma \gg \omega\epsilon$ or $\tan\theta \gg 1$. From the work of the preceding section, the attenuation and phase constants were shown to be

$$\alpha = \sqrt{\pi f\mu\sigma} \quad \textbf{(6–83)}$$

and

$$\beta = \sqrt{\pi f\mu\sigma} \quad \textbf{(6–84)}$$

Refer to Fig. 6–10 in the development that follows. Assume that a single plane wave with $\overline{E}_x$ and $\overline{H}_y$ components is propagating in the positive z-direction as shown. Assume that at $z = 0$, the phasor value of $\overline{E}_x$ is given by $\overline{E}_{x_0}$. From the transmission line analogies, the value of $\overline{E}_x$ at any value of z (for $z > 0$) is given by

$$\overline{E}_x = \overline{E}_{x_0} e^{-\gamma z} = \overline{E}_{x_0} e^{-\alpha z} e^{-j\beta z} \tag{6–85}$$

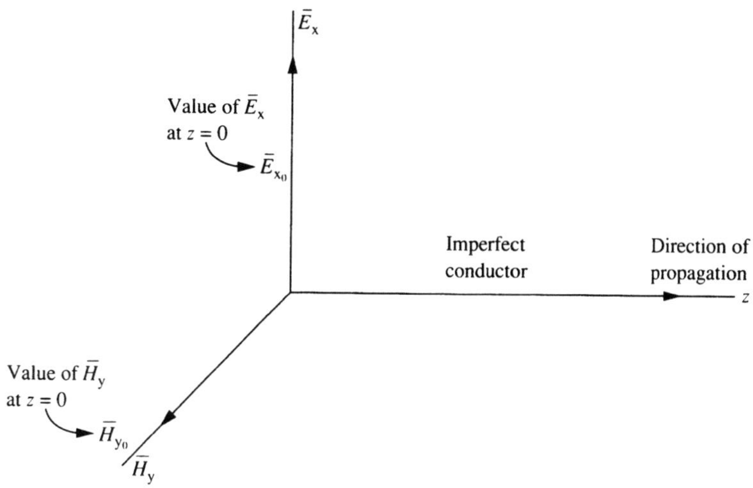

FIGURE 6–10 Plane wave propagating in imperfect conductor.

Substituting the values of α and β from (6–83) and (6–84) results in

$$\overline{E}_x = \overline{E}_{x_0} (e^{-(\sqrt{\pi f \mu \sigma})z})(e^{-j(\sqrt{\pi f \mu \sigma})z}) \tag{6–86}$$

Let E_x represent the magnitude of $\overline{E}_x$, and let E_{x_0} represent the magnitude of $\overline{E}_{x_0}$; i.e., $E_x = |\overline{E}_x|$ and $E_{x_0} = |\overline{E}_{x_0}|$. From (6–85) we have

$$E_x = E_{x_0} e^{-(\sqrt{\pi f \mu \sigma})z} \tag{6–87}$$

It is desired to represent the argument of the exponential in the form z/δ; i.e., E_x will be expressed as

$$E_x = E_{x_0} e^{-z/\delta} \tag{6–88}$$

in which

$$\delta = \frac{1}{\sqrt{\pi f \mu \sigma}} \tag{6–89}$$

The quantity δ is called the **skin depth.** (It is also called the **depth of penetration.**) It is the value of the depth at which the electric field has decayed to $1/e$ or about 36.8% of the value at the surface. Although this is a spatial variation, it is similar to the time constant of an exponential voltage or current.

From (6–89), the following properties of a conductor can be deduced:

1. The skin depth decreases as the permeability increases.
2. The skin depth decreases as the conductivity increases. (Remember that earlier it was indicated that for a perfect conductor, all the current flows on the surface, which implies that $\delta = 0$.)
3. The skin depth decreases as the frequency increases.

Property (3) has a significant effect in dealing with conductors at high frequencies. The current tends to move toward the surface, and very little flows near the center. Thus, the "ac resistance" of the conductor is effectively much greater than the "dc resistance" since the net area over which the current flows is reduced. Therefore, the measurement of the resistance of a wire with a dc ohmmeter has little bearing on what the effective resistance will be at high frequencies.

The tendency for nearly all the current to flow on or near the surface has been used to lower the cost of conducting wire designed specifically for high-frequency applications. Less expensive material with lower conductivity can be used as the core for such wire since very little current will flow near the center. However, a highly conducting surface coating can be used to lower the resistance near the surface as much as possible.

EXAMPLE 6–15

Copper has a conductivity of $\sigma = 5.8 \times 10^7$ S/m and a permeability μ_0. Calculate the skin depth at the following frequencies: (a) 60 Hz, (b) 1 MHz, and (c) 1 GHz.

Solution

At any frequency f, the skin depth δ can be expressed as

$$\delta = \frac{1}{\sqrt{\pi f \mu \sigma}} = \frac{1}{\sqrt{\pi (4\pi \times 10^{-7})(5.8 \times 10^7) f}} = \frac{0.06609}{\sqrt{f}} \qquad \textbf{(6–90)}$$

Specific values of frequency can now be substituted in (6–90).

(a) $f = 60$ Hz

$$\delta = \frac{0.06609}{\sqrt{60}} = 8.532 \times 10^{-3}\ \text{m} = 8.532\ \text{mm} \qquad \textbf{(6–91)}$$

(b) $f = 1$ MHz

$$\delta = \frac{0.06609}{\sqrt{10^6}} = 66.09 \times 10^{-6}\ \text{m} = 66.09\ \mu\text{m} \qquad \textbf{(6–92)}$$

(c) $f = 1$ GHz

$$\delta = \frac{0.06609}{\sqrt{10^9}} = 2.090 \times 10^{-6}\,\text{m} = 2.090\,\mu\text{m} \tag{6–93}$$

EXAMPLE 6–16

In this example, the difficulty of transmitting a signal to a submarine submerged under the sea will be investigated. Assume that for sea water, $\mu_r = 1$ and $\sigma = 4$ S/m. Assume that a submarine is located at a depth of 10 m below the surface. Calculate the attenuation in decibels between the surface and the submarine for the following two frequencies: (1) 1 MHz and (b) 20 kHz.

Solution

At a frequency f, the skin depth is

$$\delta = \frac{1}{\sqrt{\pi f \mu \sigma}} = \frac{1}{\sqrt{\pi(4\pi \times 10^{-7})(4)f}} = \frac{251.6}{\sqrt{f}} \tag{6–94}$$

We will now investigate the phenomenon at the two frequencies.

(a) $f = 1$ MHz

$$\delta = \frac{251.6}{\sqrt{10^6}} = 0.2516\,\text{m} \tag{6–95}$$

The electric and magnetic fields will be attenuated to $1/e$ of their surface values in a distance of less than 1 ft. The attenuation at a frequency of 1 MHz and a depth of 10 m is

$$\text{attenuation} = 8.686 \times \frac{10}{0.2516} = 345.23\,\text{dB} \tag{6–96}$$

This is an enormous attenuation, and it would be impossible to recover the signal at this frequency.

(b) $f = 20$ kHz

$$\delta = \frac{251.6}{\sqrt{20 \times 10^3}} = 1.779\,\text{m} \tag{6–97}$$

The attenuation at a frequency of 20 kHz and a depth of 10 m is

$$\text{attenuation} = 8.686 \times \frac{10}{1.779} = 48.83\,\text{dB} \tag{6–98}$$

This level of attenuation is more tolerable and would permit possible detection with a sensitive receiver.

This analysis indicates why submarine radio transmission is performed at very low frequencies, almost in the audio range. The tradeoff is that very large antenna structures are required at the transmitting station.

6–12 ■ WAVE TRANSMISSION AT DIELECTRIC BOUNDARIES

Assume that two different types of perfect dielectric materials form a boundary plane as shown in Fig. 6–11(a). (The second dimension of the plane is perpendicular to the page.) The dielectric of region 1 has an intrinsic impedance η_1, while that of region 2 has an intrinsic impedance η_2. A plane wave with power density $\mathcal{P}_1$ is assumed to be traveling from left to right, and its E and H rms values are denoted as E_1 and H_1, respectively. The regions on both sides are assumed to be infinite in length which, from a practical point of view, means that they are sufficiently long that no other effects from different boundary conditions are present. At this point in the text, consideration will be limited to **normal incidence,** i.e., to situations in which the incident wave is perpendicular to the boundary at the point of contact. Situations involving other angles will be considered later.

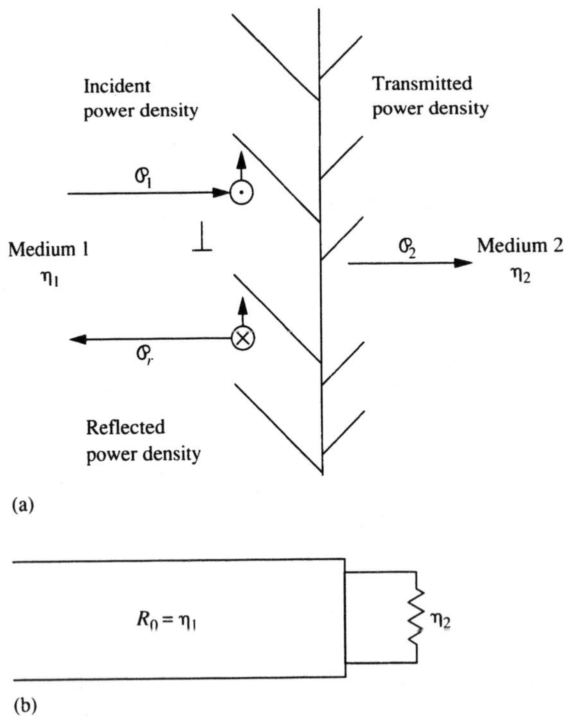

FIGURE 6–11 (a) Boundary plane between two dielectric media and (b) transmission line analogy.

The normal case can be handled easily using the transmission line analogy. Referring to Fig. 6–11(b), a fictitious transmission line with characteristic impedance η_1 is assumed to be terminated with an impedance η_2. The effects on E and H at the boundary are represented in terms of the voltage and current on the fictitious transmission line.

The reflection coefficient at the boundary will be denoted simply as Γ. (There is no need to put a subscript on it since there is only one boundary. The equivalent "generator" end is very far away in the analogy.) The value of Γ is

$$\Gamma = \frac{\eta_2 - \eta_1}{\eta_2 + \eta_1} \tag{6–99}$$

Let E_r represent the rms value of the reflected electric field, and let H_r represent the corresponding rms value for the reflected magnetic field. From the transmission line analogy, we have

$$E_r = \Gamma E_1 \tag{6–100}$$

$$H_r = -\Gamma H_1 \tag{6–101}$$

The values of E_2 and H_2 at the boundary are given by

$$E_2 = E_1 + E_r = E_1 + \Gamma E_1 = E_1(1 + \Gamma) \tag{6–102}$$

$$H_2 = H_1 + H_r = H_1 - \Gamma H_1 = H_1(1 - \Gamma) \tag{6–103}$$

Since these latter values constitute the E and H values for the power density propagating into medium 2, we can express that power density $\mathcal{P}_2$ as

$$\mathcal{P}_2 = E_2 H_2 = E_1(1 + \Gamma)H_1(1 - \Gamma) \tag{6–104a}$$

$$= E_1 H_1(1 - \Gamma^2) = \mathcal{P}_1(1 - \Gamma^2) \tag{6–104b}$$

This value is the **transmitted power,** and it is determined by multiplying the incident power by $1 - \Gamma^2$.

By power conservation, the **reflected power** is simply

$$\mathcal{P}_r = \Gamma^2 \mathcal{P}_1 \tag{6–105}$$

EXAMPLE 6–17

A plane wave propagating in air has a power density of 100 μW/m^2. The wave encounters a plane boundary at normal incidence of which the material beyond the boundary is polyethylene ($\epsilon_r = 2.25$). Determine (a) the electric field reflection coefficient, (b) the VSWR, (c) the reflected power density, and (d) the transmitted power density.

Solution

The intrinsic impedance in medium 1 is simply

$$\eta_1 = \eta_0 = 377\ \Omega \tag{6–106}$$

The intrinsic impedance in medium 2 is

$$\eta_2 = \frac{\eta_0}{\sqrt{\epsilon_r}} = \frac{377}{\sqrt{2.25}} = 251.3\ \Omega \tag{6–107}$$

A transmission line analogy can now be constructed as shown in Fig. 6–12.

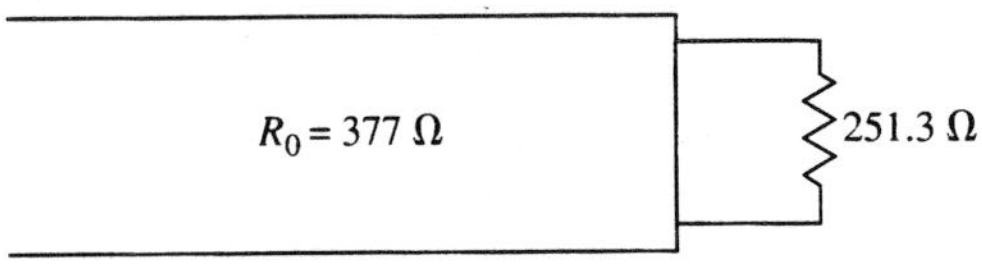

FIGURE 6–12 Transmission line analogy of Example 6–17.

(a) The electric field reflection coefficient corresponds to the voltage reflection coefficient of the transmission line analogy and is

$$\Gamma = \frac{\eta_2 - \eta_1}{\eta_2 + \eta_1} = \frac{251.3 - 377}{251.3 + 377} = -0.200 \tag{6–108}$$

(b) Since both values of impedance are purely resistive, the VSWR is simply

$$S = \frac{\eta_1}{\eta_2} = \frac{377}{251.3} = 1.5 \tag{6–109}$$

(c) The reflected power density is

$$\mathcal{P}_r = \Gamma^2 \mathcal{P}_1 = (-0.200)^2 \times 100 = 4\ \mu\text{W/m}^2 \tag{6–110}$$

(d) The transmitted power density is

$$\mathcal{P}_2 = (1 - \Gamma^2)\mathcal{P}_1 = [1 - (-0.200)^2] \times 100 = 96\ \mu\text{W/m}^2 \tag{6–111}$$

Alternatively, we could have employed the principle of conservation of power; i.e.,

$$\mathcal{P}_2 = \mathcal{P}_1 - \mathcal{P}_r = 100 - 4 = 96\ \mu\text{W/m}^2 \tag{6–112}$$

■ PROBLEMS

6–1. The voltage between two parallel plates separated by a distance of 5 mm is 200 V. Determine the electric field intensity.

6–2. The voltage between two parallel plates separated by a distance of 0.4 in is 60 V. Determine the electric field intensity.

6–3. The electric field intensity in the region between two parallel plates separated by a distance of 4 cm is 2 kV/m. Determine the voltage between the plates.

6–4. The electric field intensity in the region between two parallel plates separated by a distance of 8 mm is 200 V/mm. Determine the voltage between the plates.

6–5. A direct current of 5 A is flowing in a conductor. Determine the magnetic field intensity at a distance of 3 m from the conductor.

6–6. A direct current of 40 mA is flowing in a conductor. Determine the magnetic field intensity at a distance of 5 ft from the conductor.

6–7. For the parallel plates of Problem 6–3, determine the electric flux density if the dielectric is polyethylene ($\epsilon_r = 2.25$).

6–8. For the parallel plates of Problem 6–4, determine the electric flux density if the dielectric is air.

6–9. For the current-carrying conductor of Problem 6–5, determine the magnetic flux density at a distance of 3 m from the conductor if the medium is air.

6–10. For the current-carrying conductor of Problem 6–6, determine the magnetic flux density at a distance of 5 ft from the conductor if the medium is air.

6–11. The electric flux density normal to a rectangular surface with dimensions 8 m × 75 cm is 4 μC/m^2. Determine the value of the electric flux across the area.

6–12. The electric flux density normal to a circular surface with a diameter of 3 m is 8 μC/m^2. Determine the value of the electric flux across the area.

6–13. The magnetic flux density normal to a circular surface with a radius of 5 m is 4 nWb/m^2. Determine the value of the magnetic flux across the area.

6–14. The magnetic flux density normal to a rectangular surface with dimensions 30 cm × 60 cm is 12 nWb/m^2. Determine the value of the magnetic flux across the area.

6–15. A current of 8 A is uniformly distributed over a rectangular conductor with dimensions 5 mm × 4 mm. Determine the current density.

6–16. A current of 4 A is uniformly distributed over a circular conductor with a diameter of 3 cm. Determine the current density.

6–17. Assume that the conductivity for the conductor of Problem 6–15 is 5 MS/m. Determine the electric field intensity.

6–18. Assume that the conductivity for the conductor of Problem 6–16 is 6×10^7 S/m. Determine the electric field intensity.

6–19. The rms magnitude of the magnetic field of a plane wave in air is $H_y = 200$ μA/m. Assuming that E is in the positive x-direction, determine the following for a circular surface of diameter 50 m in the x-y plane over which the fields are constant:
(a) E_x
(b) $\mathcal{P}_z$
(c) total power transmitted through area

6–20. The rms magnitude of the electric field of a plane wave in sea water ($\epsilon_r = 80$) is $E_x = 3$ V/m. Assuming that H is in the positive y-direction, determine the following for a square surface with sides of 15 m each in the x-y plane over which the fields are constant:
(a) H_y
(b) $\mathcal{P}_z$
(c) total power transmitted through area

6–21. In a lossless dielectric medium, the rms electric and magnetic field intensities are $E_x = 18.97$ mV/m and $H_y = 158.1$ μA/m. Determine the following:
(a) intrinsic impedance
(b) power density
(c) dielectric constant

6–22. The power density of a plane wave propagating in free space is $\mathcal{P}_z = 120$ μW/m^2. Determine the following:
(a) E_x
(b) H_y

6–23. Using the values of ϵ_r and σ for sea water in Example 6–13, determine the various quantities at a frequency of 10 MHz.

6–24. Using the values of ϵ_r and σ for sea water in Example 6–13, determine the various quantities at a frequency of 1 GHz.

6–25. Fresh water has typical values of dielectric constant and conductivity given by $\epsilon_r = 80$ and $\sigma = 10^{-3}$ S/m. At a frequency of 100 MHz, determine the following quantities:
(a) intrinsic impedance
(b) propagation constant
(c) attenuation constant
(d) velocity of propagation

6–26. Repeat the analysis of Problem 6–25 for fresh water at a frequency of 1 MHz.

6–27. Using the criteria in the text and the values of $\epsilon_r = 80$ and $\sigma = 4$ S/m, classify whether sea water is a *dielectric*, a *conductor*, or a *quasi conductor* at a frequency of 900 MHz.

6–28. Repeat the analysis of Problem 6–27 at a frequency of 1 MHz.

6–29. Fresh water has typical values of dielectric constant and conductivity given by $\epsilon_r = 80$ and $\sigma = 10^{-3}$ S/m. Classify whether it is a *dielectric*, a *conductor*, or a *quasi conductor* at a frequency of 100 kHz.

6–30. Repeat the analysis of Problem 6–29 at a frequency of 30 GHz.

6–31. Repeat the analysis of Problem 6–29 at a frequency of 10 kHz.

6–32. For fresh water with parameters as given in Problem 6–29, determine the frequency at which the loss tangent is unity.

6–33. Based on the criteria in the text and the values of $\epsilon_r = 80$ and $\sigma = 4$ S/m for sea water, determine the frequency constituting the boundary between behavior as a conductor and as a quasi conductor.

6–34. Based on the criteria in the text and the values of $\epsilon_r = 80$ and $\sigma = 10^{-3}$ S/m for fresh water, determine the frequency constituting the boundary between behavior as a conductor and as a quasi conductor.

6–35. With the same assumptions as in Problem 6–33 for sea water, determine the frequency constituting the boundary between the behavior as a quasi conductor and as a dielectric.

6–36. With the same assumptions as in Problem 6–34 for fresh water, determine the frequency constituting the boundary between behavior as a quasi conductor and as a dielectric.

6–37. Based on the assumptions that $\sigma = 4$ S/m and $\mu = \mu_0$ for sea water, calculate the skin depth at a frequency of 10 kHz.

6–38. Based on the assumptions that $\sigma = 10^{-3}$ S/m and $\mu = \mu_0$ for fresh water, calculate the skin depth at a frequency of 10 kHz.

6–39. Based on the assumptions of Problem 6–37, calculate the skin depth of sea water at a frequency of 100 kHz.

6–40. Based on the assumptions of Problem 6–38, calculate the skin depth of fresh water at a frequency of 100 kHz.

6–41. A plane wave propagating in air has a power density of 20 mW/m^2. The wave encounters a plane boundary at normal incidence of which the material beyond the boundary is marble ($\epsilon_r = 8$). Determine the following:
(a) electric field reflection coefficient
(b) VSWR
(c) reflected power density
(d) transmitted power density

6–42. A plane wave propagating in a medium of mica ($\epsilon_r = 6$) has a power density of 80 μW/m^2. The wave encounters a plane boundary at normal incidence of which the material beyond the boundary is polystyrene ($\epsilon_r = 2.7$). Determine the following:
(a) electric field reflection coefficient
(b) VSWR
(c) reflected power density
(d) transmitted power density.

6–43. A plane wave propagating in air at a frequency of 1 GHz has a power density of 100 μW/m^2. It encounters a sheet of polyethylene ($\epsilon_r = 2.25$) of thickness 7.5 cm as shown in Fig. P6–43. Determine the following:
(a) power density of wave reflected back to the left of the sheet
(b) power density of wave transmitted into air on the right of the slab
(*Hint:* Use a transmission line analogy with the polyethylene sheet representing a transmission line and air on the right as the termination.)

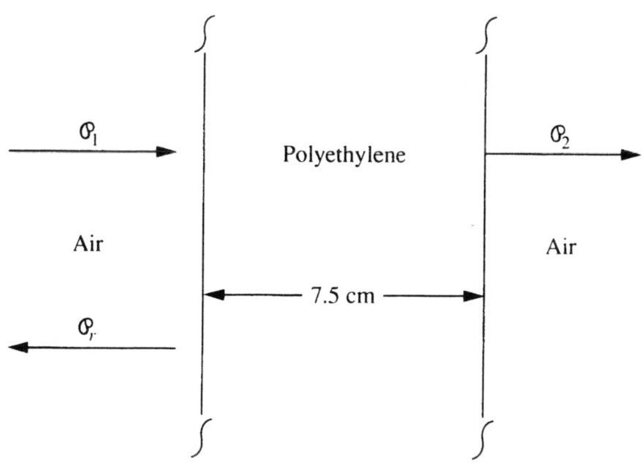

FIGURE P6–43

6–44. For the system of Problem 6–43, determine the smallest thickness (other than zero) of the sheet that would result in all the power transmitted to the air on the right.

7

High-Frequency Transmission Lines

OBJECTIVES

After completing this chapter, the reader should be able to:

1. Determine the boundary conditions for the electric and magnetic field intensities and flux densities between any two media.
2. Define the linear conduction current density $\overrightarrow{K}$ on the surface of a boundary between any two media.
3. State the boundary conditions for an interface between any two perfect dielectric media.
4. State the boundary conditions for an interface between a dielectric medium and a perfect conductor.
5. Define **electromagnetic field modes.**
6. Describe the transverse electromagnetic (TEM) mode, the transverse electric (TE) mode, and the transverse magnetic (TM) mode.
7. Determine the inductance and capacitance per unit length for a two-wire parallel transmission line.
8. Determine the characteristic impedance and velocity of propagation for a two-wire parallel transmission line.
9. Determine the inductance and capacitance per unit length for a coaxial cable.
10. Determine the characteristic impedance and velocity of propagation for a coaxial cable.
11. Describe the **stripline** and **microstrip** types of substrate transmission lines.
12. Determine the wavelength of an electromagnetic wave traveling on a *strip-type* substrate transmission line.
13. Determine the characteristic impedance, velocity of propagation, and wavelength on a *stripline* type of substrate transmission line.
14. Determine the physical dimensions of a *stripline* to obtain a specified characteristic impedance.
15. Determine the characteristic impedance, velocity of propagation, and wavelength on a *microstrip* type of substrate transmission line.
16. Determine the physical dimensions of a *microstrip* to obtain a specified characteristic impedance.

7-1 ■ BOUNDARY CONDITIONS

The electric and magnetic fields are vectors that have both magnitude and direction. At the boundary between two media, these vectors can be resolved into two components: one normal to the surface of the boundary and the other tangential to the boundary. These components are illustrated in Fig. 7–1. The two media can be characterized by three physical parameters: permittivity ϵ, permeability μ, and conductivity σ. Both the electric and magnetic fields at the boundary between the two media must satisfy Maxwell's equations in each medium.

Generalized Boundary Conditions

There are four generalized boundary conditions based on (1) the *tangential* component of the **electric field intensity** $\vec{E}_t$, (2) the *tangential* component of the **magnetic field intensity** $\vec{H}_t$, (3) the *normal* component of the **electric flux density** $\vec{D}_n$, and (4) the *normal* component of the **magnetic flux density** $\vec{B}_n$. The normal component of the electric flux density $\vec{D}_n$ in a medium is related to the normal component of the electric field intensity $\vec{E}_n$ by

$$\vec{D}_n = \epsilon \vec{E}_n \tag{7–1}$$

The normal component of the magnetic flux density $\vec{B}_n$ is related to the normal component of the magnetic field intensity $\vec{H}_n$ by

$$\vec{B}_n = \mu \vec{H}_n \tag{7–2}$$

The four generalized boundary conditions as required to satisfy Maxwell's equations are shown in Fig. 7–2.

The tangential component of the electric field intensity $\vec{E}_{t_1}$ in medium 1 at the boundary between the two media must be equal to the tangential component of the electric field intensity $\vec{E}_{t_2}$ in medium 2; i.e.,

$$\vec{E}_{t_1} = \vec{E}_{t_2} \tag{7–3}$$

This boundary condition is illustrated in Fig. 7–2(a). One of Maxwell's equations (Faraday's law) can be used to prove this boundary condition. The proofs of the generalized boundary conditions are beyond the scope of this book; however, they can be found in any advanced textbook on electromagnetic field theory.

A linear conduction current density $\vec{K}$, whose units are amperes per meter (A/m), can flow on the surface of the boundary between the two media. (The linear conduction current density is related to the conduction current density $\vec{J}$, which was defined in Chapter 6.) It can be shown by one of Maxwell's equations (Ampère's law) that the difference between the tangential component of the magnetic field intensity $\vec{H}_{t_1}$ in medium 1 and the tangential component of the magnetic field intensity $\vec{H}_{t_2}$ in medium 2 at the boundary between the two media must be equal to the linear conduction current density $\vec{K}$ on the

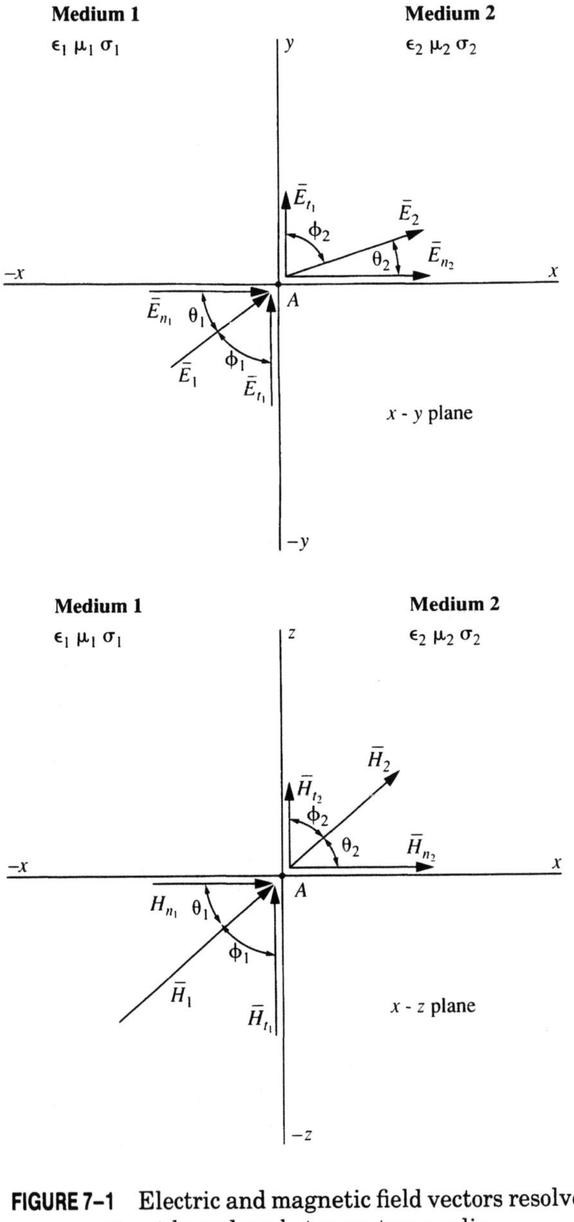

FIGURE 7–1 Electric and magnetic field vectors resolved into two normal and tangential components at boundary between two media.

surface of the boundary between the two media. In mathematical terms,

$$\vec{H}_{t_1} - \vec{H}_{t_2} = \vec{K} \tag{7–4}$$

Referring to Fig. 7–2(b), note that the linear conduction current density vector flows on the surface of the boundary, which is either into or out of the

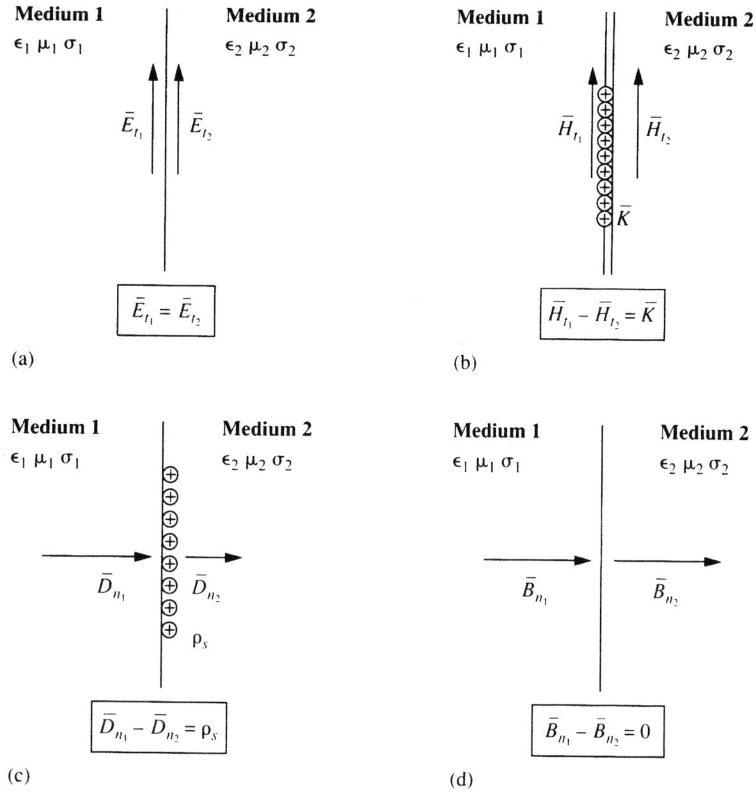

FIGURE 7–2 Generalized boundary conditions.

plane of the page. The direction of the vector is determined by one version of the "right-hand rule." If $\vec{H}_{t1} > \vec{H}_{t2}$, then $\vec{K}$ is directed into the page; and if $\vec{H}_{t1} < \vec{H}_{t2}$, then $\vec{K}$ is directed out of the page.

A surface charge density ρ_s can exist on the surface of the boundary between two media. One of Maxwell's equations (Gauss's law for the electric field) can be used to derive the third generalized boundary condition. The difference between the normal component of the electric flux density $\vec{D}_{n1}$ in medium 1 and the normal component of the electric flux density $\vec{D}_{n2}$ in medium 2 is equal to the surface charge density ρ_s on the surface of the boundary between the two media. In mathematical terms,

$$\vec{D}_{n1} - \vec{D}_{n2} = \rho_s \tag{7–5}$$

This boundary condition is illustrated in Fig. 7–2(c).

The final generalized boundary condition is shown in Fig. 7–2(d). The last of Maxwell's equations (Gauss's law for magnetic fields) can be used to derive this boundary condition. The normal component of the magnetic flux density $\vec{B}_{n1}$ in medium 1 and the normal component of the magnetic flux density $\vec{B}_{n2}$ in medium 2 at the boundary between the two media must be equal. In mathematical terms,

$$\vec{B}_{n1} = \vec{B}_{n2} \tag{7–6}$$

Dielectric-to-Dielectric Boundary Conditions

A commonly encountered situation requires the determination of the boundary conditions between two dielectrics, both of which can be assumed to behave as perfect dielectrics. As discussed in Section 6–10, a perfect dielectric is one in which the conductivity is zero. Since the conductivity is zero, there are no free electrons, and therefore both the linear conduction current density and the surface charge density must be zero. The four generalized boundary conditions reduce to the following for the boundary between perfect dielectrics:

$$\vec{E}_{t1} = \vec{E}_{t2} \tag{7–7}$$

$$\vec{H}_{t1} = \vec{H}_{t2} \tag{7–8}$$

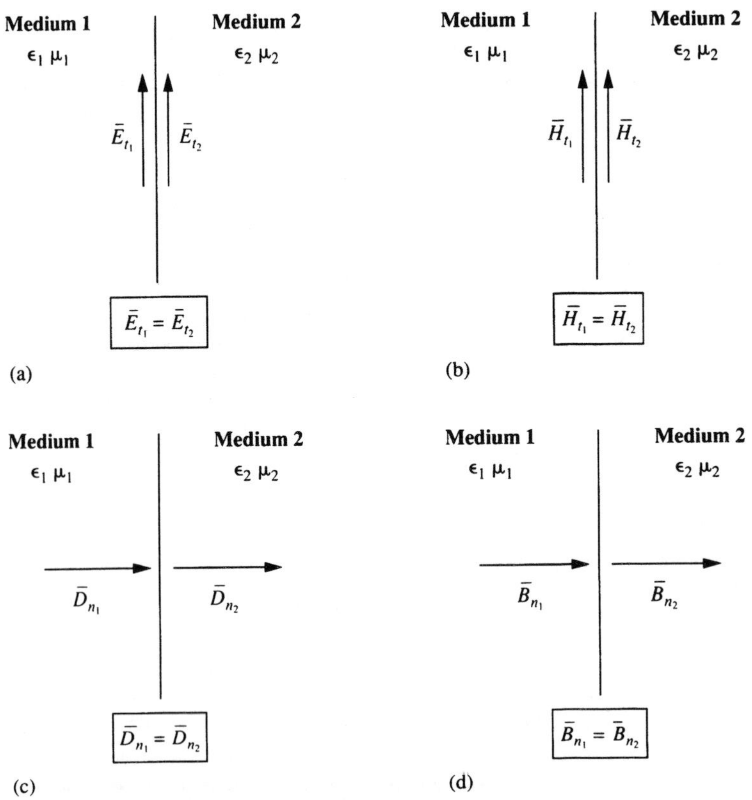

FIGURE 7–3 Dielectric-to-dielectric boundary conditions.

$$\vec{D}_{n_1} = \vec{D}_{n_2} \tag{7–9}$$
$$\vec{B}_{n_1} = \vec{B}_{n_2} \tag{7–10}$$

The four boundary conditions for the case of two perfect dielectrics are shown in Fig. 7–3.

Dielectric-to-Conductor Boundary Conditions

The boundary conditions between a dielectric and a perfect conductor represent a second situation that is encountered often. As discussed in Section 6–10, the conductivity in a perfect conductor is infinite; therefore, no fields can exist in a perfect conductor. On the surface of the perfect conductor at the boundary, both a linear conduction current density $\vec{K}$ and a surface charge density ρ_s can exist. Since $\vec{E}_t$ must be zero in the perfect conductor, (7–3) becomes

$$\vec{E}_{t_1} = 0 \tag{7–11}$$

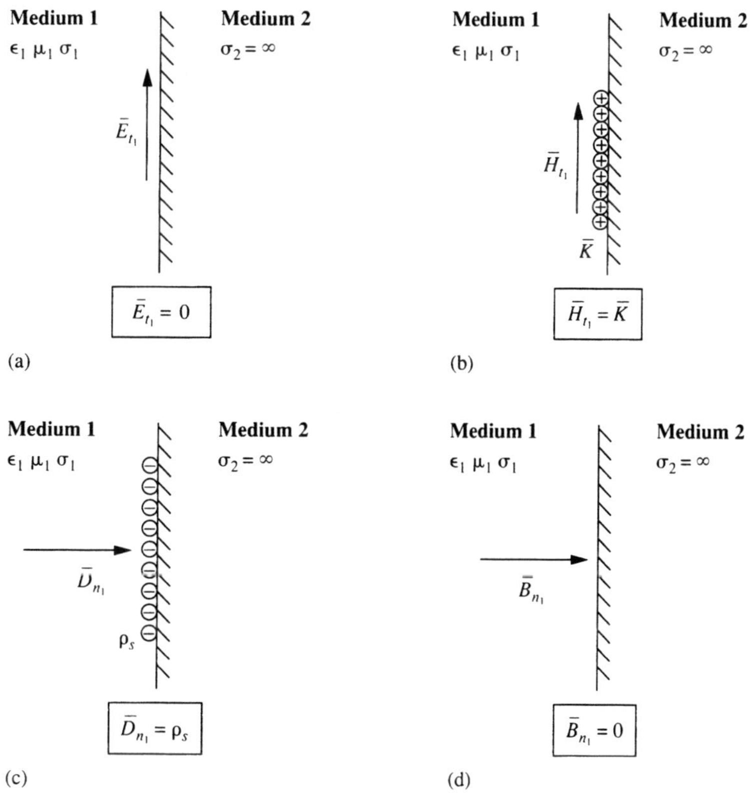

FIGURE 7–4 Dielectric-to-perfect-conductor boundary conditions.

Since $\vec{H}_t$ must be zero in the perfect conductor, (7–4) becomes

$$\vec{H}_{t1} = \vec{K} \tag{7–12}$$

Since $\vec{D}_n$ must be zero in the perfect conductor, (7–5) becomes

$$\vec{D}_{n1} = \rho_s \tag{7–13}$$

Since $\vec{B}_n$ must be zero in the perfect conductor, (7–6) becomes

$$\vec{B}_n = 0 \tag{7–14}$$

The four boundary conditions for the case of an interface between a dielectric and a perfect conductor are shown in Fig. 7–4.

EXAMPLE 7–1

Refer to the boundary between two different media shown in Fig. 7–5. An electric field intensity vector $\vec{E}_1$ in medium 1 at point A on the boundary makes an angle ϕ_1 of 60° with respect to the boundary as shown. The magnitude of $\vec{E}_1$ is 1000 V/m. The dielectric constant in medium 1 is 2.3, and that in medium 2 is 9.7. Assume that in both media, $\mu = \mu_0$. A surface charge density ρ_s with value -2.522×10^{-8} C/m² exists at point A. Determine (a) the magnitude of the electric field intensity $\vec{E}_2$ and (b) the angle ϕ_2 that it makes with respect to the boundary at point A.

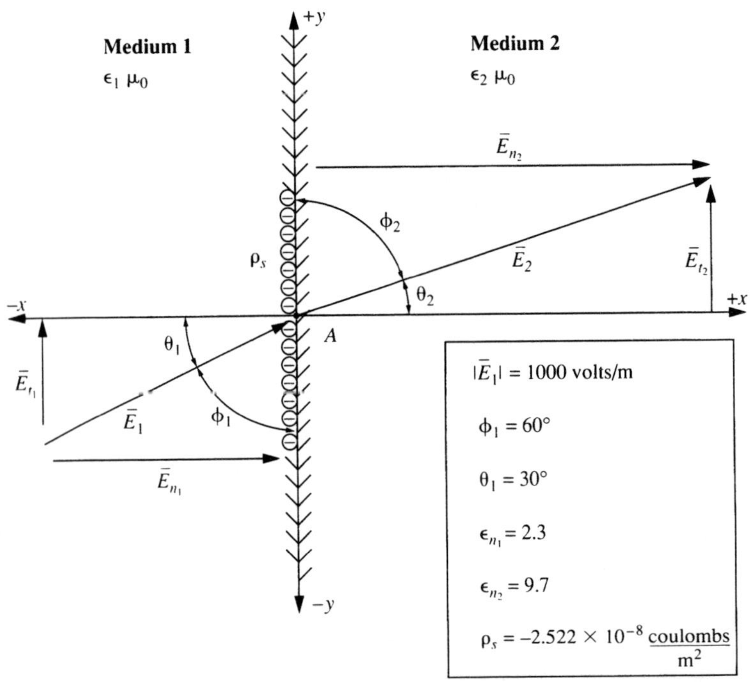

FIGURE 7–5 Boundary conditions for Example 7–1.

Solution

As shown in Fig. 7–5, the tangential and normal components of the electric field intensity can be determined as follows. The tangential component in medium 1 is

$$\vec{E}_{t_1} = \vec{E}_1 \times \cos\phi_1 = 1 \times 10^3 \times \cos 60° = 500 \text{ V/m} \tag{7–15}$$

The normal component of the electric field intensity in medium 1 is

$$\vec{E}_{n_1} = \vec{E}_1 \times \sin\phi_1 = 1 \times 10^3 \times \sin 60° = 866 \text{ V/m} \tag{7–16}$$

The tangential component of the electric field intensity in medium 2 can be determined from the boundary condition of (7–7):

$$\vec{E}_{t_2} = \vec{E}_{t_1} = 500 \text{ V/m} \tag{7–17}$$

The normal component of the electric flux density in medium 1 can be determined from (7–1):

$$\vec{D}_{n_1} = \epsilon_{r_1}\epsilon_0\vec{E}_{n_1} = (2.3)\left(\frac{1}{36\pi} \times 10^{-9}\right)(866) = 1.761 \times 10^{-8} \text{ C/m}^2 \tag{7–18}$$

The normal component of the electric flux density in medium 2 can be determined from the boundary condition of (7–5):

$$\vec{D}_{n_2} = \vec{D}_{n_1} - \rho_s = 1.761 \times 10^{-8} - (-2.522 \times 10^{-8}) = 4.283 \times 10^{-8} \text{ C/m}^2 \tag{7–19}$$

The normal component of the electric field intensity in medium 2 can be determined from (7–1):

$$\vec{E}_{n_2} = \frac{\vec{D}_{n2}}{\epsilon_{r_2}\epsilon_0} = \frac{4.283 \times 10^{-8}}{(9.7)\left(\frac{1}{36\pi} \times 10^{-9}\right)} = 500 \text{ V/m} \tag{7–20}$$

(a) The magnitude of the electric field intensity in medium 2 is

$$|\vec{E}_2| = \sqrt{|\vec{E}_{t_2}|^2 + |\vec{E}_{n_2}|^2} = \sqrt{(500)^2 + (500)^2} = 707.1 \text{ V/m} \tag{7–21}$$

(b) The angle* ϕ_2 that the electric field intensity vector makes with the boundary in Fig. 7–5 is

$$\phi_2 = \tan^{-1}\left\{\frac{|\vec{E}_{n2}|}{|\vec{E}_{t_2}|}\right\} = \tan^{-1}\left\{\frac{500}{500}\right\} = 45° \tag{7–22}$$

EXAMPLE 7–2

Refer to the boundary between different media shown in Fig. 7–6. A magnetic field intensity vector $\vec{H}_1$ in medium 1 at point A makes an angle ϕ_1 of 45° with respect to the boundary as shown. The magnitude of the magnetic field intensity vector in medium 1 is 14.14 A/m. Assume that $\mu = \mu_0$ in both media. A linear

*Angles will be expressed in degrees in this chapter and in all following chapters. Degrees are more appropriate here, while radians were more appropriate in the first six chapters.

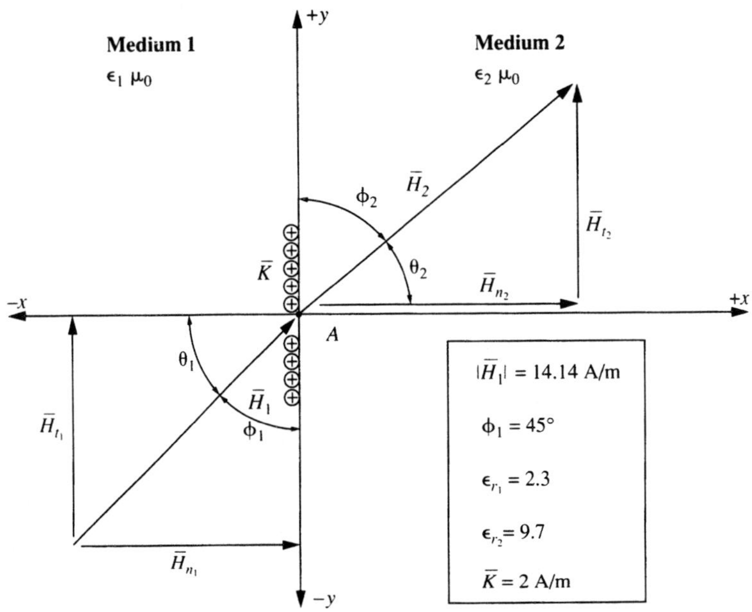

FIGURE 7–6 Boundary conditions for Example 7–2.

conduction current density $\vec{K}$ of value 2 A/m directed into the page in Fig. 7–6 exists at point A. (The tails of the arrows in Fig. 7–6 indicate that the direction of $\vec{K}$ is into the paper.) Determine (a) the magnitude of the magnetic field intensity vector in medium 2 at point A and (b) the angle ϕ_2 that it makes with respect to the boundary at point A.

Solution

As shown in Fig. 7–6, the tangential and normal components of the magnetic field intensity vector can be determined as follows. The tangential component in medium 1 is

$$\vec{H}_{t_1} = \vec{H}_1 \times \cos 45° = 14.14 \times \cos 45° = 10 \text{ A/m} \tag{7–23}$$

The normal component in medium 1 is

$$\vec{H}_{n_1} = \vec{H}_1 \times \sin 45° = 14.14 \times \sin 45° = 10 \text{ A/m} \tag{7–24}$$

The tangential component of the magnetic field intensity vector in medium 2 can be determined from boundary condition (7–4):

$$\vec{H}_{t_2} = \vec{H}_{t_1} - \vec{K} = 10 - 2 = 8 \text{ A/m} \tag{7–25}$$

The normal component of the magnetic field intensity vector in medium 2 can be determined from (7–2) and boundary condition (7–6):

$$\vec{B}_{n_1} = \mu_0 \vec{H}_{n_1} \quad \text{and} \quad \vec{B}_{n_2} = \mu_0 \vec{H}_{n_2} \tag{7–26}$$

Therefore,

$$\vec{H}_{n_2} = \vec{H}_{n_1} = 10 \text{ A/m} \tag{7–27}$$

(a) The magnitude of the magnetic field intensity vector in medium 2 is

$$|\vec{H}_2| = \sqrt{|\vec{H}_{t_2}|^2 + |\vec{H}_{n_2}|^2} = \sqrt{8^2 + 10^2} = 12.81 \text{ A/m} \tag{7–28}$$

(b) The angle ϕ_2 that the magnetic field intensity vector makes with the boundary is

$$\phi_2 = \tan^{-1}\left\{\frac{|\vec{H}_{n_2}|}{|\vec{H}_{t_2}|}\right\} = \tan^{-1}\left\{\frac{10}{8}\right\} = 51.34^\circ \tag{7–29}$$

EXAMPLE 7-3

An electric field intensity vector of 100 V/m exists at a point on the boundary between a dielectric medium and a perfect conductor. The dielectric constant of the dielectric medium is 10. The permeability of the dielectric medium is μ_0. A magnetic field intensity vector of 10 A/m exists at the same point. Determine (a) the tangential and normal components of the electric and magnetic field intensities in the dielectric medium and within the perfect conductor at the point on the boundary, (b) the surface charge density on the surface of the perfect conductor, and (c) the linear conduction current density.

Solution

The boundary conditions for the boundary between a dielectric and a perfect conductor, given in Eqs. (7–11) through (7–14), require that the tangential component of the electric field intensity vector and the normal component of the magnetic flux density vector be equal to zero at the boundary. This condition is a result of the infinite conductivity of the perfect conductor, which requires that no electromagnetic fields exist inside a perfect conductor.

(a) The only two field components at the boundary are

$$\vec{D}_{n_1} = \epsilon_r \epsilon_0 \vec{E}_{n_1} = (10)\left(\frac{1}{36\pi} \times 10^{-9}\right)(100) = 8.842 \times 10^{-9} \text{ C/m}^2 \tag{7–30}$$

$$\vec{H}_{t_1} = 10 \text{ A/m} \tag{7–31}$$

(b) The surface charge density can be determined from boundary condition (7–13):

$$\rho_s = \vec{D}_{n_1} = 8.842 \times 10^{-9} \text{ C/m}^2 \tag{7–32}$$

(c) The linear conduction current density on the surface can be determined from the boundary condition (7–12):

$$\vec{K} = \vec{H}_{t_1} = 10 \text{ A/m} \tag{7–33}$$

7–2 ■ ELECTROMAGNETIC FIELD MODES

Electromagnetic waves are transverse waves since they exist in a plane perpendicular to the direction of propagation. Also, the time-varying electric and magnetic fields must be orthogonal, i.e., oriented 90° in space with respect to each other. The electric and magnetic fields are functions of distance (x, y, and z) and time (t). There can be many solutions of Maxwell's equations that satisfy the required boundary conditions. These solutions for a particular physical configuration such as a transmission line are defined as **modes.**

The **dominant mode** is the mode that exists at the lowest frequency. Higher-order modes can also exist on the transmission line; however, it is desirable to have a frequency region where only the dominant mode can propagate.

There are three types of transverse electromagnetic modes: (a) the **T**rans**v**erse **E**lectro**M**agnetic **(TEM)** mode, (b) the **T**ransverse **E**lectric **(TE)** mode, and (c) the **T**ransverse **M**agnetic **(TM)** mode. Figure 7–7 illustrates the electric and magnetic fields in a TEM mode on a coaxial transmission line. Both a cross-sectional (transverse) view and a side view taken through the center of the transverse view at AA' are shown.

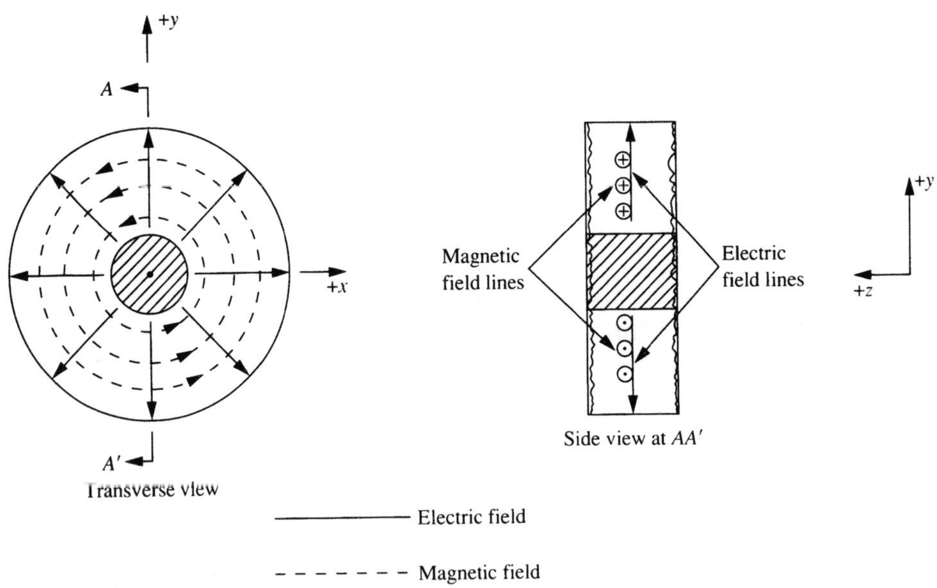

FIGURE 7–7 Transverse electromagnetic modes on a coaxial transmission line.

Figure 7–8 illustrates the dominant TE_{10} mode and the dominant TM_{11} mode for a rectangular waveguide. As in the TEM modes shown in Fig. 7–7, both views are shown. The coaxial transmission line will be described later in this chapter. The rectangular waveguide will be covered in Chapter 8.

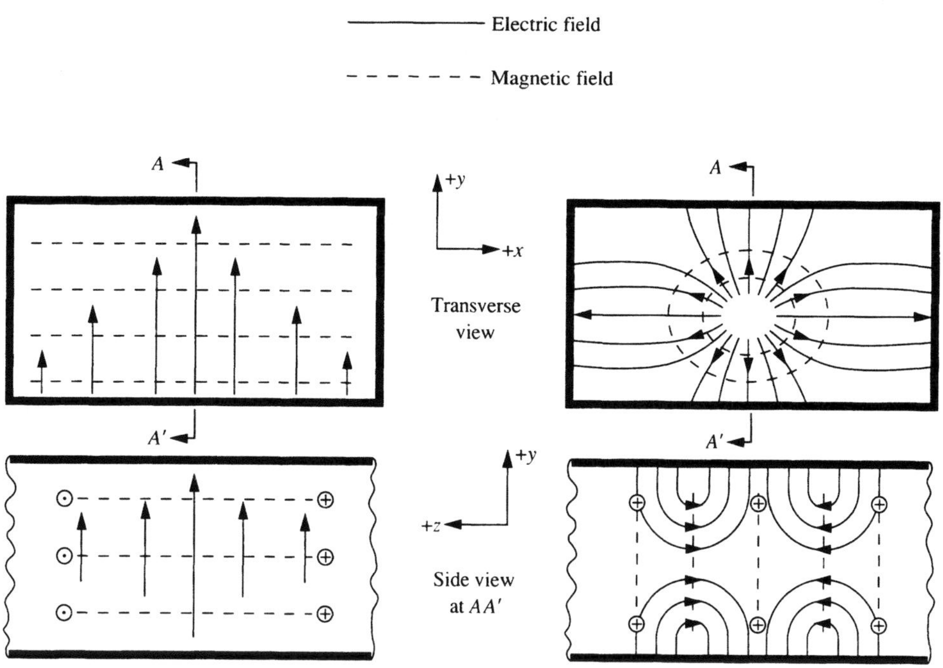

(a) Transverse electric (TE_{10} mode)

(b) Transverse magnetic (TM_{11} mode)

FIGURE 7–8 Transverse electromagnetic modes in rectangular waveguide.

The TEM mode is one in which variations of the electric and magnetic fields as a function of distance are restricted to the transverse (x-y) plane at any fixed instant of time. There will be no variation of $\vec{E}$ and $\vec{H}$ in the propagation (z-) direction, as shown in the side view of Fig. 7–7(a). The TE mode is one in which variation of the electric field as a function of distance is restricted to the transverse (x-y) plane. However, the magnetic field is a function of distance in all three directions (x, y, and z), as illustrated in the two views of the TE_{10} mode in Fig. 7–8(a). The TM mode is one in which variation of the magnetic field as a function of distance is restricted to the transverse (x-y) plane. However, the electric field is a function of distance in all three directions (x, y, and z), as illustrated in the two views of the TM_{11} mode in Fig. 7–8(b).

7–3 ■ TWO-WIRE PARALLEL TRANSMISSION LINE

The two-wire parallel transmission line is one of the oldest types of transmission lines used in the transmission of electrical energy. Such lines are used for power transmission, telephone lines, and television signals. The two-wire parallel line is a balanced transmission line and is usually limited to frequencies below 500 MHz due to radiation losses. A cross-sectional drawing of the two-wire parallel transmission is shown in Fig. 7–9(a). Let d represent the diameter of each of the wires, and D represent the spacing between the centers of each conductor. The dominant mode for the two-wire parallel line is the TEM mode, as shown in Fig. 7–9(b).

The formulas for the inductance and capacitance can be derived from the basic definitions of Section 1–5. The inductance per unit length is

$$L = \frac{\mu}{\pi} \ln \frac{2D}{d} \text{ H/m} \tag{7–34}$$

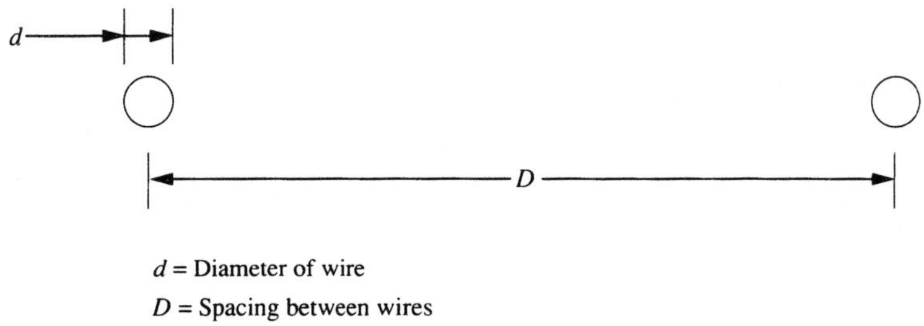

(a) Cross-sectional view

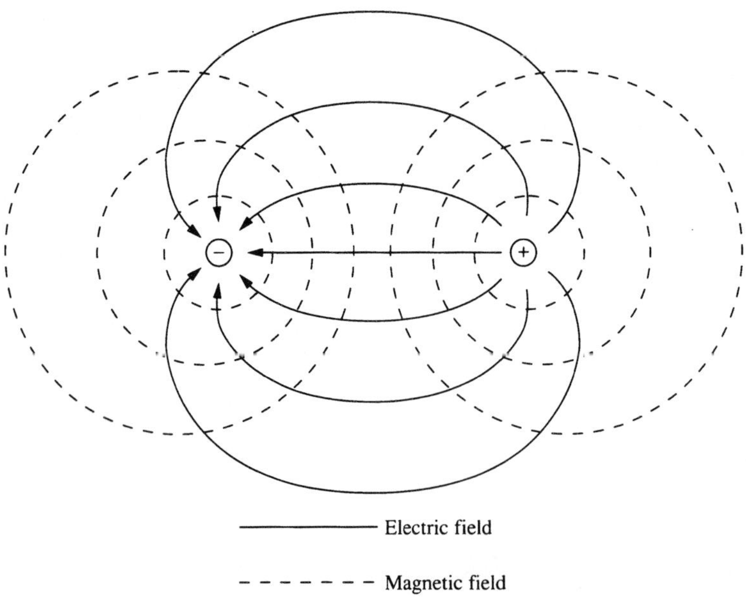

(b) Electromagnetic fields (TEM mode)

FIGURE 7–9 Two-wire parallel transmission line.

Since $\mu_r = 1$ for nonmagnetic materials, and since the permeability of free space is $\mu_0 = 4\pi \times 10^{-7}$ H/m, (7–34) reduces to

$$L = 4 \times 10^{-7} \ln \frac{2D}{d} \text{ H/m} \tag{7–35}$$

The capacitance per unit length is

$$C = \frac{\pi\epsilon}{\ln \frac{2D}{d}} \text{ F/m} \tag{7–36}$$

Since $\epsilon = \epsilon_r \epsilon_0$ and $\epsilon_0 = \frac{1}{36\pi} \times 10^{-9}$ F/m, then (7–36) reduces to

$$C = 27.78 \times 10^{-12} \frac{\epsilon_r}{\ln \frac{2D}{d}} \text{ F/m} \tag{7–37}$$

The characteristic impedance R_0 of any lossless transmission line is

$$R_0 = \sqrt{\frac{L}{C}}\ \Omega \tag{7–38}$$

The characteristic impedance of a two-wire parallel transmission line is determined by substituting (7–34) and (7–36) into (7–38):

$$R_0 = \frac{1}{\pi}\sqrt{\frac{\mu}{\epsilon}} \ln \frac{2D}{d}\ \Omega \tag{7–39}$$

Since $\mu_r = 1$ for nonmagnetic materials, we obtain, after substituting the values for ϵ_0 and μ_0 into (7–39),

$$R_0 = \frac{120}{\sqrt{\epsilon_r}} \ln \frac{2D}{d}\ \Omega \tag{7–40}$$

The velocity of propagation on any lossless transmission line is

$$\nu = \frac{1}{\sqrt{LC}} \tag{7–41}$$

Substituting (7–34) and (7–36) into (7–41), we obtain the velocity of propagation for a two-wire parallel transmission line:

$$\nu = \frac{1}{\sqrt{\epsilon\mu}} = \frac{1}{\sqrt{\epsilon_r \epsilon_0 \mu_r \mu_0}} = \frac{c}{\sqrt{\epsilon_r}} = \frac{3 \times 10^8}{\sqrt{\epsilon_r}} \text{ m/s} \tag{7–42}$$

where

$$c = \frac{1}{\sqrt{\epsilon_0 \mu_0}} = 3 \times 10^8 \text{ m/s} \tag{7–43}$$

For most applications of parallel lines, the medium surrounding the conductors is air, in which $\epsilon_r = 1$ is assumed. Therefore, the velocity of propagation is the speed of light.

EXAMPLE 7–4

A two-wire parallel transmission line has the following physical dimensions:

Diameter of each conductor (d) = 0.0644 cm
Spacing between the two conductors (D) = 0.3922 cm
Dielectric constant of medium (ϵ_r) = 1

Determine (a) the inductance per unit length, (b) the capacitance per unit length, (c) the characteristic impedance, and (d) the velocity of propagation.

Solution

Since $\ln(2D/d)$ is required for the first three parameters, it will be determined first:

$$\ln\frac{2D}{d} = \ln\left[\frac{(2)(0.3922)}{0.0644}\right] = 2.50 \tag{7–44}$$

(a) The inductance per unit length is determined from (7–35):

$$L = 4 \times 10^{-7} \ln\frac{2D}{d} = (4 \times 10^{-7})(2.50) = 1\ \mu\text{H/m} \tag{7–45}$$

(b) The capacitance per unit length is determined from (7–37):

$$C = 27.78 \times 10^{-12}\frac{\epsilon_r}{\ln\dfrac{2D}{d}} \tag{7–46a}$$

$$= \frac{27.78 \times 10^{-12}}{2.5} = 11.11\ \text{pF/m} \tag{7–46b}$$

(c) The characteristic impedance is determined from (7–40):

$$R_0 = \frac{120}{\sqrt{\epsilon_r}}\ln\frac{2D}{d} = (120)(2.5) = 300\ \Omega \tag{7–47}$$

(d) The velocity of propagation is determined from (7–42):

$$\nu = \frac{3 \times 10^8}{\sqrt{\epsilon_r}} = 3 \times 10^8\ \text{m/s} \tag{7–48}$$

7–4 ■ COAXIAL TRANSMISSION LINE

The coaxial transmission line is the most widely used transmission line for high-frequency applications. A cross-sectional drawing of a coaxial cable is shown in Fig. 7–10(a). It consists of an inner conductor with diameter d, which is surrounded by a concentric cylinder of an insulating material with a dielectric constant ϵ_r. The outer conductor is a concentric cylinder with inner diameter D. The coaxial cable is an unbalanced transmission line since the outer conductor is normally operated at ground potential. The dominant mode in a coaxial cable is the TEM mode; however, higher-order modes can exist.

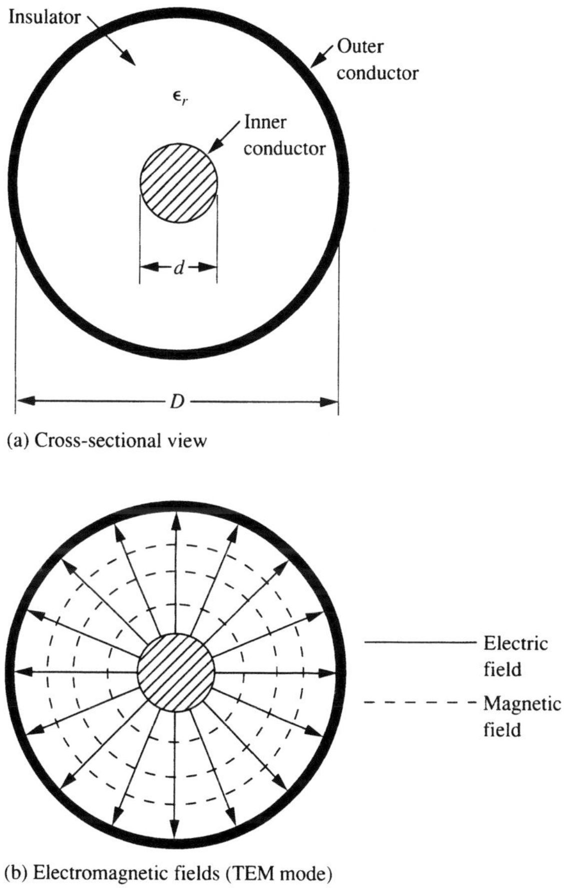

FIGURE 7–10 Coaxial transmission line.

The TEM mode pattern for the coaxial cable is shown in Fig. 7–10(b). As shown in the mode patterns, coaxial cables are nonradiating since the electromagnetic fields are confined in the insulator between the inner and outer conductors. The recent development of precision connectors for smaller-diameter coaxial cables has increased the frequency that coaxial cables can operate to above 40 GHz.

The formulas for the inductance and capacitance of coaxial cable can be determined from Maxwell's equations and the basic definitions of Chapter 1. The inductance per unit length is

$$L = \frac{\mu}{2\pi} \ln \frac{D}{d} \text{ H/m} \tag{7–49}$$

Since $\mu_r = 1$ for nonmagnetic materials, and since the permeability of free space is $\mu_0 = 4\pi \times 10^{-7}$ H/m, (7–49) reduces to

$$L = 2 \times 10^{-7} \ln \frac{D}{d} \text{ H/m} \tag{7–50}$$

The capacitance per unit length is

$$C = \frac{2\pi\epsilon}{\ln(D/d)} \text{ F/m} \tag{7–51}$$

Since $\epsilon = \epsilon_r\epsilon_0$, where ϵ_r is the relative permittivity (dielectric constant) and $\epsilon_0 = 1/(36\pi) \times 10^{-9}$ F/m, then (7–51) reduces to

$$C = 55.56 \times 10^{-12} \frac{\epsilon_r}{\ln(D/d)} \text{ F/m} \tag{7–52}$$

The characteristic impedance of a coaxial cable can be obtained by substituting (7–49) and (7–51) into (7–38):

$$R_0 = \frac{1}{2\pi}\sqrt{\frac{\mu}{\epsilon}} \ln \frac{D}{d} \ \Omega \tag{7–53}$$

Since $\mu_r = 1$ for nonmagnetic materials, and substituting the values for ϵ_0 and μ_0 into (7–53), we obtain

$$R_0 = \frac{60}{\sqrt{\epsilon_r}} \ln\left\{\frac{D}{d}\right\} \Omega \tag{7–54}$$

The velocity of propagation for the coaxial cable can be obtained by substituting (7–49) and (7–51) into (7–41):

$$\nu = \frac{1}{\sqrt{\mu\epsilon}} = \frac{c}{\sqrt{\mu_r\epsilon_r}} \text{ m/s} \tag{7–55}$$

Since $c = \dfrac{1}{\sqrt{\mu_0\epsilon_0}}$ and $\mu_r = 1$ for coaxial cables, the velocity of propagation for coaxial cable becomes

$$\nu = \frac{3 \times 10^8}{\sqrt{\epsilon_r}} \text{ m/s} \tag{7–56}$$

The wavelength λ of the electromagnetic signals traveling on any transmission line can be determined from the following equation:

$$\lambda = \frac{\text{velocity of propagation}}{\text{frequency}} = \frac{\nu}{f} \tag{7–57}$$

EXAMPLE 7–5

A coaxial cable has the following physical dimensions:

Diameter of inner conductor (d) = 0.29 cm

Inner diameter of outer conductor (D) = 1.05 cm

Polyethylene dielectric (ϵ_r) = 2.3

Determine (a) the inductance per unit length, (b) the capacitance per unit length, (c) the characteristic impedance, and (d) the velocity of propagation.

Solution

Since $\ln(D/d)$ is required for the first three parameters, it will be determined first:

$$\ln\frac{D}{d} = \ln\frac{1.05}{0.29} = 1.287 \tag{7–58}$$

(a) The inductance per unit length is determined from (7–50):

$$L = 2 \times 10^{-7} \ln\frac{D}{d} \tag{7–59a}$$

$$= (2 \times 10^{-7})(1.287) = 257.4 \text{ nH/m} \tag{7–59b}$$

(b) The capacitance per unit length is determined from (7–52):

$$C = 55.56 \times 10^{-12} \frac{\epsilon_r}{\ln(D/d)} \tag{7–60a}$$

$$= \frac{(55.56 \times 10^{-12})(2.3)}{1.287} = 99.29 \text{ pF/m} \tag{7–60b}$$

(c) The characteristic impedance is determined from (7–54):

$$R_0 = \frac{60}{\sqrt{\epsilon_r}} \ln\frac{D}{d} = \frac{60}{\sqrt{2.3}}(1.287) = 50.92\ \Omega \tag{7–61}$$

(d) The velocity of propagation is determined from (7–56):

$$\nu = \frac{3 \times 10^8}{\sqrt{\epsilon_r}} = \frac{3 \times 10^8}{\sqrt{2.3}} = 1.978 \times 10^8 \text{ m/s} \tag{7–62}$$

7–5 ■ STRIP-TYPE SUBSTRATE TRANSMISSION LINE

A type of transmission line widely used at frequencies from 100 MHz up to approximately 100 GHz is the **strip-type substrate transmission line.** This type of transmission line consists of one or two dielectric substrates that have been plated with a conductor such as copper to form a ground plane on one side and the strip-type conductor or circuit element on the other side. Computer-aided design programs are used to design the circuits and transmission line components. Photo-etching techniques are used in the manufacturing, which results in economical, repeatable, and accurate components in the RF and microwave frequency regions.

There are two popular types of strip-type substrate lines: (a) a symmetrical type commonly referred to as **stripline** and (b) an asymmetrical type commonly referred to as **microstrip.** Stripline is commonly used for passive transmission line components, while microstrip is used for both passive and active components.

Stripline

Stripline consists of two dielectric substrates that have been plated with a conductor. Figure 7–11(a) shows a cross-sectional view of the stripline structure. The lower substrate has a metallic ground plane on the lower surface of the dielectric substrate. The low-loss dielectric substrate has a dielectric constant ϵ_r. The metallic center conductor is photo-etched on the upper surface of the lower substrate. The width of the center conductor is w, and the thickness of the metallic conductor is t. Usually the thicknesses of the metallic center conductor and the metallic ground planes are the same. The thickness of each dielectric substrate is $b/2$. The dominant mode for the stripline is a TEM mode which is shown in Fig. 7–11(b). The advantage of the stripline configuration is similar to that of the coaxial cable: The fields are essentially confined within the transmission line, and there are no significant radiation losses. The major disadvantages of stripline are that (a) the circuit is not accessible during development for adjustments and tuning and (b) it is difficult to mount discrete and active components such as transistors, diodes, circulators, chip resistors, and chip capacitors.

The analytical solution for the characteristic impedance of stripline using Maxwell's equations in closed form is difficult due to complicated physical structure and boundary conditions. A combination of analytical and empirical techniques has led to the following equations for the characteristic impedance of stripline. These equations, along with many stripline design equations, can be found in Howe.[†] The equations are divided into a **high-impedance region** and a **low-impedance region** determined by the ratio of w to $(b - t)$. The impedance of a stripline is inversely proportional to the ratio of the width w

[†]H. Howe, Jr., *Stripline Circuit Design* (Dedham, MA: Artech House, 1974).

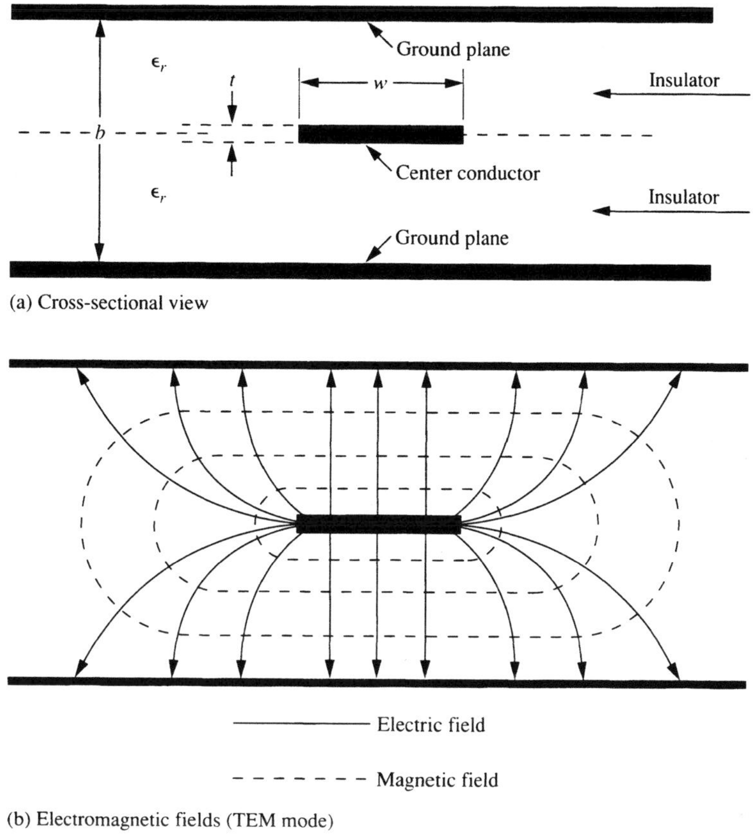

FIGURE 7–11 Stripline transmission line.

of the inner conductor to the distance b between the two ground planes. In the high-impedance region, the characteristic impedance can be computed by determining the diameter of an equivalent circular inner conductor.

Design Equations

High-Impedance Region

$$\frac{w}{b-t} \leq 0.35 \quad \text{and} \quad \frac{t}{b} \leq 0.25 \tag{7–63}$$

The characteristic impedance of a stripline in the high-impedance region is

$$Z_0 = \frac{60}{\sqrt{\epsilon_r}} \ln\left[\frac{4b}{\pi d}\right] \Omega \tag{7–64}$$

where d is the diameter of a circular conductor equivalent to the rectangular center conductor of the stripline with width w and thickness t.

$$d = \frac{w}{2}\left\{1 + \frac{t\left[1 + \ln\left(\frac{4\pi w}{t}\right) + 0.51\pi\left(\frac{t}{w}\right)^2\right]}{\pi w}\right\} \quad (7\text{–}65)$$

Low-Impedance Region

$$\frac{w}{b - t} > 0.35 \quad (7\text{–}66)$$

The characteristic impedance of a stripline in the low-impedance region is

$$Z_0 = \frac{94.15}{\sqrt{\epsilon_r}\left(\frac{w}{bA} + B\right)}\ \Omega \quad (7\text{–}67)$$

where

$$A = 1 - \frac{t}{b} \quad (7\text{–}68)$$

and

$$B = \frac{1}{\pi}\left[\frac{2}{A}\ln\left(1 + \frac{1}{A}\right) - \left(\frac{1}{A} - 1\right)\ln\left(\frac{1}{A^2} - 1\right)\right] \quad (7\text{–}69)$$

The velocity of propagation for the stripline transmission line is

$$\nu = \frac{c}{\sqrt{\epsilon_r}}\ \text{m/s} \quad (7\text{–}70)$$

The wavelength of the electromagnetic signal on the stripline transmission line is

$$\lambda = \frac{\nu}{f} = \frac{c}{\sqrt{\epsilon_r} f}\ \text{m} \quad (7\text{–}71)$$

A graphical design aid for determining the width of the conductor for a stripline design as a function of the characteristic impedance, the dielectric constant ϵ_r, and the ratio of conductor thickness t to dielectric insulator thickness b is presented in Fig. 7–12. These curves are solutions to Eqs. (7–64) through (7–69) for both the high-impedance and low-impedance regions. The curves of the product of the square root of the dielectric constant times the characteristic impedance $\sqrt{\epsilon_r}Z_0$ are a function of the conductor width to dielectric thickness ratio w/b for commonly available thickness to dielectric thickness ratios t/b. However, many computer-aided design tools are available today for designing stripline circuits and transmission lines.

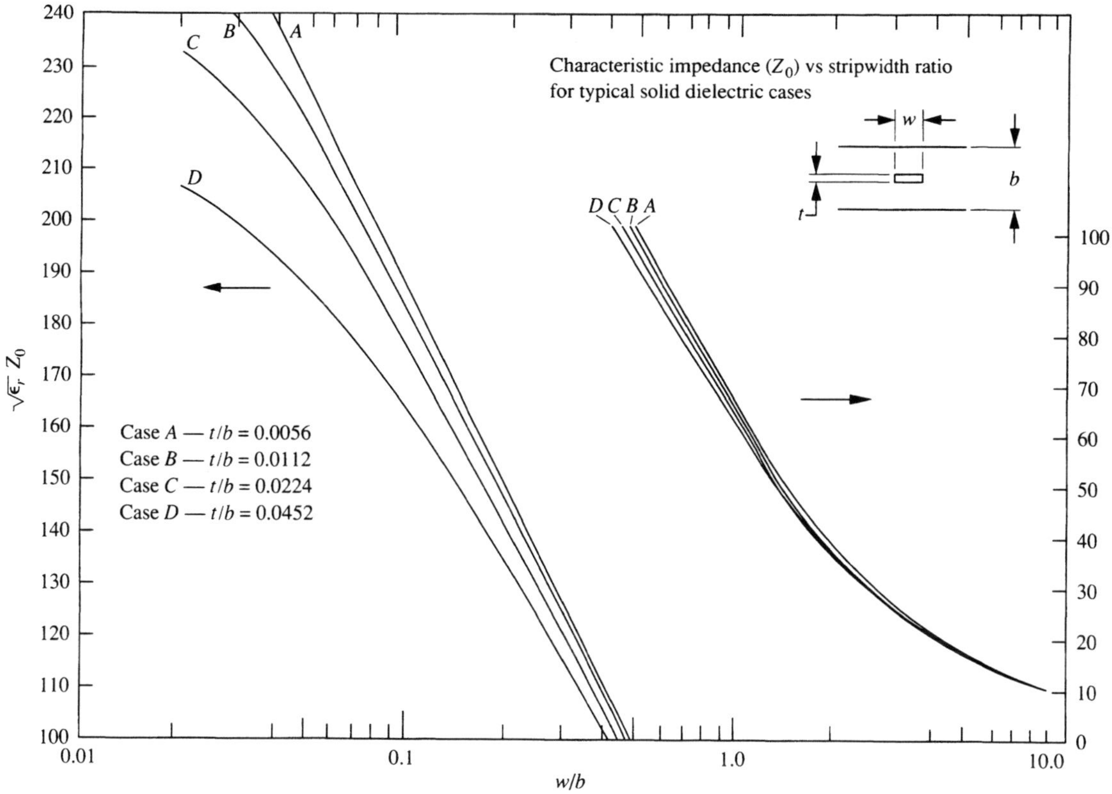

FIGURE 7–12 Characteristic impedance with stripline using commonly available t/b ratios. Adapted from H. Howe, Jr., *Stripline Circuit Design* (Dedham, MA: Artech House, 1974).

EXAMPLE 7–6

A stripline transmission line is fabricated from polyolefin printed circuit board which is plated with 1-oz copper. The distance between the ground planes b is 0.3175 cm. The thickness t of the 1-oz copper is 0.003556 cm. The width w of the conductor is 0.1 cm. Polyolefin has a dielectric constant of 2.32. Determine for this stripline (a) the characteristic impedance and (b) the velocity of propagation.

Solution

The first step is to determine which set of equations to use—those for the high-impedance region or those for the low-impedance region. Solving for the

$w/(b-t)$ and t/b factors yields

$$\frac{w}{b-t} = \frac{0.1}{0.3175 - 0.003556} = 0.319 \tag{7–72}$$

$$\frac{t}{b} = \frac{0.003556}{0.3175} = 0.0112 \tag{7–73}$$

The $w/(b-t)$ and t/b factors satisfy the criteria of (7–63) for the high-impedance region. The diameter of the equivalent circular conductor d is determined from (7–65):

$$d = \frac{w}{2}\left\{1 + \frac{t\left[1 + \ln\left(\frac{4\pi w}{t}\right) + 0.51\pi\left(\frac{t}{w}\right)^2\right]}{\pi w}\right\} \tag{7–74a}$$

$$= \frac{0.1}{2}\left\{1 + \frac{(0.003556)\left[1 + \ln\left(\frac{4\pi(0.1)}{0.003556}\right) + 0.51\pi\left(\frac{0.003556}{0.1}\right)^2\right]}{\pi(0.1)}\right\} \tag{7–74b}$$

$$= 0.0539 \tag{7–74c}$$

(a) The characteristic impedance is determined from (7–64):

$$Z_0 = \frac{60}{\sqrt{\epsilon_r}}\ln\left[\frac{4b}{\pi d}\right] \tag{7–75a}$$

$$= \frac{60}{\sqrt{2.32}}\ln\left[\frac{(4)(0.3175)}{\pi(0.0539)}\right] = 79.38\ \Omega \tag{7–75b}$$

(b) The velocity of propagation is determined from (7–70):

$$\nu = \frac{c}{\sqrt{\epsilon_r}} = \frac{3 \times 10^8}{\sqrt{2.32}} = 1.970 \times 10^8\ \text{m/s} \tag{7–76}$$

EXAMPLE 7–7

A stripline transmission line is fabricated from the same material as in Example 7–6. However, the width w of the conductor is 1.790 cm. Determine the characteristic impedance of this stripline. (Since the material is the same as in Example 7–6, the velocity of propagation will be the same.)

Solution

The first step is to determine which set of equations to use—those for the high-impedance region or those for the low-impedance region. Solving for the $w/(b-t)$ and t/b factors yields

$$\frac{w}{b-t} = \frac{1.790}{0.3175 - 0.003556} = 5.702 \tag{7–77}$$

$$\frac{t}{b} = \frac{0.003556}{0.3175} = 0.0112 \qquad \textbf{(7–78)}$$

The $w/(b-t)$ factor satisfies the criteria of (7–66) for the low-impedance region. First, we solve (7–68) for the A factor:

$$A = 1 - \frac{t}{b} = 1 - \frac{0.003556}{0.3175} = 0.9888 \qquad \textbf{(7–79)}$$

We then solve (7–70) for the B factor:

$$B = \frac{1}{\pi}\left[\frac{2}{A}\ln\left(1 + \frac{1}{A}\right) - \left(\frac{1}{A} - 1\right)\ln\left(\frac{1}{A^2} - 1\right)\right] \qquad \textbf{(7–80a)}$$

$$= \frac{1}{\pi}\left[\frac{2}{0.9888}\ln\left(1 + \frac{1}{0.9888}\right) - \left(\frac{1}{0.9888} - 1\right)\ln\left(\frac{1}{(0.9888)^2} - 1\right)\right] \qquad \textbf{(7–80b)}$$

$$= 0.4635 \qquad \textbf{(7–80c)}$$

The characteristic impedance is determined from (7–67):

$$Z_0 = \frac{94.15}{\sqrt{\epsilon_r}\left(\dfrac{w}{bA} + B\right)} \qquad \textbf{(7–81a)}$$

$$= \frac{94.15}{\sqrt{2.32}\left(\dfrac{1.790}{0.3175 \times 0.9888} + 0.4635\right)} = 10.03\ \Omega \qquad \textbf{(7–81b)}$$

Microstrip

Microstrip consists of a dielectric substrate with a metallic ground plane on the lower surface and a metallic conductor photo-etched on the upper surface to form microstrip transmission lines and circuits. Figure 7–13(a) shows a cross-sectional view of the microstrip structure. The low-loss dielectric substrate has a relative permittivity of ϵ_r. The width w of the metallic conductor, the thickness t of the metallic conductor, and the thickness of the dielectric substrate h determine the height of the conductor above the ground plane. The dominant mode for the microstrip is a quasi-TEM mode which is shown in Fig. 7–13(b). As shown in this figure, the electric and magnetic fields of the quasi-TEM mode exist both in the air above the microstrip line and in the dielectric substrate between the conductor and the ground plane. Therefore, to determine the characteristic impedance and velocity of propagation, one must determine an effective dielectric constant, which is a function of the width-to-height ratio (w/h).

Design and analysis equations to achieve accuracies better than 1% must be divided into two regions based on the width-to-height ratio (w/h). Analysis equations are used to determine the characteristic impedance and effective dielectric constant when the physical characteristics of width, height, and

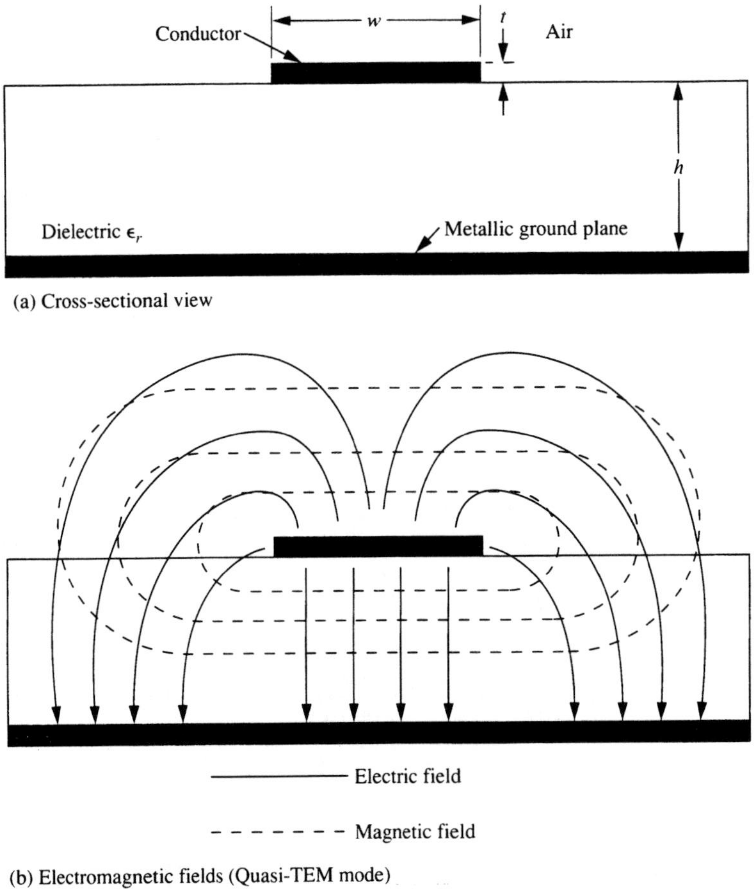

FIGURE 7–13 Microstrip transmission line.

dielectric constant are known. Design equations are used to determine the physical characteristics of the microstrip line to obtain the desired characteristic impedance. Separate design and analysis equations are required because the characteristic impedance is a function of both the w/h ratio and the effective dielectric constant. The effective dielectric constant is also a function of the w/h ratio. Hammerstadt[‡] has presented revised design and analysis equations for microstrip characteristic impedance and effective dielectric constant that have accuracies better than 1%. These equations are valid for dielectric constants less than 16 and for w/h ratios from 0.05 to 20. They assume that the conductor thickness t is small compared to the dielectric thickness h and that

[‡]E. O. Hammerstadt, "Equations for Microstrip Circuit Design," *Proceedings of the Fifth European Microwave Conference* (Microwave Exhibitions & Publishers, 1975) pp. 268–72.

t can be neglected. These restrictions are satisfied for the majority of all design requirements.

Analysis Equations

High-Impedance Region ($w/h < 1$)

The characteristic impedance of microstrip in the high-impedance region is

$$Z_0 = \frac{60}{\sqrt{\epsilon_{\text{eff}}}} \ln\left(\frac{8h}{w} + \frac{w}{4h}\right) \Omega \tag{7–82}$$

The effective dielectric constant in the high-impedance region is

$$\epsilon_{\text{eff}} = \frac{\epsilon_r + 1}{2} + \frac{\epsilon_r - 1}{2}\left[\frac{1}{\sqrt{1 + \dfrac{12h}{w}}} + 0.04\left(1 - \frac{w}{h}\right)^2\right] \tag{7–83}$$

Low-Impedance Region ($w/h \geq 1$)

The characteristic impedance of microstrip in the low-impedance region is

$$Z_0 = \frac{120\pi}{\sqrt{\epsilon_{\text{eff}}}}\left\{\frac{1}{\left[\dfrac{w}{h} + 1.393 + 0.667 \ln\left(1.444 + \dfrac{w}{h}\right)\right]}\right\} \Omega \tag{7–84}$$

The effective dielectric constant in the low-impedance region is

$$\epsilon_{\text{eff}} = \frac{\epsilon_r + 1}{2} + \frac{\epsilon_r - 1}{2}\left(\frac{1}{\sqrt{1 + 12\dfrac{h}{w}}}\right) \tag{7–85}$$

The velocity of propagation for both the high- and low-impedance regions is

$$\nu = \frac{c}{\sqrt{\epsilon_{\text{eff}}}} \text{ m/s} \tag{7–86}$$

The wavelength of the electromagnetic signals on the microstrip transmission line is

$$\lambda = \frac{c}{\sqrt{\epsilon_{\text{eff}}} f} \text{ m} \tag{7–87}$$

Figure 7–14 presents a graphical design aid that shows the characteristic impedance as a function of the width-to-height ratio w/h for various dielectric constants ϵ_r. Also, the effective dielectric constant ϵ_{eff} is plotted as a function of w/h in the figure. These curves can be used to estimate the w/h ratio required to design a specified characteristic impedance. This ratio is required so that the correct design equations, Eqs. (7–88) through (7–91), can be used to deter-

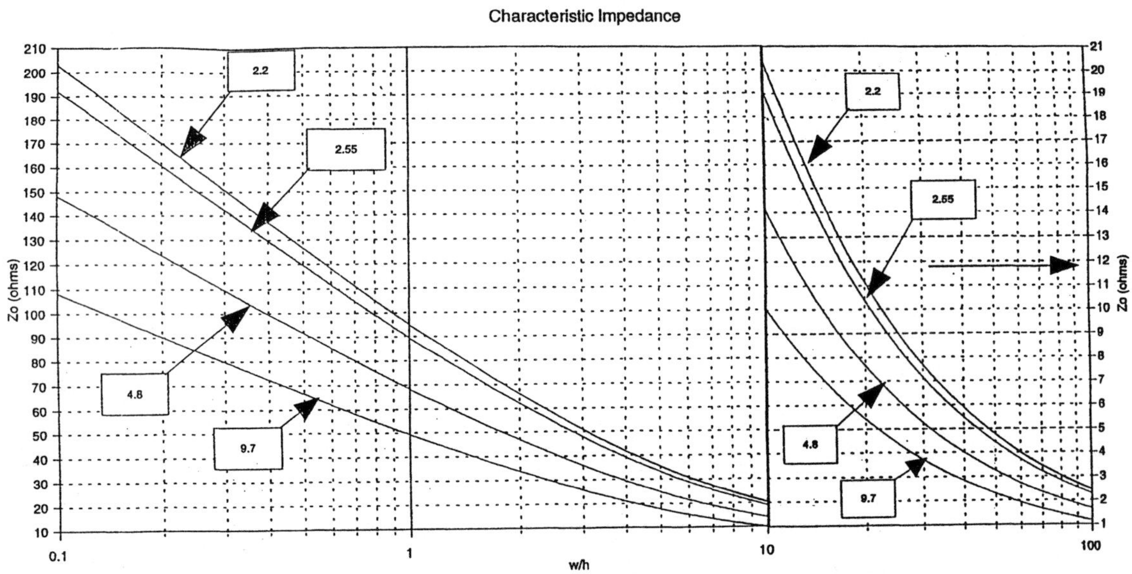

(a)

FIGURE 7–14 (a) Characteristic impedance of a microstrip line for various dielectric constants (ϵ_r).

mine the width of the microstrip as a function of the dielectric constant of the substrate and specified characteristic impedance.

Design Equations. Note that the regions for the design equations are divided at $w/h = 2$, while the analysis equations are divided at $w/h = 1$.

High-Impedance Region ($w/h < 2$)

The following equation is used to determine the width of the microstrip required to obtain a specified characteristic impedance for a given substrate where the dielectric constant ϵ_r and the substrate thickness h are known:

$$\frac{w}{h} = \frac{8}{e^A - \dfrac{2}{e^A}} \tag{7–88}$$

where

$$A = \frac{Z_0}{60}\sqrt{\frac{\epsilon_r + 1}{2}} + \frac{\epsilon_r - 1}{\epsilon_r + 1}\left(0.23 + \frac{0.11}{\epsilon_r}\right) \tag{7–89}$$

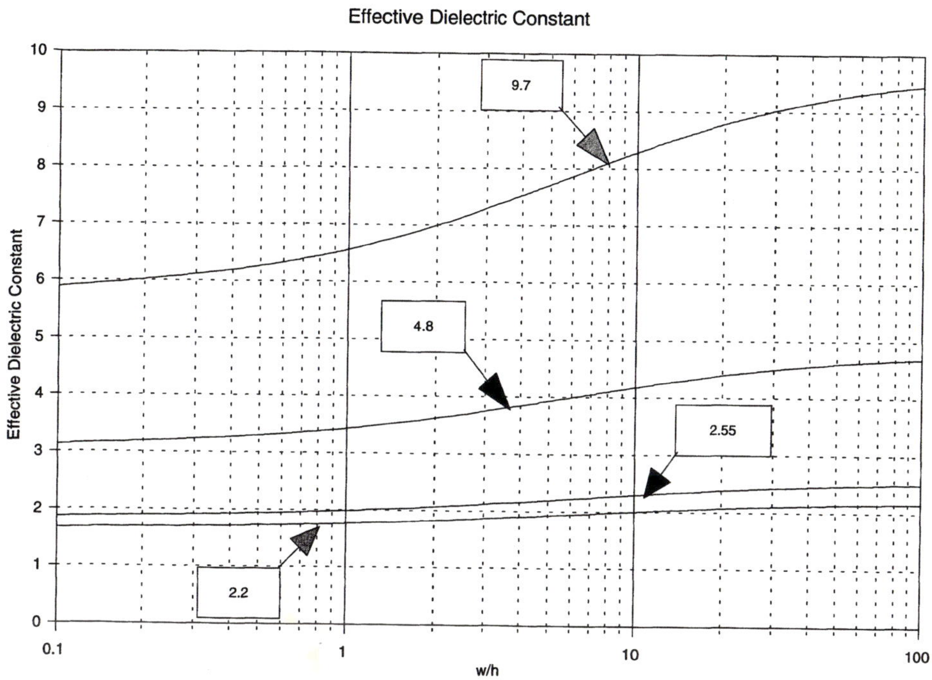

(b)

FIGURE 7–14 (b) Effective dielectric constant of a microstrip line for various dielectric constants (ϵ_r).

Low-Impedance Region ($w/h \geq 2$)
The following equation is used to determine the width of the microstrip required to obtain a specified characteristic impedance for a given substrate where the dielectric constant ϵ_r and the substrate thickness h are known:

$$\frac{w}{h} = \frac{2}{\pi}\left\{B - 1 - \ln(2B - 1) + \frac{\epsilon_r - 1}{2\epsilon_r}\left[\ln(B - 1) + 0.39 - \frac{0.61}{\epsilon_r}\right]\right\} \quad \textbf{(7–90)}$$

where

$$B = \frac{120\pi^2}{2Z_0\sqrt{\epsilon_r}} \qquad \textbf{(7–91)}$$

Many computer-aided design programs are available for the microwave engineer or technologist to use in the design of microstrip circuits and transmission lines.

EXAMPLE 7–8

Determine the characteristic impedance and the effective dielectric constant for a microstrip transmission line fabricated on an alumina substrate (ϵ_r = 9.7). The w/h ratio is 5.0.

Solution

Since the w/h ratio > 1, the low-impedance analysis equations (7–84) and (7–85) are used. First, the effective dielectric constant is determined from (7–85):

$$\epsilon_{\text{eff}} = \frac{\epsilon_r + 1}{2} + \frac{\epsilon_r - 1}{2}\left(\frac{1}{\sqrt{1 + 12\dfrac{h}{w}}}\right) \qquad \textbf{(7–92a)}$$

$$= \frac{9.7 + 1}{2} + \frac{9.7 - 1}{2}\left(\frac{1}{\sqrt{1 + 12 \times \left(\dfrac{1}{5}\right)}}\right) \qquad \textbf{(7–92b)}$$

$$= 7.709 \qquad \textbf{(7–92c)}$$

The characteristic impedance is determined from (7–84):

$$Z_0 = \frac{120\pi}{\sqrt{\epsilon_{\text{eff}}}}\left\{\frac{1}{\left[\dfrac{w}{h} + 1.393 + 0.667\ln\left(1.444 + \dfrac{w}{h}\right)\right]}\right\} \qquad \textbf{(7–93a)}$$

$$= \frac{120\pi}{\sqrt{7.709}}\left\{\frac{1}{5 + 1.393 + 0.667\ln(1.444 + 5)}\right\} \qquad \textbf{(7–93b)}$$

$$= 17.78\ \Omega \qquad \textbf{(7–93c)}$$

EXAMPLE 7–9

Determine the characteristic impedance and the effective dielectric constant for a microstrip transmission line fabricated on an alumina substrate (ϵ_r = 9.7). The w/h ratio is 0.5.

Solution

Since the w/h ratio < 1, the high-impedance analysis equations (7–82) and (7–83) are used. First, the effective dielectric constant is determined from (7–83):

$$\epsilon_{\text{eff}} = \frac{\epsilon_r + 1}{2} + \frac{\epsilon_r - 1}{2}\left[\frac{1}{\sqrt{1 + \frac{12h}{w}}} + 0.04\left(1 - \frac{w}{h}\right)^2\right] \tag{7–94a}$$

$$= \frac{9.7 + 1}{2} + \frac{9.7 - 1}{2}\left[\frac{1}{\sqrt{1 + 12 \times \left(\frac{1}{0.5}\right)}} + 0.04(1 - 0.5)^2\right] \tag{7–94b}$$

$$= 6.264 \tag{7–94c}$$

The characteristic impedance is determined from (7–82):

$$Z_0 = \frac{60}{\sqrt{\epsilon_{\text{eff}}}}\ln\left(\frac{8h}{w} + \frac{w}{4h}\right) \tag{7–95a}$$

$$= \frac{60}{\sqrt{6.264}}\ln\left[8 \times \left(\frac{1}{0.5}\right) + \frac{0.5}{4}\right] \tag{7–95b}$$

$$= 66.65\ \Omega \tag{7–95c}$$

EXAMPLE 7–10

Determine the w/h ratio required to fabricate a microstrip transmission line with a characteristic impedance of 100 Ω on an alumina substrate ($\epsilon_r = 9.7$).

Solution

The first step is to use Fig. 7–14 to estimate a value for w/h. From Fig. 7–14, to obtain a characteristic impedance of 100 Ω with a substrate dielectric constant of 9.7, the ratio is between 0.1 and 0.2. Therefore, the high-impedance design equations (7–88) and (7–89) will be used. The value of the A factor is

$$A = \frac{Z_0}{60}\sqrt{\frac{\epsilon_r + 1}{2}} + \frac{\epsilon_r - 1}{\epsilon_r + 1}\left(0.23 + \frac{0.11}{\epsilon_r}\right) \tag{7–96a}$$

$$= \frac{100}{60}\sqrt{\frac{9.7 + 1}{2}} + \frac{9.7 - 1}{9.7 + 1}\left(0.23 + \frac{0.11}{9.7}\right) \tag{7–96b}$$

$$= 4.051 \tag{7–96c}$$

The w/h ratio as determined from (7–88) is

$$\frac{w}{h} = \frac{8}{e^A - \frac{2}{e^A}} = \frac{8}{e^{4.051} - \frac{2}{e^{4.051}}} = 0.1393 \tag{7–97}$$

EXAMPLE 7–11

Determine the velocity of propagation for the microstrip in Example 7–10.

Solution

The velocity of propagation is determined from (7–86). The effective dielectric constant must first be determined from (7–83) since $w/h < 1$, which requires the high-impedance analysis equations. The effective dielectric constant is

$$\epsilon_{\text{eff}} = \frac{\epsilon_r + 1}{2} + \frac{\epsilon_r - 1}{2}\left[\frac{1}{\sqrt{1 + \frac{12h}{w}}} + 0.04\left(1 - \frac{w}{h}\right)^2\right] \tag{7–98a}$$

$$= \frac{9.7 + 1}{2} + \frac{9.7 - 1}{2}\left[\frac{1}{\sqrt{1 + (12) \times \left(\frac{1}{0.1393}\right)}} + 0.04(1 - 0.1393)^2\right] \tag{7–98b}$$

$$= 5.945 \tag{7–98c}$$

The velocity of propagation is

$$\nu = \frac{c}{\sqrt{\epsilon_{\text{eff}}}} = \frac{3 \times 10^8}{\sqrt{5.945}} = 1.230 \times 10^8 \text{ m/s} \tag{7–99}$$

EXAMPLE 7–12

Determine the w/h ratio required to fabricate a microstrip transmission line with a characteristic impedance of 10 Ω on an alumina substrate ($\epsilon_r = 9.7$).

Solution

The first step is to use Fig. 7–14 to estimate a value for w/h. From Fig. 7–14, to obtain a characteristic impedance of 10 Ω with a substrate dielectric constant of 9.7, the ratio is approximately 10. Therefore, the low-impedance design equations (7–90) and (7–91) will be used. The value of the B factor is

$$B = \frac{120\pi^2}{2Z_0\sqrt{\epsilon_r}} = \frac{120\pi^2}{(2)(10)\sqrt{9.7}} = 19.014 \tag{7–100}$$

Let $B - 1 = 19.014 - 1 = 18.014$, and let $2B - 1 = 38.027 - 1 = 37.027$. Then

$$\frac{w}{h} = \frac{2}{\pi}\left\{B - 1 - \ln(2B - 1) + \frac{\epsilon_r - 1}{2\epsilon_r}\left[\ln(B - 1) + 0.39 - \frac{0.61}{\epsilon_r}\right]\right\} \tag{7–101a}$$

$$= \frac{2}{\pi}\left\{18.014 - \ln(37.027) + \frac{9.7 - 1}{(2) \times (9.7)}\left[\ln(18.014) + 0.39 - \frac{0.61}{9.7}\right]\right\} \tag{7–101b}$$

$$= 10.09 \tag{7–101c}$$

PROBLEMS

7–1. Refer to the boundary between two different media shown in Fig. 7–1. An electric field intensity vector $\vec{E}_1$ in medium 1 at point A on the boundary makes an angle ϕ_1 of 30° with respect to the boundary as shown. The magnitude of $\vec{E}_1$ is 500 V/m. The dielectric constant in medium 1 is 1.0, and that in medium 2 is 2.8. Assume that in both media, $\mu = \mu_0$. A surface charge density ρ_s with value 4.0×10^{-8} C/m^2 exists at point A. Determine the magnitude of the electric field intensity $\vec{E}_2$ and the angle ϕ_2 that it makes with the boundary at point A.

7–2. Refer to the boundary between two different media shown in Fig. 7–1. An electric field intensity vector $\vec{E}_1$ in medium 1 at point A on the boundary makes an angle ϕ_1 of 45° with respect to the boundary as shown. The magnitude of $\vec{E}_1$ is 1200 V/m. The dielectric constant in medium 1 is 2.3, and that in medium 2 is 1.0. Assume that in both media, $\mu = \mu_0$. A surface charge density ρ_s with value -3.0×10^{-8} C/m^2 exists at point A. Determine the magnitude of the electric field intensity $\vec{E}_2$ and the angle ϕ_2 that it makes with the boundary at point A.

7–3. In Fig. 7–1, a magnetic field intensity vector $\vec{H}_1$ in medium 1 at point A on the boundary between media 1 and 2 makes an angle ϕ_1 of 75° to the boundary as shown. The magnitude of the magnetic field intensity vector in medium 1 is 10.0 A/m. The permeability of both media 1 and 2 is the permeability of free space, $4\pi \times 10^{-7}$ H/m. A linear conduction current density $\vec{K}$ of 4 A/m directed into the paper in Fig. 7–2(b) exists at point A. The tails of the arrows in Fig. 7–2(b) indicate that the direction of $\vec{K}$ is into the paper. Determine the magnitude of the magnetic field intensity vector in medium 2 at point A and the angle ϕ_2 that it makes with the boundary at point A.

7–4. In Fig. 7–1, a magnetic field intensity vector $\vec{H}_1$ in medium 1 at point A on the boundary between media 1 and 2 makes an angle ϕ_1 of 30° to the boundary as shown. The magnitude of the magnetic field intensity vector in medium 1 is 5.0 A/m. The permeability of both media 1 and 2 is the permeability of free space, $4\pi \times 10^{-7}$ H/m. A linear conduction current density $\vec{K}$ of 2 A/m directed into the paper in Fig. 7–2(b) exists at point A. The tails of the arrows in Fig. 7–2(b) indicate that the direction of $\vec{K}$ is into the paper. Determine the magnitude of the magnetic field intensity vector in medium 2 at point A and the angle ϕ_2 that it makes with the boundary at point A.

7–5. An electric field intensity vector of 200 V/m exists at a point on the boundary between a dielectric medium and a perfect conductor. The dielectric constant of the dielectric medium is 3. The permeability of the dielectric medium is μ_0. A magnetic field intensity vector of 16 A/m exists at the same point. Determine the following:

(a) tangential and normal components of the electric and magnetic field intensities in the dielectric medium and within the perfect conductor at the point on the boundary

(b) surface charge density on the surface of the perfect conductor

(c) linear conduction current density

7–6. An electric field intensity vector of 150 V/m exists at a point on the boundary between a dielectric medium and a perfect conductor. The dielectric constant of the dielectric medium is 2. The permeability of the dielectric medium is μ_0. A magnetic field intensity vector of 12 A/m exists at the same point. Determine the following:

(a) tangential and normal components of the electric and magnetic field intensities in the dielectric medium and within the perfect conductor at the point on the boundary
(b) surface charge density on the surface of the perfect conductor
(c) linear conduction current density

7–7. A two-wire parallel transmission line has the following physical dimensions:

Diameter of each conductor (d) = 0.2 cm

Spacing between the two conductors (D) = 4.0 cm

Dielectric constant of medium (ϵ_r) = 2.3

Determine the following:

(a) inductance per unit length
(b) capacitance per unit length
(c) characteristic impedance
(d) velocity of propagation

7–8. An open two-wire parallel transmission line has the following physical dimensions:

Diameter of each conductor (d) = 0.4 in

Spacing between the two conductors (D) = 8.0 in

Dielectric constant of medium (ϵ_r) = 1.0

Determine the following:

(a) inductance per unit length
(b) capacitance per unit length
(c) characteristic impedance
(d) velocity of propagation

7–9. A coaxial cable has the following physical dimensions:

Diameter of inner conductor (d) = 0.1 in

Inner diameter of outer conductor (D) = 0.4 in

Polystyrene dielectric (ϵ_r) = 2.7

Determine the following:

(a) inductance per unit length
(b) capacitance per unit length
(c) characteristic impedance
(d) velocity of propagation

7–10. A coaxial cable has the following physical dimensions:

Diameter of inner conductor (d) = 5 mm

Inner diameter of outer conductor (D) = 3 cm

Polyethylene dielectric (ϵ_r) = 2.3

Determine the following:

(a) inductance per unit length

(b) capacitance per unit length
(c) characteristic impedance
(d) velocity of propagation

7–11. Determine the following for a coaxial cable that has a nominal velocity of propagation of 66% of electromagnetic propagation in free space and a characteristic impedance of 50 Ω:
(a) inductance per unit length
(b) capacitance per unit length
(c) dielectric constant of the insulating material between the inner and outer conductors

7–12. Determine the following for a coaxial cable that has a nominal velocity of propagation of 66% of electromagnetic propagation in free space and a characteristic impedance of 72 Ω:
(a) inductance per unit length
(b) capacitance per unit length
(c) dielectric constant of the insulating material between the inner and outer conductors

7–13. A stripline transmission line is fabricated from polyolefin printed circuit board that is plated with 1-oz copper. The distance b between the ground planes is 0.3175 cm. The thickness t of the 1-oz copper is 0.003556 cm. The width w of the conductor is 0.05 cm. Polyolefin has a dielectric constant of 2.32. Determine the following for this stripline:
(a) characteristic impedance
(b) velocity of propagation

7–14. A stripline transmission line is fabricated from polyolefin printed circuit board that is plated with 1-oz copper. The distance b between the ground planes is 0.3175 cm. The thickness t of the 1-oz copper is 0.003556 cm. The width w of the conductor is 0.08 cm. Polyolefin has a dielectric constant of 2.32. Determine the following for this stripline:
(a) characteristic impedance
(b) velocity of propagation

7–15. A stripline transmission line is fabricated from polyolefin printed circuit board that is plated with 1-oz copper. The distance b between the ground planes is 0.3175 cm. The thickness t of the 1-oz copper is 0.003556 cm. The width w of the conductor is 1.5 cm. Polyolefin has a dielectric constant of 2.32. Determine the following for this stripline:
(a) characteristic impedance
(b) velocity of propagation

7–16. A stripline transmission line is fabricated from polyolefin printed circuit board that is plated with 1-oz copper. The distance b between the ground planes is 0.3175 cm. The thickness t of the 1-oz copper is 0.003556 cm. The width w of the conductor is 2.0 cm. Polyolefin has a dielectric constant of 2.32. Determine the following for this stripline:
(a) characteristic impedance
(b) velocity of propagation

7–17. Determine the characteristic impedance and the effective dielectric constant for a microstrip transmission line fabricated on an alumina substrate (ϵ_r = 9.7). The w/h ratio is 3.0.

7–18. Determine the characteristic impedance and the effective dielectric constant for a microstrip transmission line fabricated on a polyolefin substrate (ϵ_r = 2.3). The w/h ratio is 4.0.

7–19. Determine the characteristic impedance and the effective dielectric constant for a microstrip transmission line fabricated on an alumina substrate (ϵ_r = 9.7). The w/h ratio is 0.2.

7–20. Determine the characteristic impedance and the effective dielectric constant for a microstrip transmission line fabricated on a polyolefin substrate (ϵ_r = 2.3). The w/h ratio is 0.4.

7–21. Determine the w/h ratio required to fabricate a microstrip transmission line with a characteristic impedance of 150 Ω on an alumina substrate (ϵ_r = 2.55)

7–22. Determine the w/h ratio required to fabricate a microstrip transmission line with a characteristic impedance of 100 Ω on a polyolefin substrate (ϵ_r = 2.3).

7–23. Determine the w/h ratio required to fabricate a microstrip transmission line with a characteristic impedance of 20 Ω on a polyolefin substrate (ϵ_r = 2.3).

7–24. Determine the w/h ratio required to fabricate a microstrip transmission line with a characteristic impedance of 5 Ω on an alumina substrate (ϵ_r = 9.7).

7–25. Determine the velocity of propagation for the microstrip in Problem 7–21.

7–26. Determine the velocity of propagation for the microstrip in Problem 7–22.

7–27. Determine the velocity of propagation for the microstrip in Problem 7–23.

7–28. Determine the velocity of propagation for the microstrip in Problem 7–24.

8

Waveguides

OBJECTIVES

After completing this chapter, the reader should be able to:

1. Define **waveguides** and their application to transmission of electromagnetic energy.
2. Describe the propagation of electromagnetic waves within a waveguide.
3. Define the **transverse electric (TE) mode** and the **transverse magnetic (TM) mode** within a waveguide.
4. Define the **dominant mode.**
5. Describe the rectangular waveguide structure.
6. Define the **cutoff wavelength** and the **cutoff frequency** and determine them for any mode.
7. Define and determine the **guide wavelength.**
8. Define and determine the **phase velocity** and the **group velocity.**
9. Explain the difference between **characteristic, intrinsic,** and **wave impedances.**
10. Determine the wave impedance of a rectangular waveguide for any mode.
11. Determine the waveguide losses for an air-filled rectangular waveguide.

8–1 ■ INTRODUCTION

The transmission of electromagnetic signals can be accomplished by three distinct methods. In the first method, **transmission lines** can be used to conduct voltage and current waves on a pair of conductors to transmit an electromagnetic signal from one physical location to another location. This method was presented in the first five chapters of the text. There are electric and magnetic fields surrounding the conductors of the transmission line which were produced by these voltages and currents.

A second method, which was discussed in Chapter 6, is to use an **antenna** which launches electric and magnetic fields in free space. These fields propagate through free space typically at the speed of light as described in Section 6–8 to a receiving antenna. This antenna converts the electric and magnetic fields to voltage and current waves on a transmission line. Antennas will be discussed in more detail in Chapter 9.

A third method is to use an antenna to launch an electric and magnetic field into a waveguiding structure commonly referred to as a **waveguide.** Waveguides, which are commonly used in the microwave frequency range from 1 GHz to 300 GHz, are found in several different forms. They normally consist of metallic hollow structures that have either rectangular or circular cross sections. However, several other types of cross sections have been used for specific applications. The electric and magnetic fields travel through the waveguide and must satisfy Maxwell's equations and the associated boundary conditions, as discussed in Chapter 7, at the metallic walls of the waveguide structure. The most common form of waveguide in the microwave region is the rectangular waveguide, in which the height is approximately equal to one-half the width. This type of waveguide will be presented in this chapter. Detailed discussions of other waveguide types can be found in advanced books on microwave engineering.

A different form of waveguide which is used in the optical frequency region is the **fiber optic waveguide.** This type of waveguide uses two concentric cylinders of dielectric materials that have different indices of refraction. The boundary between the two materials is used to guide the propagation of electromagnetic signals in the optical frequency region. This form of waveguide will be presented in Chapter 12 on fiber optic technology.

8–2 ■ RECTANGULAR WAVEGUIDES

Waveguide structures are high-pass filters with a cutoff frequency. Frequencies below the cutoff frequency cannot propagate, and they experience a significant attenuation. All frequencies above the cutoff frequency can propagate through the waveguide structure. There are one or more electromagnetic modes that satisfy the boundary conditions for each frequency that can exist in a waveguide structure. It is undesirable to operate a waveguiding structure at a frequency where more than one mode can exist. The electromagnetic mode that exists at the lowest frequency range above the cutoff frequency is defined as the **dominant mode.** It is desirable to design a waveguide structure in which there is a frequency range above the cutoff frequency where only one dominant mode exists.

There are two types of modes that can exist in a waveguide structure: the **transverse electric (TE) mode** and the **transverse magnetic (TM) mode.** The TE and TM modes were discussed in Chapter 7. The square waveguide in which the height equals the width does not have a frequency range where only one mode exists. However, the rectangular waveguide does have a 2 : 1 frequency range where only one mode can exist.

The rectangular waveguide structure is one in which the height is equal to one-half the width. A sketch of the rectangular waveguide structure is shown in Fig. 8–1. The waveguide consists of metallic walls that surround the dielectric medium in which the electric and magnetic fields propagate. Normally, the rectangular waveguides are filled with air. However, some waveguides are filled with a dielectric material to reduce the physical size of the waveguide for a specific operating frequency.

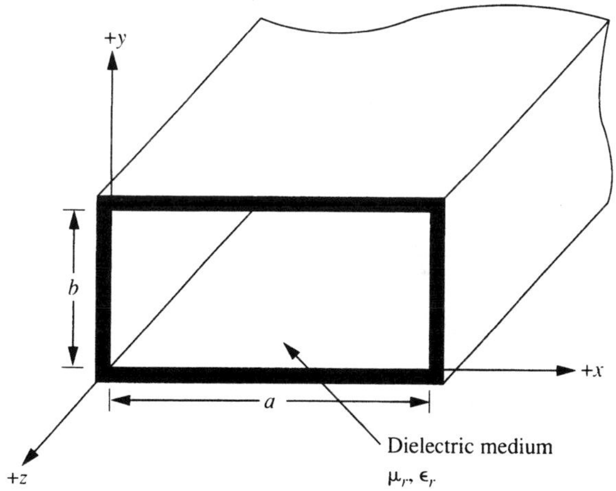

FIGURE 8–1 Rectangular waveguide structure.

The inner width dimension is denoted as a, and the inner height dimension is denoted as b as shown in Fig. 8–1. As will be shown in a later section, the cutoff wavelength for the rectangular waveguide is

$$\lambda_c = 2a \tag{8–1}$$

where

λ_c = cutoff wavelength
a = inner width dimension

The cutoff frequency for an air-filled waveguide is

$$f_c = \frac{c}{\lambda_c} \tag{8–2}$$

where

f_c = cutoff frequency in hertz
c = velocity of light (3×10^8 m/s)

The dominant mode for the rectangular waveguide is the TE_{10} mode.

A common rectangular waveguide is the X-band WR–90 waveguide. The recommended operating range is 8.2 to 12.4 GHz. The cutoff frequency is 6.560 GHz. The cutoff frequency for the next nearest mode is 13.12 GHz. The inner dimensions are a = 22.860 mm (0.900 in) and b = 10.160 mm (0.400 in). The outer dimensions are 25.40 mm by 12.70 mm, or 1 in by 0.5 in.

EXAMPLE 8–1

A WR–28 rectangular waveguide has an operating frequency range of 26.4 to 40.0 GHz. The inner dimensions are 7.112 mm by 3.556 mm. Determine the cutoff frequency for the dominant mode.

Solution

The cutoff wavelength is determined from (8–1):

$$\lambda_c = 2a = 2 \times (7.112 \times 10^{-3}) \tag{8–3a}$$

$$= 1.4224 \times 10^{-2}\ \text{m} = 1.4224\ \text{cm} \tag{8–3b}$$

The cutoff frequency is determined from (8–2):

$$f_c = \frac{c}{\lambda_c} = \frac{3 \times 10^8}{1.4224 \times 10^{-2}} = 21.091\ \text{GHz} \tag{8–4}$$

8–3 ■ WAVEGUIDE MODES

As discussed in the preceding section, the two types of electromagnetic modes that can exist in a waveguide structure are the transverse electric (TE) and the transverse magnetic (TM) modes. The **TE mode** is one in which at any instant of time, variations in the electric field with position (x, y, z) exist only in the x-y plane as shown in Fig. 8–2(a). The **TM mode** is one in which at any instant of time, variations in the magnetic field exist only in the x-y plane as shown in Fig. 8–2(b).

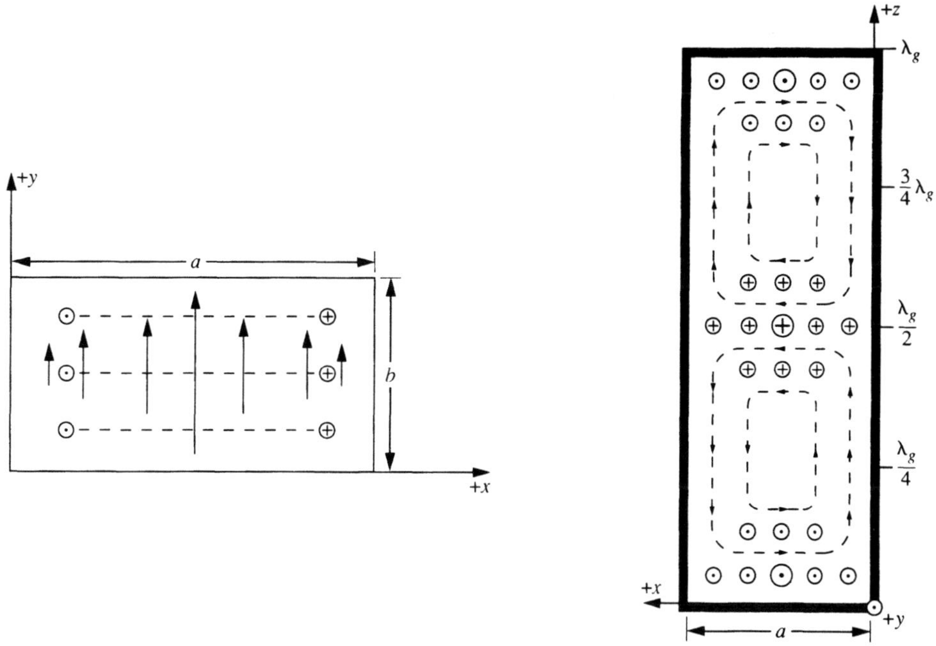

(a) TE_{10} mode

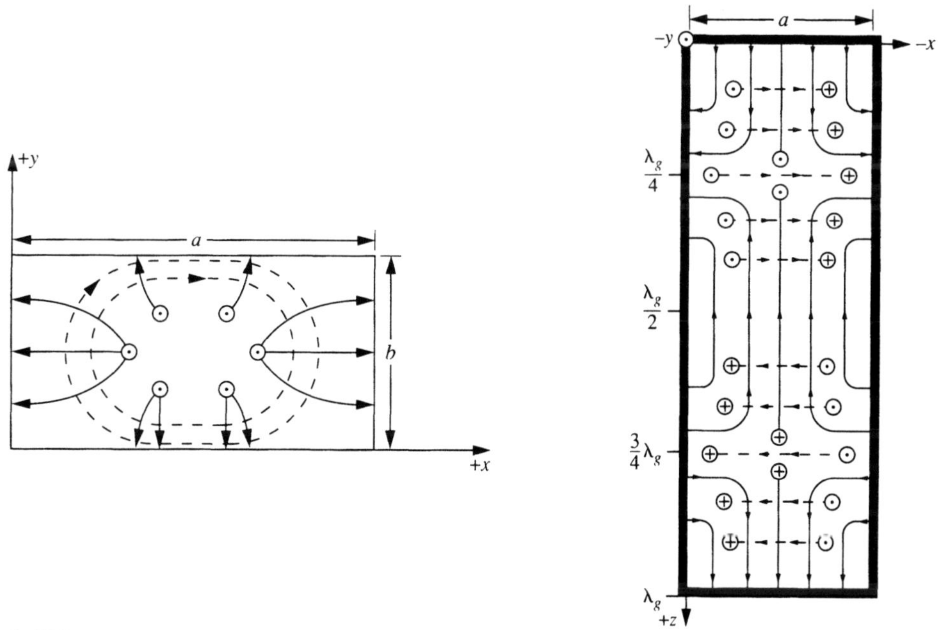

(b) TM_{11} mode

⊙ Out of page ——————— *E* lines

⊕ Into page - - - - - - - *H* lines

FIGURE 8–2 Waveguide modes.

The different modes that can exist are designated by two numeric subscripts such as TE_{10} or TM_{11}. The first subscript defines the number of half-wave variations of the transverse field in the a dimension of the waveguide. The second subscript defines the number of half-wave variations of the transverse field in the b dimension. The TE_{10} mode has one half-wavelength variation of the electric field in the a dimension, as shown in Fig. 8–2(b), and no variations in the b dimension.

The walls of the waveguides can be considered as nearly perfect conductors. Therefore, the boundary conditions require that the electric field be normal, i.e., perpendicular, to the waveguide walls. The magnetic fields must be tangen-

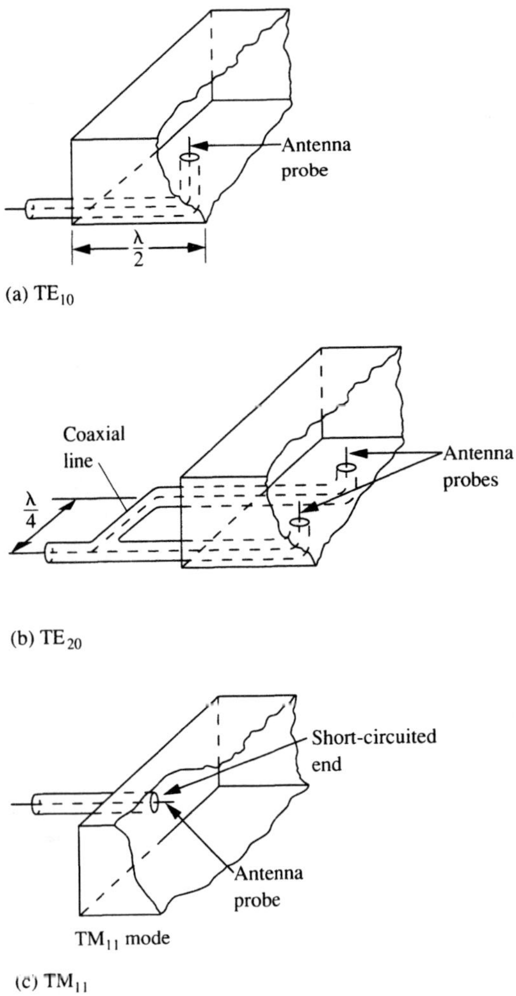

FIGURE 8–3 Methods to launch different waveguide modes.

tial, i.e., parallel, to the waveguide walls. Because of these boundary conditions, a zero subscript can exist in the TE mode but not in the TM mode. For example, the TE_{10}, TE_{01}, TE_{20}, and so on, modes can exist in a rectangular waveguide, but only the TM_{11}, TM_{12}, TM_{21}, and so on, modes can exist.

The desired mode is obtained in a rectangular waveguide by the configuration of the launching devices that generate the electromagnetic fields. Figure 8–3 illustrates how the TE_{10}, TE_{20}, and TM_{11} modes are launched in a rectangular waveguide. These launching devices are, in fact, antennas. At the receiving end of the waveguide, a similar launching device, i.e., a receiving antenna, must be used to convert the electromagnetic fields within the waveguide back to voltage and currents on a transmission line. Therefore, one must know which waveguide mode was used to launch the electromagnetic fields at the transmitting end of the waveguide. If more than one mode can exist at a particular frequency in the waveguide, then any discontinuities, such as bends, joints, and waveguide components, can cause electromagnetic energy to be transferred from one mode to another. This will result in an additional loss in the waveguide since it will not be recovered in the receiving end.

8-4 ■ CUTOFF WAVELENGTH

The cutoff wavelength for the TE_{10} mode was given in (8–1). The cutoff wavelength for any mode is the longest wavelength or lowest frequency that can exist in a rectangular waveguide for that particular waveguide mode. The cutoff wavelength for any generalized mode TE_{mn} or TM_{mn} is

$$\lambda_c = \frac{1}{\sqrt{\left(\frac{m}{2a}\right)^2 + \left(\frac{n}{2b}\right)^2}} \qquad \textbf{(8–5)}$$

where

a = width of the rectangular waveguide

b = height of the rectangular waveguide

m = number of half-wave variations in a

n = number of half-wave variations in b

When $m = 1$ and $n = 0$, (8–5) reduces to (8–1).

EXAMPLE 8–2

Determine the cutoff wavelengths and the cutoff frequencies for (a) the TE_{01}, (b) the TE_{20}, and (c) the TM_{11} modes in a WR–90 rectangular waveguide with air dielectric. The inner dimensions are $a = 22.860$ mm and $b = 10.160$ mm. Compare these frequencies with the specified operating frequency range of 8.2 to 12.4 GHz.

Solution

The cutoff wavelength can be determined from (8–5):

$$\lambda_c = \frac{1}{\sqrt{\left(\frac{m}{2a}\right)^2 + \left(\frac{n}{2b}\right)^2}} \tag{8–6a}$$

$$= \frac{1}{\sqrt{\left(\frac{m}{2(22.860 \times 10^{-3})}\right)^2 + \left(\frac{n}{2(10.160 \times 10^{-3})}\right)^2}} \tag{8–6b}$$

The cutoff frequency can be determined from (8–2):

$$f_c = \frac{c}{\lambda_c} \tag{8–7}$$

(a) Substituting values for m and n into (8–6) and solving for the TE_{01} mode, we have

$$\lambda_c = \frac{1}{\sqrt{\left(\frac{0}{2(22.860 \times 10^{-3})}\right)^2 + \left(\frac{1}{2(10.160 \times 10^{-3})}\right)^2}} \tag{8–8a}$$

$$= \frac{1}{\sqrt{\left(\frac{1}{2.032 \times 10^{-2}}\right)^2}} \tag{8–8b}$$

$$= 2.032 \times 10^{-2}\ \text{m} = 2.032\ \text{cm} \tag{8–8c}$$

The cutoff frequency for the TE_{01} mode is

$$f_c = \frac{3 \times 10^8}{2.032 \times 10^{-2}} = 14.76\ \text{GHz} \tag{8–9}$$

(b) Substituting values for m and n into (8–6) and solving for the TE_{20} mode, we have

$$\lambda_c = \frac{1}{\sqrt{\left(\frac{2}{2(22.860 \times 10^{-3})}\right)^2 + \left(\frac{0}{2(10.160 \times 10^{-3})}\right)^2}} \tag{8–10a}$$

$$= \frac{1}{\sqrt{\left(\frac{2}{4.572 \times 10^{-2}}\right)^2}} \tag{8–10b}$$

$$= 2.286 \times 10^{-2}\ \text{m} = 2.286\ \text{cm} \tag{8–10c}$$

The cutoff frequency for the TE_{20} mode is

$$f_c = \frac{3 \times 10^8}{2.286 \times 10^{-2}} = 13.12\ \text{GHz} \tag{8–11}$$

(c) Substituting values for m and n into (8–5) and solving for the TM_{11} mode, we have

$$\lambda_c = \frac{1}{\sqrt{\left(\frac{1}{2(22.860 \times 10^{-3})}\right)^2 + \left(\frac{1}{2(10.160 \times 10^{-3})}\right)^2}} \quad \textbf{(8–12a)}$$

$$= \frac{1}{\sqrt{\left(\frac{1}{4.572 \times 10^{-2}}\right)^2 + \left(\frac{1}{2.032 \times 10^{-2}}\right)^2}} \quad \textbf{(8–12b)}$$

$$= 1.857 \times 10^{-2}\,\text{m} = 1.857\,\text{cm} \quad \textbf{(8–12c)}$$

The cutoff frequency for the TM_{11} mode is

$$f_c = \frac{3 \times 10^8}{1.8569 \times 10^{-2}} = 16.16\,\text{GHz} \quad \textbf{(8–13)}$$

The lowest cutoff frequency, except in the dominant TE_{10} mode, is 13.12 GHz for the TE_{20} mode. However, this is above the upper recommended frequency range of 12.4 GHz for the WR–90 rectangular waveguide. Therefore, if the waveguide is operating within the recommended frequency range, only one waveguide mode can exist.

8–5 ■ WAVE PROPAGATION IN A RECTANGULAR WAVEGUIDE

The direction of propagation of electromagnetic radiation, i.e., the Poynting vector as described in Section 6–8, is perpendicular to the plane of the electric field intensity vector and the magnetic field intensity vector [see Fig. 6–6(b)]. For the electromagnetic radiation to propagate directly down the waveguide in the z-direction, it would require that both the electric field and the magnetic field exist in the x-y plane, which is the TEM mode. Because of the boundary conditions within a waveguide, this condition cannot exist. Therefore, the direction of propagation makes an angle θ with the $-z$-axis as shown in Fig. 8–4. The Poynting vector travels across the waveguide and is reflected off the metallic walls back to the other wall. Thus, the propagation of electromagnetic radiation follows a zigzag path through the waveguide.

The angle θ is a function of the frequency of the electromagnetic radiation. As the frequency approaches the cutoff frequency, θ approaches 90°. At the cutoff frequency, $\theta = 90°$, and the propagation is across the waveguide and does not travel down the waveguide. The angle θ can be determined from

$$\theta = \sin^{-1}\frac{\lambda}{\lambda_c} = \sin^{-1}\frac{f_c}{f} \quad \textbf{(8–14)}$$

Refer to (8–15) in the next section for the derivation of (8–14).

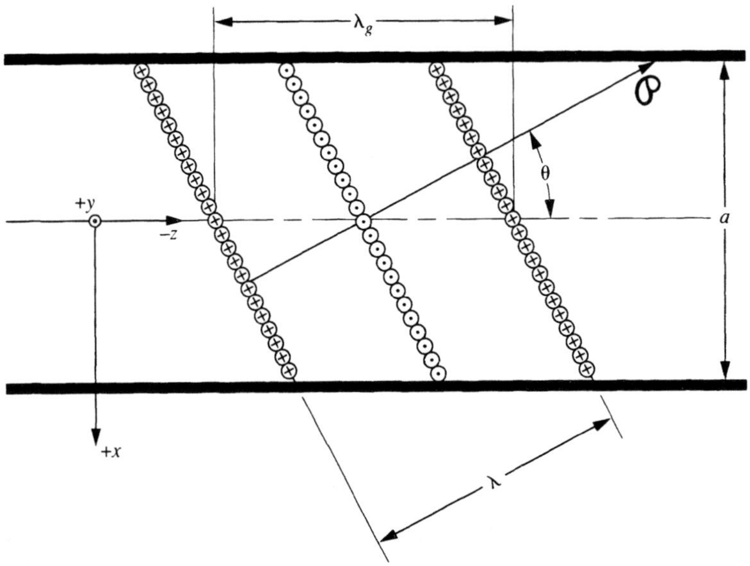

FIGURE 8–4 Wave propagation in a rectangular waveguide.

8–6 ■ GUIDE WAVELENGTH

The **guide wavelength** λ_g is the distance that a point of constant phase such as zero degrees travels during the time for one period of the sinusoidal wave along the negative z-direction of the waveguide. Assuming an air-filled waveguide and referring to Fig. 8–4, the free-space wavelength λ is the distance traveled across the waveguide where the direction of travel makes an angle θ with the negative z-direction of the waveguide. As shown in Fig. 8–4, the guide wavelength is always longer than the free-space wavelength. When the frequency approaches the cutoff frequency, the guide wavelength becomes very large and approaches infinity at the cutoff frequency. When the frequency is much greater than the cutoff frequency, the guide wavelength approaches the free-space wavelength.

An equation for the guide wavelength as a function of the free-space wavelength and the cutoff wavelength can be derived as follows. Refer to Fig. 8–5, and assume that an electromagnetic wave is propagating in a waveguide. The direction of propagation makes an angle θ with the negative z-direction of the waveguide. The heavy solid line xx' represents the maximum point on the sinusoidal function, and the heavy dashed line yy' represents the minimum point as shown in view AA'. The point O is at zero on the sinusoidal function in view AA' and also at the midpoint of the waveguide on the $-z$-axis. Referring to triangle OBC and substituting (8–1), we have

$$\sin\theta = \frac{\lambda/4}{a/2} = \frac{\lambda}{2a} = \frac{\lambda}{\lambda_c} \tag{8–15}$$

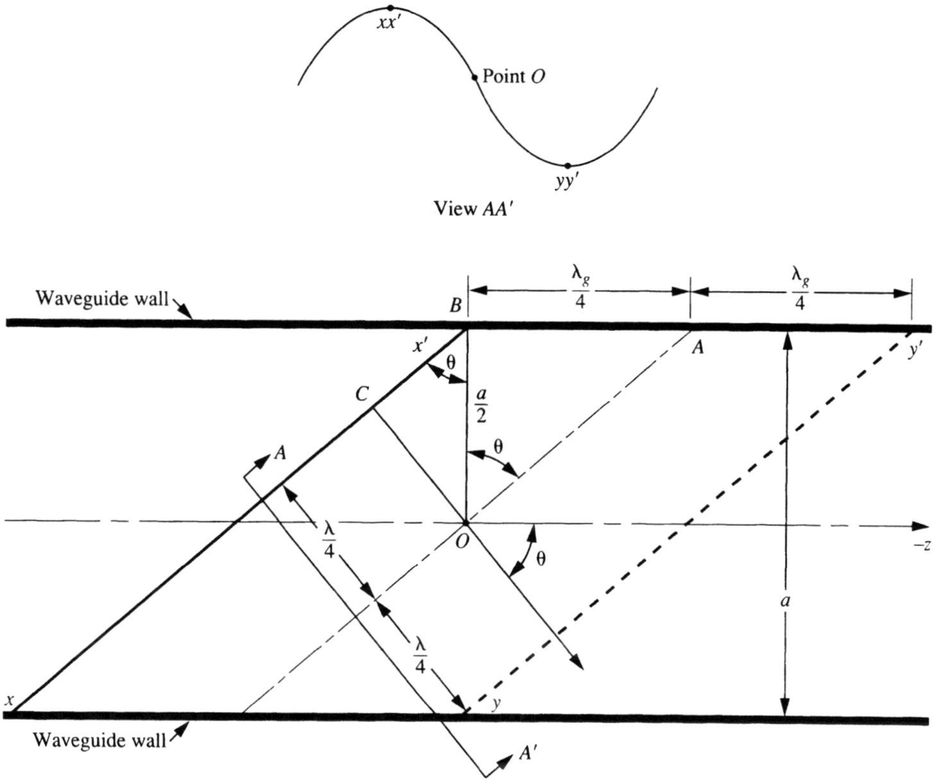

FIGURE 8–5 Guide wavelength.

From triangle OAB and substituting (8–1), we have

$$\tan\theta = \frac{\lambda_g/4}{a/2} = \frac{\lambda_g}{2a} = \frac{\lambda_g}{\lambda_c} \tag{8–16}$$

Since $\tan\theta = \sin\theta/\cos\theta$, we obtain

$$\cos\theta = \frac{\lambda/\lambda_c}{\lambda_g/\lambda_c} = \frac{\lambda}{\lambda_g} \tag{8–17}$$

Since $\sin^2\theta + \cos^2\theta = 1$, and from (8–15) and (8–16), we obtain

$$\lambda_g = \frac{\lambda}{\sqrt{1-(\lambda/\lambda_c)^2}} \tag{8–18}$$

EXAMPLE 8–3

Determine the guide wavelength in a WR–90 waveguide with an air dielectric at both ends of the recommended frequency range of 8.2 and 12.4 GHz. The cutoff wavelength is 45.72 mm. Compare the guide wavelength with the free-space wavelength at each frequency.

Solution

The free-space wavelength at 8.2 GHz is

$$\lambda = \frac{c}{f} = \frac{3 \times 10^8}{8.2 \times 10^9} = 3.659 \times 10^{-2}\ \text{m} = 3.659\ \text{cm} \tag{8–19}$$

The free-space wavelength at 12.4 GHz is

$$\lambda = \frac{c}{f} = \frac{3 \times 10^8}{12.4 \times 10^9} = 2.419 \times 10^{-2}\ \text{m} = 2.419\ \text{cm} \tag{8–20}$$

Substituting the results from (8–19) and (8–20) into (8–18), we find that the guide wavelength for 8.2 GHz is

$$\lambda_g = \frac{\lambda}{\sqrt{1 - \left(\frac{\lambda}{\lambda_c}\right)^2}} = \frac{3.659 \times 10^{-2}}{\sqrt{1 - \left(\frac{3.659 \times 10^{-2}}{4.572 \times 10^{-2}}\right)^2}} \tag{8–21a}$$

$$= 6.102 \times 10^{-2}\ \text{m} = 6.102\ \text{cm} \tag{8–21b}$$

The guide wavelength for 12.4 GHz is

$$\lambda_g = \frac{\lambda}{\sqrt{1 - \left(\frac{\lambda}{\lambda_c}\right)^2}} = \frac{2.419 \times 10^{-2}}{\sqrt{1 - \left(\frac{2.419 \times 10^{-2}}{4.572 \times 10^{-2}}\right)^2}} \tag{8–22a}$$

$$= 2.851 \times 10^{-2}\ \text{m} = 2.851\ \text{cm} \tag{8–22b}$$

Since 8.2 GHz is closer to the cutoff frequency of 6.562 GHz, the guide wavelength of 6.102 cm is further away from the free-space wavelength of 3.659 cm. At 12.4 GHz, the guide wavelength of 2.851 cm is closer to the free-space wavelength of 2.419 cm.

8–7 ■ PHASE VELOCITY AND GROUP VELOCITY

There are two velocities associated with the propagation of electromagnetic radiation within a waveguiding structure: (a) the **phase velocity,** which is the velocity at which a constant phase point on the sinusoidal wave travels down the waveguiding structure; and (b) the **group velocity,** which is the velocity that the energy is moving through the waveguide.

As was shown in Figs. 8–4 and 8–5, the guide wavelength is the distance traveled down the waveguide by a constant phase point in one cycle of the electromagnetic radiation. Since the period or time required to travel through one cycle is the reciprocal of the frequency, then the phase velocity is the guide wavelength multiplied by the cyclic frequency, or

$$\nu_p = \lambda_g f \tag{8–23}$$

where

ν_p = phase velocity

The Poynting vector $\mathcal{P}$ is the power of the electromagnetic radiation propagating through the waveguide. As shown in Fig. 8–4, the Poynting vector travels at the speed of light across the waveguide, making an angle θ with the $-z$-axis. The group velocity is the component of the velocity of the Poynting vector which is in the negative z-direction or down the waveguide toward the receiving end. If the waveguide is air filled, then the velocity of the Poynting vector is the speed of light. Figure 8–6 shows the relationship of the velocity of the Poynting vector and the group velocity. Using Fig. 8–6 and Eq. (8–17), we find the group velocity:

$$\nu_g = c \times \cos\theta = c \times \frac{\lambda}{\lambda_g} \qquad \textbf{(8–24)}$$

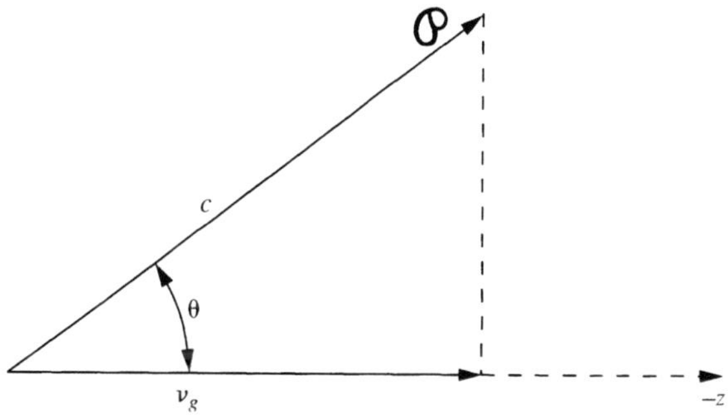

FIGURE 8–6 Group velocity.

Substituting (8–23) and $\lambda = c/f$ into (8–24) yields

$$\nu_g = c \times \frac{c/f}{\nu_p/f} = \frac{c^2}{\nu_p} \qquad \textbf{(8–25)}$$

Rearranging (8–25), we obtain

$$\nu_g \nu_p = c^2 \qquad \textbf{(8–26)}$$

Substituting (8–24) into (8–26) yields

$$\nu_p = c \times \frac{\lambda_g}{\lambda} \qquad \textbf{(8–27)}$$

Equations to determine the phase and group velocity as a function of the free-space wavelength and cutoff wavelength can be obtained by substituting

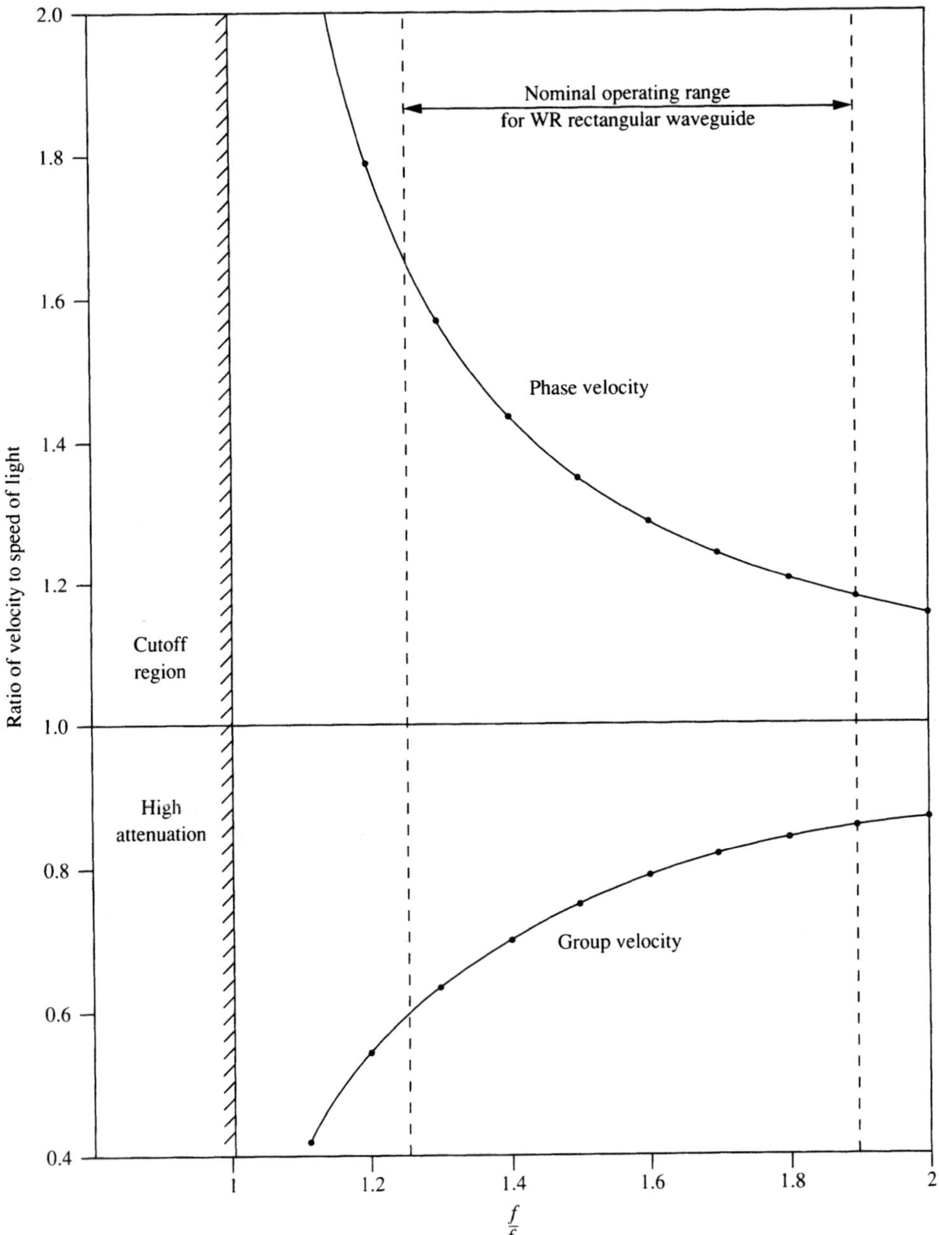

FIGURE 8–7 Phase velocity and group velocity.

(8–18) into (8–24) and (8–27):

$$\nu_g = c \times \sqrt{1 - \left(\frac{\lambda}{\lambda_c}\right)^2} \tag{8–28}$$

$$\nu_p = \frac{c}{\sqrt{1 - \left(\frac{\lambda}{\lambda_c}\right)^2}} \tag{8–29}$$

Substituting $\lambda = c/f$ into (8–28) and (8–29), we have

$$\nu_g = c \times \sqrt{1 - \left(\frac{f_c}{f}\right)^2} \tag{8–30}$$

$$\nu_p = \frac{c}{\sqrt{1 - \left(\frac{f_c}{f}\right)^2}} \tag{8–31}$$

Equations (8–30) and (8–31) have been plotted in Fig. 8–7 as a function of the ratio of the frequency of the electromagnetic radiation to the cutoff frequency. As can be seen from the curves, the group velocity approaches zero as the frequency approaches the cutoff frequency, while the phase velocity approaches infinity. When the frequency is much greater than the cutoff frequency, both the group velocity and the phase velocity approach the velocity of light. The nominal operating frequency range for a WR rectangular waveguide is shown in Fig. 8–7 as two vertical dashed lines. The lower limit is at an f/f_c ratio of 1.25 where the group velocity is approximately 60% of free-space velocity. The upper frequency limit is set at an f/f_c ratio of 1.9 where the group velocity is approximately 85% of free-space velocity. The upper limit is set by the fact that for a rectangular waveguide, the cutoff frequency for the next mode (TE_{20}) is at an f/f_c ratio of 2.

EXAMPLE 8–4

Determine the phase and group velocities of the electromagnetic waves propagating in a WR–90 waveguide at the two ends of the operating frequency range, i.e., at 8.2 and 12.4 GHz. The cutoff wavelength is 4.572 cm.

Solution

Both the phase and group velocity equations (8–30) and (8–31) use a common factor which will first be evaluated at 8.2 and 12.4 GHz. The cutoff frequency was determined in Section 8–2 to be 6.56 GHz. Solving for the factor at 8.2 GHz yields

$$\sqrt{1 - \left(\frac{f_c}{f}\right)^2} = \sqrt{1 - \left(\frac{6.56 \times 10^9}{8.2 \times 10^9}\right)^2} = 0.600 \tag{8–32}$$

Solving for the factor at 12.4 GHz, we obtain

$$\sqrt{1-\left(\frac{f_c}{f}\right)^2} = \sqrt{1-\left(\frac{6.56 \times 10^9}{12.4 \times 10^9}\right)^2} = 0.8486 \tag{8–33}$$

The group velocity at 8.2 GHz can be obtained from (8–30):

$$\nu_g = c \times \sqrt{1-\left(\frac{f_c}{f}\right)^2} = (3 \times 10^8) \times 0.600 = 1.8 \times 10^8 \text{ m/s} \tag{8–34}$$

The group velocity at 12.4 GHz is

$$\nu_g = c \times \sqrt{1-\left(\frac{f_c}{f}\right)^2} = (3 \times 10^8) \times 0.8486 = 2.546 \times 10^8 \text{ m/s} \tag{8–35}$$

The phase velocity at 8.2 GHz can be obtained from (8–31):

$$\nu_p = \frac{c}{\sqrt{1-\left(\frac{f_c}{f}\right)^2}} = \frac{3 \times 10^8}{0.600} = 5 \times 10^8 \text{ m/s} \tag{8–36}$$

The phase velocity at 12.4 GHz is

$$\nu_p = \frac{c}{\sqrt{1-\left(\frac{f_c}{f}\right)^2}} = \frac{3 \times 10^8}{0.8486} = 3.535 \times 10^8 \text{ m/s} \tag{8–37}$$

8–8 ■ WAVEGUIDE IMPEDANCE

The characteristic impedance of a transmission line is defined as the ratio of the forward traveling voltage wave $\overline{V}^+$ to the forward traveling current wave $\overline{I}^+$. For a two-conductor transmission line, this ratio can be uniquely defined as the voltage between the conductors divided by the conduction current. The characteristic impedance is a function of only the physical dimensions of the two-conductor transmission line and the permeability and permittivity of the medium surrounding the transmission line.

The intrinsic impedance η of free space is defined as the ratio of the electric field intensity vector to the magnetic field intensity vector, both of which lie in a plane transverse to the direction of propagation. The electric and magnetic fields have components only in the transverse plane. The intrinsic impedance can be determined from

$$\eta = \sqrt{\frac{\mu}{\epsilon}} = \sqrt{\frac{\mu_0\mu_r}{\epsilon_0\epsilon_r}} \tag{8–38}$$

For free space, $\eta = 377\ \Omega$ as indicated in Chapter 6.

The propagation on a two-conductor transmission line or in free space is in a TEM mode. The propagation within the waveguide is in either the TE or the TM mode. The voltage at any point along the waveguide varies from zero at the edge to a maximum at the center for the TE_{10} mode. The conduction current flows on the waveguide surface. These currents for the TE_{10} mode in a rectangular waveguide are illustrated in Fig. 8–8. As shown, the conduction current at any point along the waveguide varies with position in the transverse plane.

Defining a waveguide impedance as a transmission line characteristic impedance is not possible. Some authors have defined a quasi-characteristic impedance for the TE_{10} mode by defining it as the ratio of the voltage at the centerline divided by the conduction current along the waveguide at the centerline. Other definitions based on power and centerline voltage or power and conduction current have been proposed.

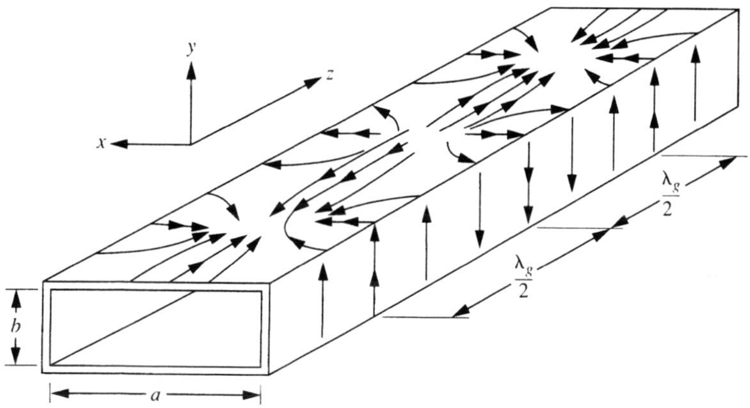

FIGURE 8–8 Conduction currents in rectangular guide for the TE_{10} mode. (P. A. Rizzi, *Microwave Engineering, Passive Circuits* (Englewood Cliffs, NJ: Prentice-Hall, 1988) p. 204, Fig. 5–19. Used with permission.)

A different waveguide impedance concept based on the definition of the intrinsic impedance in free space is that of the **wave impedance.** The wave impedance is defined as the ratio of the transverse component of the electric field to the transverse component of the magnetic field at any point along the waveguide. The magnetic field in the TE mode has components in both the transverse and longitudal directions, while in the TM mode, the electric field has components in both directions. Due to this fact, the wave impedances for the TE and TM modes are not the same as a function of frequency. The wave impedance $Z_{W_{TE_{mn}}}$ for the TE_{mn} mode is

$$Z_{W_{TE_{mn}}} = \frac{\eta}{\sqrt{1 - \left(\frac{\lambda}{\lambda_{c_{mn}}}\right)^2}}\ \Omega \qquad \textbf{(8–39)}$$

where

η = intrinsic impedance of medium within the waveguide
λ = wavelength
$\lambda_{c_{mn}}$ = cutoff wavelength for mn mode from (8–5)

The wave impedance for the TM_{mn} mode is

$$Z_{W_{TM_{mn}}} = \eta\sqrt{1 - \left(\frac{\lambda}{\lambda_{c_{mn}}}\right)^2}\ \Omega \tag{8–40}$$

The wave impedance for both modes approaches the intrinsic impedance when the frequency is much greater than the cutoff frequency. When the frequency approaches the cutoff frequency, the wave impedance for the TE mode approaches infinity, while the wave impedance for the TM mode approaches zero.

EXAMPLE 8–5

Determine the wave impedance for the TE_{10} mode in an air dielectric WR–90 waveguide at an operating frequency of 8.2 GHz.
The cutoff wavelength is 4.572 cm.

Solution

The wave impedance for the TE_{10} mode is determined from (8–39) by substituting the cutoff wavelength, $m = 1$, and $n = 0$:

$$Z_{W_{TE_{10}}} = \frac{\eta}{\sqrt{1 - \left(\frac{\lambda}{\lambda_{c_{10}}}\right)^2}} \tag{8–41a}$$

$$= \frac{\eta}{\sqrt{1 - \left(\frac{\lambda}{4.572 \times 10^{-2}}\right)^2}}\ \Omega \tag{8–41b}$$

The wavelength λ at 8.2 GHz is 3.659 cm, and since WR–90 is air filled, then $\eta = 377\ \Omega$. Substituting into (8–41), we obtain

$$Z_{W_{TE_{10}}} = \frac{377}{\sqrt{1 - \left(\frac{3.659 \times 10^{-2}}{4.572 \times 10^{-2}}\right)^2}} = 628.6\ \Omega \tag{8–42}$$

EXAMPLE 8–6

Determine the wave impedance for the TM_{11} mode in a WR–90 waveguide at a frequency that is 50% greater than the cutoff frequency.

Solution

First determine the cutoff wavelength for the TM_{11} mode in the WR–90 waveguide from (8–5). For the WR–90 waveguide, $a = 22.860$ mm and $b = 10.160$ mm. Since $m = 1$ and $n = 1$, then

$$\lambda_c = \frac{1}{\sqrt{\left(\frac{m}{2a}\right)^2 + \left(\frac{n}{2b}\right)^2}} \tag{8–43a}$$

$$= \frac{1}{\sqrt{\left(\frac{1}{2(2.286 \times 10^{-2})}\right)^2 + \left(\frac{1}{2(1.016 \times 10^{-2})}\right)^2}} \tag{8–43b}$$

$$= 1.857 \times 10^{-2}\,\text{m} = 1.857\,\text{cm} \tag{8–43c}$$

The cutoff frequency is

$$f_c = \frac{c}{\lambda_c} = \frac{3 \times 10^8}{1.857 \times 10^{-2}} = 16.16\,\text{GHz} \tag{8–44}$$

The frequency that is 50% greater than the cutoff frequency is 24.24 GHz. The wave impedance for the TM_{11} mode is determined from (8–40):

$$Z_{W_{TM_{11}}} = \eta\sqrt{1 - \left(\frac{\lambda}{\lambda_{c_{11}}}\right)^2}\ \Omega \tag{8–45}$$

The wavelength at 24.24 GHz is 1.238 cm. Substituting this and the cutoff wavelength into (8–40) along with $\eta = 377\ \Omega$, since the WR–90 waveguide is air filled, yields

$$Z_{W_{TM_{11}}} = 377\sqrt{1 - \left(\frac{1.238 \times 10^{-2}}{1.857 \times 10^{-2}}\right)^2} = 281.0\,\Omega \tag{8–46}$$

8–9 ■ WAVEGUIDE LOSSES

The dissipative losses within a rectangular waveguide consist of dissipative losses associated with the conduction current, losses within the metallic walls, and the dielectric losses in the insulator material within the waveguide. Since dielectric-filled waveguides are not very common and have not been addressed in this textbook, the dielectric losses in an air-filled waveguide can be considered equal to zero.

The attenuation constant due to the dissipative loss in the metallic walls of the waveguide for the TE_{10} mode as determined by Ramo* is

$$\alpha_c = \frac{R_s}{b\eta_0} \frac{1 + \frac{2b}{a}\left(\frac{f_c}{f}\right)^2}{\sqrt{1 - \left(\frac{f_c}{f}\right)^2}} \text{ Np/length} \tag{8–47}$$

where

R_s = sheet resistivity in ohms/m^2

η_0 = intrinsic impedance of free space (377 Ω)

Thus,

$$R_s = \frac{1}{\sigma \delta_s} \tag{8–48}$$

where σ is the conductivity of the metallic walls in S/m and the skin depth is

$$\delta_s = \frac{1}{\sqrt{\pi f \mu_r \mu_0 \sigma}} \tag{8–49}$$

where

f = frequency

μ_r = relative permeability (typically $\mu_r = 1$)

μ_0 = permeability of free space ($4\pi \times 10^{-7}$ H/m)

Today, WR–90 waveguide is constructed from aluminum which has a conductivity of 3.54×10^7 S/m; WR–62 and WR–12 waveguide are often plated with silver which has a conductivity of 6.12×10^7 S/m.

Substituting (8–49) and (8–48) into (8–47), setting η equal to 377 Ω, and converting to dB/m yields

$$\alpha_c = (4.578 \times 10^{-5}) \sqrt{\frac{f}{\sigma}\frac{1}{b}} \left[\frac{1 + \frac{2b}{a}\left(\frac{f_c}{f}\right)^2}{\sqrt{1 - \left(\frac{f_c}{f}\right)^2}} \right] \text{ dB/m} \tag{8–50}$$

Figure 8–9 provides a plot of the attenuation in dB/m for a WR–90 aluminum rectangular waveguide over the recommended frequency range of 8.2 to 12.4 GHz. Example 8–7 provides the calculations at the ends of the operating frequency region.

*S. Ramo, J. R. Whinnery, and T. Van Duzer, *Fields and Waves in Communication Electronics* (New York: Wiley, 1965).

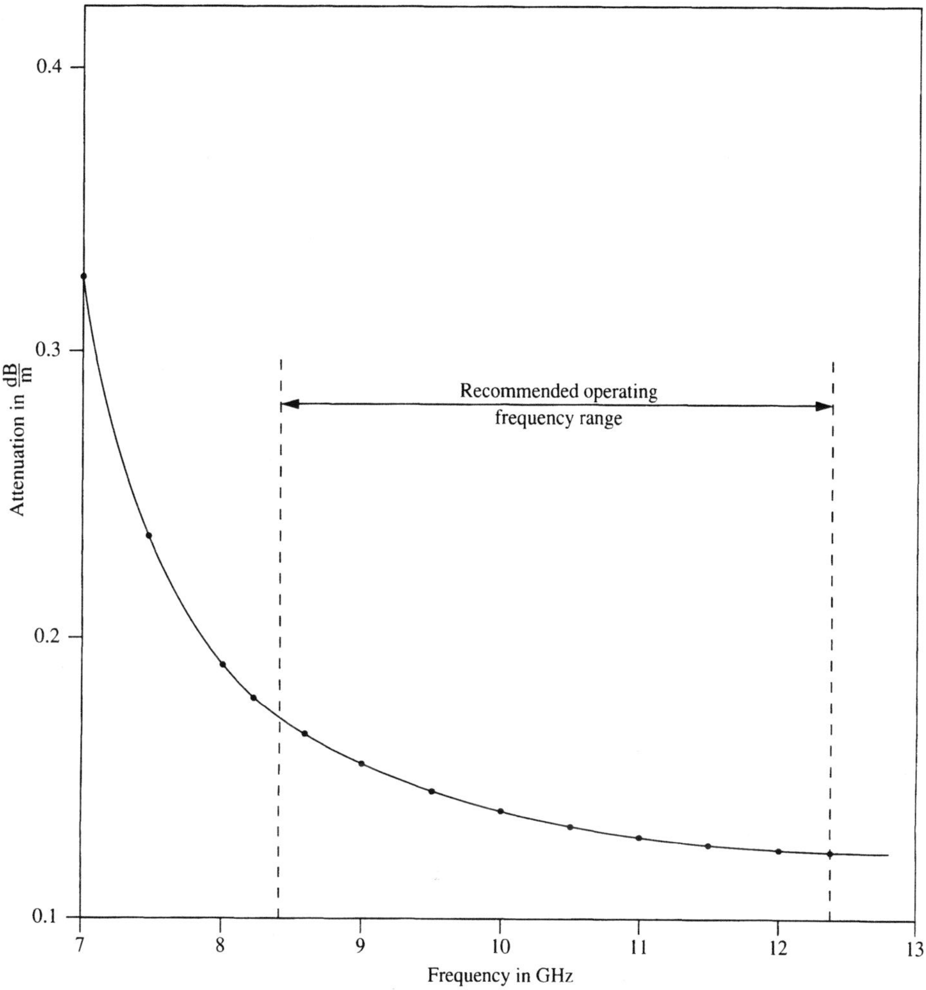

FIGURE 8–9 Attenuation in a WR–90 aluminum waveguide.

EXAMPLE 8–7

Determine the attenuation in dB/m for a WR–90 aluminum waveguide at both ends of the operating frequency range which is 8.2 to 12.4 GHz. The width a is 2.286 cm, and the height b is 1.016 cm. The cutoff frequency is 6.562 GHz, and the conductivity of aluminum is 3.54×10^7 S/m.

Solution

The attenuation can be determined from (8–50) by substituting values for a, b, f_c, and σ:

$$\alpha_c = (4.578 \times 10^{-5})\sqrt{\frac{f}{\sigma}}\frac{1}{b}\left[\frac{1+\frac{2b}{a}\left(\frac{f_c}{f}\right)^2}{\sqrt{1-\left(\frac{f_c}{f}\right)^2}}\right] \text{dB/m} \tag{8–51a}$$

$$= (7.573 \times 10^{-7})\sqrt{f}\left[\frac{1+0.8889\left(\frac{6.562 \times 10^9}{f}\right)^2}{\sqrt{1-\left(\frac{6.562 \times 10^9}{f}\right)^2}}\right] \text{dB/m} \tag{8–51b}$$

The attenuation at 8.2 GHz is

$$\alpha_c = (7.573 \times 10^{-7})\sqrt{8.2 \times 10^9}\left[\frac{1+0.8889\left(\frac{6.562 \times 10^9}{8.2 \times 10^9}\right)^2}{\sqrt{1-\left(\frac{6.562 \times 10^9}{8.2 \times 10^9}\right)^2}}\right] \tag{8–52a}$$

$$= 0.1795 \text{ dB/m} \tag{8–52b}$$

The attenuation at 12.4 GHz is

$$\alpha_c = (7.573 \times 10^{-7})\sqrt{12.4 \times 10^9}\left[\frac{1+0.8889\left(\frac{6.562 \times 10^9}{12.4 \times 10^9}\right)^2}{\sqrt{1-\left(\frac{6.562 \times 10^9}{12.4 \times 10^9}\right)^2}}\right] \tag{8–53a}$$

$$= 0.1241 \text{ dB/m} \tag{8–53b}$$

■ PROBLEMS

8–1. A WR–62 rectangular waveguide has an operating frequency range of 11.9 to 18.0 GHz. The inner dimensions are 15.799 mm by 7.899 mm. Determine the cutoff frequency for the dominant mode.

8–2. A WR–12 rectangular waveguide has an operating frequency range of 60.5 to 91.9 GHz. The inner dimensions are 3.099 mm by 1.549 mm. Determine the cutoff frequency for the dominant mode.

8–3. For the WR–62 waveguide in Problem 8–1, determine the cutoff wavelength and the cutoff frequency for the TE_{01}, TE_{20}, and TM_{11} modes.

8–4. For the WR–12 waveguide in Problem 8–2, determine the cutoff wavelength and the cutoff frequency for the TE_{01}, TE_{20}, and TM_{11} modes.

8–5. For the WR–62 waveguide in Problem 8–1, determine the guide wavelength at each end of the operating frequency range. Compare these values with the free-space wavelength.

8–6. For the WR–12 waveguide in Problem 8–2, determine the guide wavelength at each end of the operating frequency range. Compare these values with the free-space wavelength.

8–7. For the WR–62 waveguide in Problem 8–1, determine the phase and group velocities at each end of the operating frequency range.

8–8. For the WR–12 waveguide in Problem 8–2, determine the phase and group velocities at each end of the operating frequency range.

8–9. Determine the wave impedance for the TE_{10} mode in the WR–62 waveguide in Problem 8–1 at the two ends of the operating frequency range.

8–10. Determine the wave impedance for the TE_{10} mode in the WR–12 waveguide in Problem 8–2 at the two ends of the operating frequency range.

8–11. Determine the wave impedance for the TM_{11} mode in the WR–62 waveguide of Problem 8–1 at a frequency that is 50% greater than the cutoff frequency.

8–12. Determine the wave impedance for the TM_{11} mode in the WR–12 waveguide of Problem 8–2 at a frequency that is 50% greater than the cutoff frequency.

8–13. Determine the attenuation in dB/m for the silver-plated WR–62 waveguide of Problem 8–1 at the two ends of the operating frequency region. The conductivity of silver is 6.12×10^7 S/m.

8–14. Determine the attenuation in dB/m for the silver-plated WR–12 waveguide of Problem 8–2 at the two ends of the operating frequency region. The conductivity of silver is 6.12×10^7 S/m.

9

Antenna Technology

OBJECTIVES

After completing this chapter, the reader should be able to:

1. Describe an **antenna** and its functions.
2. Describe an **isotropic antenna** and its characteristics.
3. Define antenna **directivity** and **gain.**
4. Describe antenna radiation patterns.
5. Define **radiation intensity.**
6. Describe the **near field** and **far field** regions surrounding an antenna and determine the distance to the boundary between the two regions.
7. Define the **3-dB beamwidth, sidelobes** and **front-to-back ratio** of an antenna.
8. Define **antenna efficiency** and **illumination efficiency.**
9. Define and determine the **effective area** of an antenna.
10. Describe the **capture area** of an antenna.
11. Define **polarization** and describe the differences between **linear, elliptical,** and **circular polarization.**
12. Define **antenna impedance** and **radiation resistance.**
13. Describe the difference between **self-impedance** and **mutual impedance** of an antenna.
14. Determine the **radiation resistance** of an antenna.
15. Describe a **half-wave dipole.**
16. Describe a **quarter-wave monopole.**
17. Discuss the functions of a **ground plane.**
18. Define an **image antenna.**
19. Describe a **waveguide horn antenna.**
20. Describe a **parabolic reflector antenna.**
21. Determine the **effective area, isotropic gain,** and **3-dB beamwidth** of a parabolic reflector antenna.

9–1 ■ INTRODUCTION

An antenna is a device that provides the transition from a guided electromagnetic wave on a transmission line to an electromagnetic wave propagating in free space. An antenna is used both at the transmitting end or source of electromagnetic radiation and at the receiving end, where the radiation is converted from an electromagnetic wave propagating in free space back to an electromagnetic wave being guided by a transmission line. An antenna performs equally well as either a transmitting antenna or a receiving antenna. Therefore, one can test a transmitting antenna as a receiving antenna, or vice versa. The only difference is that the voltages, currents, and power levels in a transmitting antenna are usually significantly higher than those in a receiving antenna.

The antenna serves two basic functions. First, it is an impedance-matching device which matches the characteristic impedance of the transmission line to the intrinsic impedance of free space. This matching is necessary to prevent unwanted reflections back to the source or load. Second, the antenna is designed to direct the electromagnetic radiation in the desired direction.

In some antenna applications, it is desired to transmit or receive from all directions over the sphere surrounding the antenna. This type of antenna is referred to as an **isotropic** antenna. In other applications, it is desirable to concentrate the radiation to one particular direction. Here the design provides the antenna with high **directivity** or the ability to concentrate the radiation in the desired direction.

The term **antenna gain** refers to the ratio of the power density at a particular location from an antenna with directivity to the power density from an isotropic antenna radiating the same power. Therefore, the gain of an antenna is not like the gain of an amplifier; rather, it comes from the ability of the antenna to concentrate the radiation in one direction, i.e., the antenna's directivity.

The directional characteristics of an antenna are described by a **radiation pattern.** A radiation pattern provides the radiation intensity from an antenna as a function of the direction of electromagnetic radiation.

The radiation intensity is used instead of power density because it is independent of the distance from the antenna. Therefore, an antenna's radiation pattern can be used to predict the power density received by a receiving antenna or the power transmitted by a transmitting antenna at any distance from the antenna. The power density $\mathcal{P}$ is defined by

$$\mathcal{P} = \frac{P_t}{4\pi r^2} \tag{9–1}$$

where

P_t = power transmitted by the antenna

r = distance from the antenna

Multiplying the power density of (9–1) by r^2 results in the transmitted power P_t divided by 4π. There are 4π solid angles or **steradians** (sr) in a sphere.

Therefore, we define the **radiation intensity** U as

$$U = r^2 \mathcal{P} = r^2 \left(\frac{P_t}{4\pi r^2} \right) = \frac{P_t}{4\pi} \text{ W/solid angle (sr)} \tag{9–2}$$

Therefore, the radiation intensity is independent of distance from the antenna.

The purpose of this chapter is to describe the basic, important fundamental concepts of antennas and their application to free-space propagation of electromagnetic energy. The theoretical analysis and design of antennas are covered in many advanced textbooks on antenna theory. Kraus* is an excellent book that presents the basic theory of antennas with emphasis on their engineering applications.

A description of several common antenna types—such as dipoles, vertical antennas above a ground plane, horns, and parabolic reflector antennas—will be presented.

EXAMPLE 9–1

The power density 10 km from a transmitting antenna is $\mathcal{P} = 0.06\ \mu\text{W/m}^2$. Determine the radiation intensity.

Solution

The radiation intensity can be determined from (9–2) as follows:

$$U = r^2 \times \mathcal{P} \text{ W/sr} = (10^4)^2 \times (6 \times 10^{-8}) = 6 \text{ W/sr} \tag{9–3}$$

EXAMPLE 9–2

The radiation intensity from a transmitting antenna is $U = 50$ W/sr. Determine the power density at a receiving antenna located 25 km from the transmitting antenna.

Solution

Rearranging equation (9–2), we find that the power density is

$$\mathcal{P} = \frac{U}{r^2} = \frac{50}{(2.5 \times 10^4)^2} \tag{9–4a}$$

$$= 8 \times 10^{-8} \text{ W/m}^2 = 0.08\ \mu\text{W/m}^2 \tag{9–4b}$$

9–2 ■ ELECTROMAGNETIC RADIATION FROM AN ANTENNA

Time-varying voltages and currents in an antenna produce time-varying electric and magnetic fields that travel radially away from the antenna at a velocity determined by the medium in which the electromagnetic fields are propagating.

*John D. Kraus, *Antennas* (New York: McGraw-Hill, 1950).

There are two distinct regions of electric and magnetic fields surrounding an antenna. These two regions, referred to as the **near field** and the **far field,** are defined by the distance from the antenna as a function of the frequency of the electromagnetic radiation and the size of the antenna. The fields in the far field are transverse fields; i.e., the electric and magnetic field intensities are transverse to the direction of propagation. This condition, which was introduced in Chapter 6, is referred to as **plane wave propagation.**

Both transverse and radial electric and magnetic field intensities exist in the near field region. The boundary between the near and far fields is an arbitrary sphere surrounding the antenna whose radius R_{ff} is

$$R_{ff} = \frac{2D^2}{\lambda} \tag{9–5}$$

where λ is the wavelength of radiation and D is the physical linear dimension of the antenna. The radiation patterns that describe the radiation intensity of the antenna as a function of angle are usually patterns for the far field.

EXAMPLE 9–3

Determine the distance from a 100-MHz half-wavelength dipole to the boundary between the near field and the far field.

Solution

The reader can verify that the wavelength λ at 100 MHz is 3 m. Therefore, the length of the dipole is 1.5 m. The distance to the far field can be determined from (9–5):

$$R_{ff} = \frac{2D^2}{\lambda} = \frac{2(1.5)^2}{3} = 1.5 \text{ m} \tag{9–6}$$

For a dipole, the distance to the far field is the length of the dipole.

EXAMPLE 9–4

A parabolic reflector antenna with a diameter of 18.3 m operates at 2.3 GHz for space communications. Determine the far field distance for this antenna.

Solution

The reader can verify that the wavelength at 2.3 GHz is 0.13 m. The distance to the far field can be determined from (9–5):

$$R_{ff} = \frac{2D^2}{\lambda} = \frac{2(18.3)^2}{0.13} = 5.15 \times 10^3 \text{ m} \tag{9–7}$$

As shown in this example, the far field distance for a high-gain antenna is a significant distance from the antenna. The measurement of the far field radiation patterns for a large antenna operating at a high frequency can be a difficult task.

9–3 ■ RADIATION PATTERNS

The radiation pattern is a three-dimensional plot of the relative strength or radiation intensity of an antenna as a function of the spherical coordinates θ and ϕ. A three-dimensional radiation pattern for an idealized high-gain antenna is shown in Fig. 9–1. Since it is difficult to present three-dimensional information, typically the radiation patterns are shown as a pair of two-dimensional plots. The first plot shows the radiation intensity as a function of angle in the plane of the electric field intensity vector. Referring to Fig. 9–1, the E plane is the y-z plane. A plot of the radiation intensity as a function of angle in the y-z plane is the E plane radiation pattern.

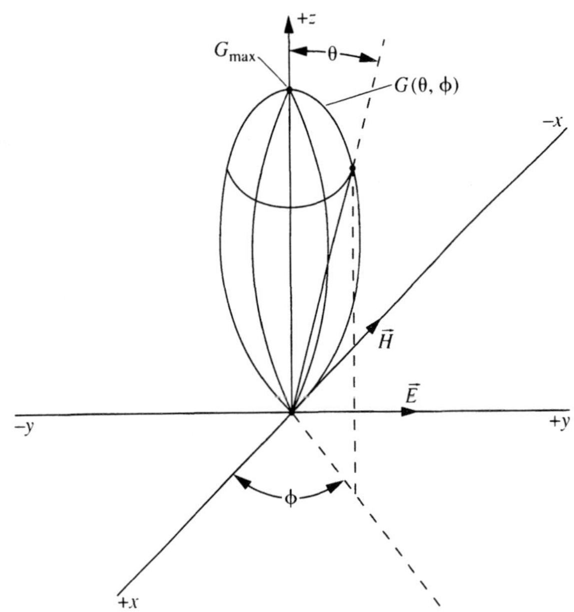

FIGURE 9–1 Three-dimensional antenna radiation pattern.

A typical E plane radiation pattern for a half-wave dipole is shown in Fig. 9–2(a). The second plot shows the radiation intensity as a function of angle in the plane of the magnetic field intensity vector. Referring to Fig. 9–1, the H plane is the x-z plane. A plot of the radiation intensity as a function of angle in the x-z plane is the H plane radiation pattern. A typical H plane radiation pattern for a half-wave dipole is shown in Fig. 9–2(b). The E plane and H plane radiation patterns are orthogonal (90°) to each other and are referred to as the **principal plane patterns.** Normally these are the only two patterns taken in an antenna measurement facility to describe the radiation patterns of an antenna.

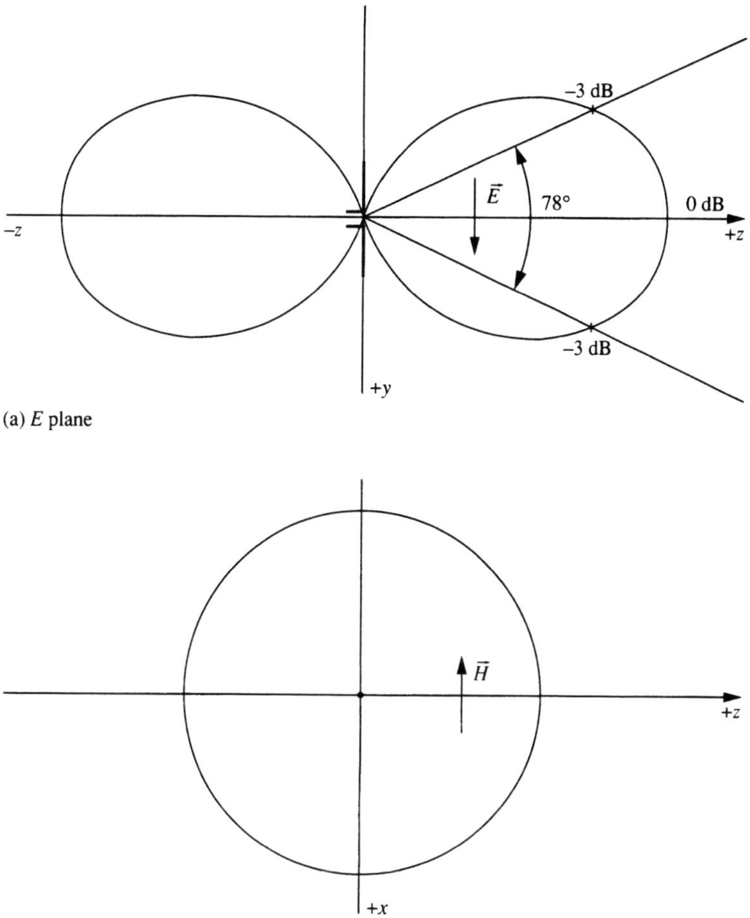

(a) *E* plane

(b) *H* plane

FIGURE 9–2 E plane and H plane radiation patterns of a half-wave dipole.

The radiation patterns are usually normalized to the maximum gain G_{max} by dividing the gain as a function of angle $G(\theta, \phi)$ by the maximum gain to obtain the normalized gain

$$g(\theta, \phi) = \frac{G(\theta, \phi)}{G_{max}} \qquad \textbf{(9–8)}$$

Therefore, the normalized maximum gain is 1, and the gain at other angles is less than 1. Since antenna patterns cover a significant dynamic range, typically from 1 down to 10^{-4} or less, antenna radiation patterns are normally plotted in decibels on a linear scale, usually on a polar plot. However, some applications require the use of a rectangular plot. The normalized decibel gain is

$$g_{dB}(\theta, \phi) = 10 \log_{10}[g(\theta, \phi)] \qquad \textbf{(9–9)}$$

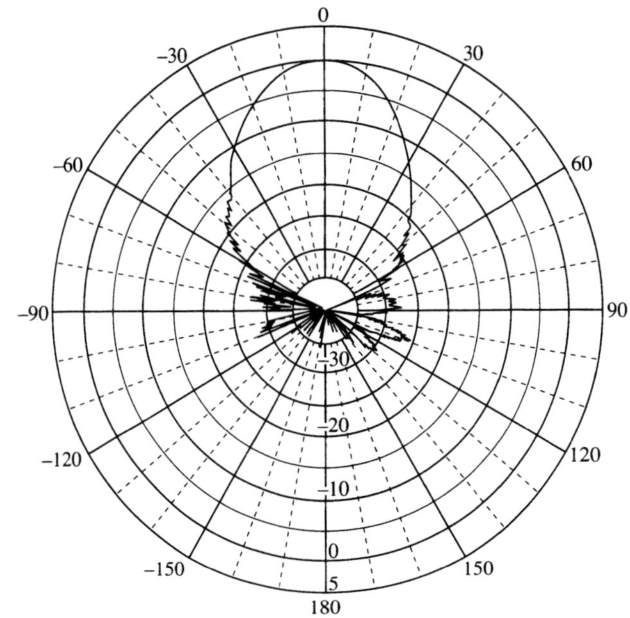

(a) Polar

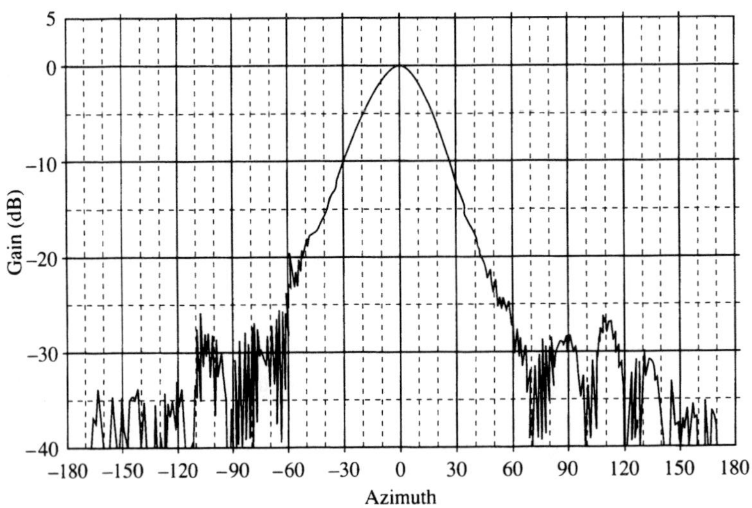

(b) Rectangular

FIGURE 9–3 Polar and rectangular patterns for a standard-gain horn antenna.

Figure 9–3 illustrates two normalized antenna radiation patterns of a standard-gain horn antenna. A polar plot is shown in (a), and a rectangular plot is shown in (b).

9–4 ■ ANTENNA BEAMWIDTHS AND SIDELOBES

An idealized antenna would have a radiation pattern whose normalized gain is 1 over the desired angular beamwidth and 0 at all other angles. A realistic antenna radiation pattern is shown in Fig. 9–4. The **antenna beamwidth** is defined as the included angle between the −3-dB points on the normalized-gain pattern. Therefore, the power gain at the edges of the antenna beam is $\frac{1}{2}$ of the on-axis, or maximum gain. The **main lobe** is the antenna beam defined between the first null on either side of the maximum-gain angle. Typically for high-gain antennas, the null-to-null beamwidth is 2.5 times the 3-dB beamwidth.

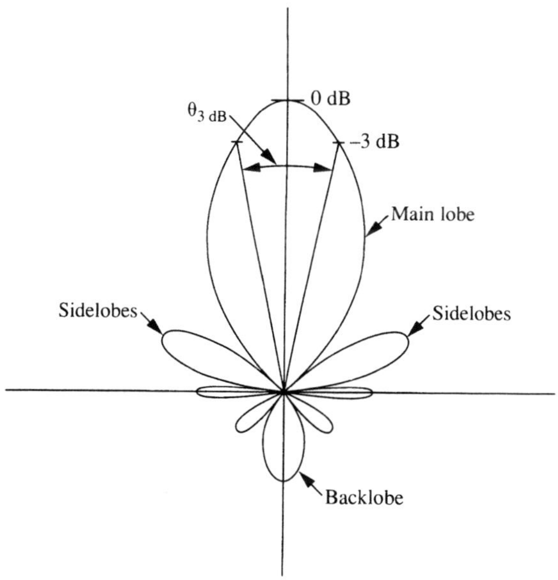

FIGURE 9–4 Realistic antenna radiation pattern.

An antenna radiates power in undesired directions. The radiation pattern has **sidelobes** as shown in Fig. 9–4. The level of the sidelobes determines how much power is radiated in undesired directions; or, if the antenna is a receiving antenna, it determines the level that undesired signals can be received through the sidelobes.

Another undesired part of the radiation pattern is the **backlobe**. A quality factor important in some antenna applications is the **front-to-back ratio.** As shown in Fig. 9–2, the front-to-back ratio of a dipole is 1 or 0 dB. A dipole cannot tell if the signal is coming from the front or back of the antenna.

9–5 ■ DIRECTIVITY AND ANTENNA GAIN

There are two commonly employed terms used to describe the radiation characteristics of an antenna: **directivity** and **antenna gain.** Directivity is a characteristic of the radiation pattern of an ideal, lossless antenna while the antenna gain includes the ohmic losses of the antenna physical structure.

Directivity

The **directivity** D of an antenna is defined from the radiation pattern as

$$D = \frac{U_{\max}}{U_0} = \frac{\text{maximum radiation intensity}}{\text{average radiation intensity}} \tag{9–10}$$

Multiplying the numerator and denominator by 4π and dividing both by r^2 gives

$$D = \frac{4\pi U_m}{4\pi U_0} = \frac{4\pi U_m}{P_t} = \frac{U_m}{\frac{P_t}{4\pi}} \tag{9–11a}$$

$$= \frac{\frac{U_m}{r^2}}{\frac{P_t}{4\pi r^2}} = \frac{\mathcal{P}_{\max}}{\frac{P_t}{4\pi r^2}} \tag{9–11b}$$

Therefore, the directivity is

$$D = \frac{\text{maximum power density}}{\text{power density from an isotropic antenna}} \tag{9–12}$$

The directivity does not take into account the ohmic losses in the antenna since it is based on the radiation patterns only.

Antenna Gain

Antenna gain is defined as the ratio of the maximum radiation intensity $U_{\max}$ to the maximum radiation intensity U_{ref} from a reference antenna with the same power input to the antenna:

$$G = \frac{U_{\max}}{U_{\text{ref}}} \tag{9–13}$$

The difference between directivity and gain is that directivity is referenced to the power radiated by the antenna, while gain is referenced to the power delivered by the transmission line to the antenna. Therefore, gain is always less than or equal to directivity, the difference being the power dissipated in the antenna ohmic losses.

Normally, antenna gain is expressed as a power ratio and is usually specified in decibels, where

$$G_{\text{dB}} = 10 \log_{10} G \tag{9–14}$$

The value of gain depends on the gain of the reference antenna. It is important to know what reference has been used for the antenna gain. Two of the common references are as follows:

1. a lossless isotropic antenna, in which the radiation intensity is uniform over the sphere surrounding the antenna, i.e., all 4π steradians
2. a reference dipole

The lossless isotropic antenna is a theoretical concept and has never been achieved, while the dipole is readily available.

Antenna measurements of gain are usually referenced to a standard dipole for low-gain antennas or to a standard-gain horn for higher-gain antennas. Accurate theoretical calculations of the gain referenced to a lossless isotropic antenna are possible for both the standard dipole and the standard-gain horn. The gain of a standard half-wavelength dipole is 1.643 or 2.16 dBi. The **dBi** designation on antenna gain is used to designate that the gain is referenced to a lossless isotropic antenna.

Antenna Efficiency

The **antenna efficiency** η_a is a measure of the power dissipated in the ohmic losses of the antenna. It is the ratio of the total power radiated from the antenna to the power delivered to the antenna from the transmission line. The antenna efficiency can be determined from the directivity and gain of an antenna as follows:

$$\eta_a = \frac{G}{D} \tag{9–15}$$

where D and G are the numeric values of directivity and gain.

EXAMPLE 9–5

An antenna is transmitting 200 W of power. The maximum power density at a distance of 10 km is 3.184 mW/m^2. Determine the directivity of the antenna.

Solution

The power density from an isotropic antenna can be determined from (9–1) as

$$\mathcal{P} = \frac{P_t}{4\pi r^2} = \frac{200}{4\pi(10 \times 10^3)^2} = 0.1592\ \mu\text{W/m}^2 \tag{9–16}$$

The directivity can be determined from (9–12) as

$$D = \frac{\text{maximum power density}}{\text{power density from an isotropic antenna}} \tag{9–17a}$$

$$= \frac{3.184 \times 10^{-3}}{0.1592 \times 10^{-6}} = 20{,}000 \tag{9–17b}$$

The directivity is normally expressed in dB relative to an isotropic antenna, i.e., in dBi. Thus,

$$D = 10 \log(20{,}000) = 43.01 \text{ dBi} \tag{9–18}$$

EXAMPLE 9–6

An antenna with a directivity of +16 dBi is transmitting a power of 1 kW. Determine the maximum power density at a distance of 50 km from the antenna.

Solution

Rearranging (9–11) yields

$$\mathcal{P}_{\max} = D\left(\frac{P_t}{4\pi r^2}\right) \tag{9–19}$$

Next, converting the directivity to a factor, we can determine the maximum power density as follows:

$$G = 10^{16\text{ dBi}/10} = 10^{1.6} = 39.81 \tag{9–20}$$

Substituting values into (9–19) gives

$$\mathcal{P}_{\max} = (39.81)\left[\frac{1000}{4\pi(50 \times 10^3)^2}\right] = 1.267\ \mu\text{W/m}^2 \tag{9–21}$$

EXAMPLE 9–7

An antenna has an antenna efficiency η_a of 95%. Given that the directivity is +33 dBi, determine the antenna gain in dBi.

Solution

The antenna efficiency is the ratio of the antenna gain to the directivity. Equation (9–15) can be used to determine the antenna gain, but the directivity must first be expressed as a ratio. Therefore,

$$D = 10^{33/10} = 1995.26 \tag{9–22}$$

The antenna gain is

$$G = \eta_a D = (0.95)(1995.26) = 1895.50 \quad \textbf{(9–23)}$$

The antenna gain in decibels referenced to an isotropic antenna is

$$G_{\text{dBi}} = 10\log(1895.50) = 32.78\text{ dBi} \quad \textbf{(9–24)}$$

9–6 ■ EFFECTIVE AREA OF AN ANTENNA

An electromagnetic wave propagating in free space must be collected by an antenna. The antenna performs the function of converting the power density of electromagnetic radiation incident on the antenna into a voltage and current flowing in the transmission line that connects the antenna to a load. Thus, the receiving antenna converts the power density incident on the antenna to a power delivered to the load. The antenna has an **effective area** by which the power density in watts per unit of area is multiplied by to obtain power in watts delivered to the load. The effective area is also referred to as the **capture area** of the antenna.

The effective area is *not* necessarily the physical area of the antenna, although it may be close to the physical area for certain types, e.g., for the parabolic reflector antenna. It can be shown that the effective area A_e is related to the gain of an antenna, where the gain is referenced to an isotropic antenna. This gain will be designated by G_i. The effective area is

$$A_e = \frac{\lambda^2}{4\pi} G_i \quad \textbf{(9–25)}$$

EXAMPLE 9–8

Determine the effective area of a 5-GHz antenna whose isotropic gain G_i is +15 dBi.

Solution

The effective area of an antenna can be determined from (9–25). The wavelength of a 5-GHz antenna is

$$\lambda = \frac{c}{f} = \frac{3 \times 10^8}{5 \times 10^9} = 0.06\text{ m} = 6\text{ cm} \quad \textbf{(9–26)}$$

The antenna gain ratio is

$$G_i = 10^{G_{\text{dBi}}/10} = 10^{1.5} = 31.623 \quad \textbf{(9–27)}$$

Substituting (9–26) and (9–27) into (9–25) yields

$$A_e = \frac{\lambda^2}{4\pi} G_i \quad \textbf{(9–28a)}$$

$$= \frac{(6 \times 10^{-2})^2}{4\pi}(31.623) = 9.059 \times 10^{-3}\text{ m}^2 \quad \textbf{(9–28b)}$$

9–7 ■ POLARIZATION

The **polarization** of an electromagnetic wave propagating in free space is the orientation of the electric field intensity vector relative to the surface of the earth. There are two basic types of polarization: **linear** and **elliptical.** In linear polarization, the electric vector does not change orientation as it travels away from an observer. In elliptical polarization, the electric vector rotates as it travels away from an observer, and the tip of the electric vector traces an ellipse.

Linear polarization can be divided into two classifications, **vertical** and **horizontal,** both of which are illustrated in Fig. 9–5. In horizontal polarization, shown in Fig. 9–5(a), the electric vector is *parallel* to the earth's surface. In vertical polarization, shown in Fig. 9–5(b), the electric vector is *perpendicular* to the earth's surface.

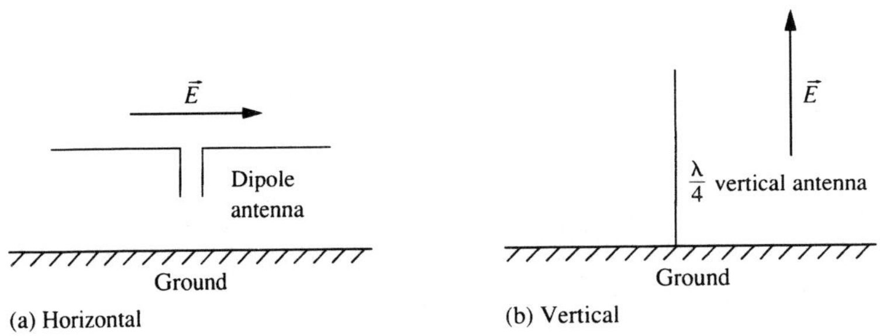

FIGURE 9–5 Linear polarization.

Circular polarization, the more common form of elliptical polarization, occurs when the tip of the rotating electric vector traces a circle in space as it travels away from the observer. Two classifications of circular polarization exist: **right-hand circular** and **left-hand circular.** Figure 9–6 illustrates the two forms of circular polarization. In right-hand circular polarization, the electric vector rotates in a *clockwise* direction as the electric vector travels away from the observer, as shown in Fig. 9–6(a). In left-hand circular polarization, the electric vector rotates in a *counterclockwise* direction, as shown in Fig. 9–6(b).

An antenna transmits vertical, horizontal, right-hand (RH) circular, or left-hand (LH) circular polarization, depending on the antenna design and orientation. There is a significant loss of approximately 30 dB if the transmitting antenna has one form of linear polarization and the receiving antenna has the other linear form. This loss also occurs if one antenna is circular polarized with one sense (RH or LH) and the receiving antenna is circular polarized with the other sense (LH or RH). This characteristic is used to provide polarization diversity in communication systems. Two channels of information can be sent over the same physical path in which one uses vertical polarization while the other uses horizontal polarization.

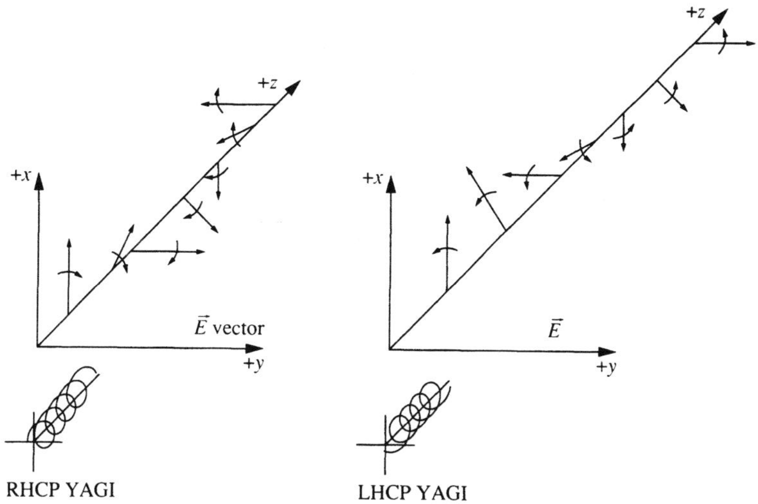

(a) Right-hand (b) Left-hand

FIGURE 9–6 Circular polarization.

A linear polarized transmitting antenna can be combined with a circular polarized receiving antenna, or vice versa, and there will be only a 3-dB loss. This combination is used where the orientation of a linear polarized antenna can change significantly, as on a missile or spacecraft.

9–8 ■ ANTENNA IMPEDANCE AND RADIATION RESISTANCE

Antenna impedance is the impedance of an antenna measured at its input terminals. It is the ratio of the phasor voltage at the feed point (input) of the antenna divided by the phasor current flowing into the antenna at the feed point. This impedance is the load on the transmission line connecting the transmitter or receiver to the antenna. It is a function of frequency and can be complex with a real (resistive) part and an imaginary (reactive) part. An RF impedance bridge connected to an antenna at the feed point (input) of the antenna would measure this impedance.

If an antenna is isolated from ground and any other surrounding objects, this impedance is the **self-impedance** of the antenna. It usually consists of a **self-resistance** and a **self-reactance.** However, at resonance, the reactive part of the antenna impedance approaches zero, and the antenna impedance has only a real, i.e., resistive, component. When other antennas, or objects, or ground is near the antenna, the currents flowing in these objects have an influence on the antenna impedance. Then the antenna impedance is determined both by the self-impedance of the antenna and by a **mutual impedance** between the antenna and the nearby objects. Further information on the self-impedance and mutual impedance of different antenna configurations can be found in any advanced text on antenna theory.

An antenna radiates electromagnetic energy. The amount of this energy leaving a sphere surrounding the antenna per unit of time is the power radiated by the antenna, P_{rad}. The power radiated from the antenna can be associated with the power absorbed by a resistor. This resistance, called the **radiation resistance** R_{rad} of the antenna, can be determined from

$$R_{rad} = \frac{P_{rad}}{I^2} \tag{9–29}$$

where I is the rms value of the antenna current magnitude at the input terminals of the antenna. The radiation resistance is the real part of the complex antenna impedance of a lossless antenna. If the antenna has ohmic losses that dissipate power as heat, this antenna ohmic loss is usually represented by an additional resistance in series with the radiation resistance, and the sum of the two resistances is equal to the real part of the antenna impedance.

EXAMPLE 9–9

An antenna has an rms current of 3 A flowing into the antenna, and it is transmitting 1 kW of power. Determine the radiation resistance of that antenna.

Solution

The radiation resistance can be determined from (9–29):

$$R_{rad} = \frac{P_{rad}}{I^2} = \frac{1000}{3^2} = 111.1\ \Omega \tag{9–30}$$

9–9 ■ DIPOLES

One of the most commonly used antennas is the **half-wave dipole,** formed from a two-wire parallel transmission line as shown in Fig. 9–7(a). Starting with an open-ended line, which has a voltage maximum at the open end and a voltage minimum $\lambda/4$ back from the open end, the two conductors are bent 90° from the transmission line as illustrated in Fig. 9–7(b). The theoretical length of the antenna is $\lambda/2$. The diameter d of the wires is much smaller than the length, and the spacing D at the feed point must be small compared to the length. A voltage minimum and a current maximum occur at the feed point as shown in Fig. 9–7(c). The antenna impedance is a minimum at the feed point. Dividing the voltage by the current, the antenna impedance for the dipole is $73 + j42.5\ \Omega$. The reactive component can be eliminated by tuning the antenna, which is accomplished by shortening the length from 0.5λ to

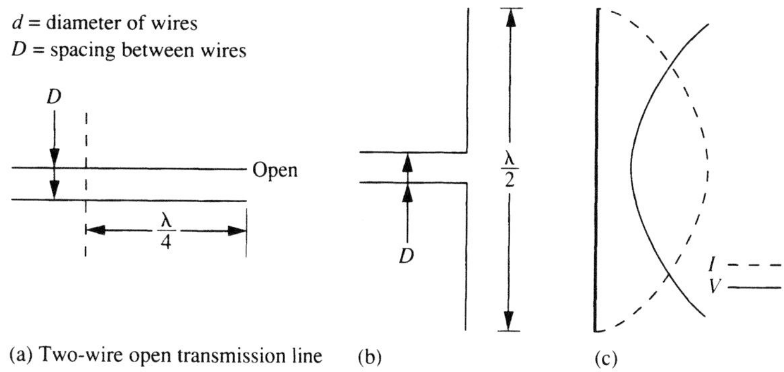

FIGURE 9–7 λ/2 dipole.

0.475λ, corresponding to approximately 95% of the theoretical length. The antenna impedance is approximately 73 Ω resistive, which for a lossless dipole is the radiation resistance of the dipole.

The dipole is a balanced antenna and must be fed by a balanced transmission line (see Section 1–4). Since the most common transmission line is an unbalanced coaxial cable, a device called a **balun** must be used to properly connect a coaxial cable to a dipole. A 75-Ω coaxial cable is commonly available, which matches closely the 73-Ω input impedance of the dipole, but a balun is required. There are several types of baluns available for this purpose.

The E plane and H plane radiation patterns were shown in Fig. 9–2 for the λ/2 dipole. The E plane pattern is like a doughnut with two maxima broadside to the dipole and a null at both ends of the dipole. The 3-dB beamwidth is 78°. The isotropic power gain (directivity) for a lossless λ/2 dipole is 1.643 or 2.16 dBi. As shown in Fig. 9–2(a), the polarization of the dipole is parallel to the dipole. The H plane pattern is illustrated in Fig. 9–2(b) and is a uniform circular pattern with a constant gain for an angle of 360° about the dipole.

The effective area of a dipole can be determined from the isotropic gain G_i and (9–25):

$$A_e = \frac{\lambda^2}{4\pi} G_i = \frac{\lambda^2}{4\pi} 1.643 = 0.13\lambda^2 \tag{9–31}$$

A variation of the λ/2 dipole is the **folded dipole,** which is shown in Fig. 9–8. The most common balanced transmission line used at RF frequencies is the 300-Ω twin lead which is used extensively in the TV and FM industries. A folded dipole is constructed from a λ/2 length of 300-Ω twin lead. The feed point is at the middle of one of the conductors. The ends of the two conductors are connected. The combination of the 73-Ω self-impedance of the dipole and the mutual impedance from the parallel conductor connected at both ends of the dipole increases the antenna impedance of the folded dipole to 280 Ω. Therefore, the folded dipole is a balanced 280-Ω antenna, which closely matches

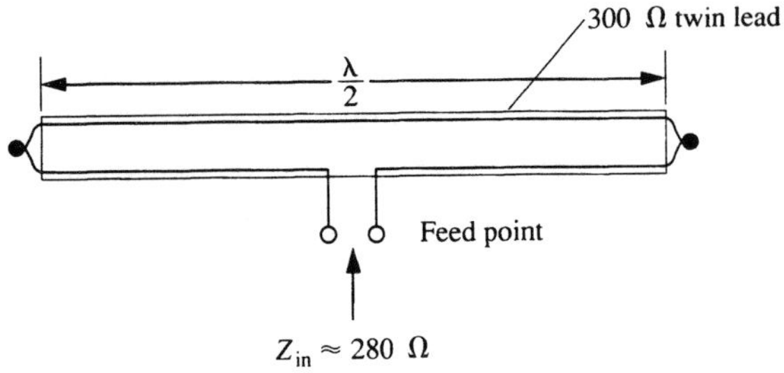

FIGURE 9–8 Folded λ/2 dipole.

the 300-Ω twin-lead balanced transmission line. A balun or an impedance-matching transformer is not required.

EXAMPLE 9–10

Determine the effective area of a dipole at 98 MHz.

Solution

The effective area of a dipole can be determined from (9–31). The wavelength at 98 MHz is

$$\lambda = \frac{c}{f} = \frac{3 \times 10^8}{98 \times 10^6} = 3.061 \text{ m} \tag{9–32}$$

Substituting the wavelength into (9–31) yields

$$A_e = 0.13\lambda^2 = 0.13(3.061)^2 = 1.218 \text{ m}^2 \tag{9–33}$$

9–10 ■ ANTENNAS ABOVE A GROUND PLANE

A **ground plane** is a uniform surface beneath an antenna which will be considered as a perfect conductor. In actual situations, a good plane can be constructed from good conductors, or, at some frequencies, the earth acts as a good ground plane. As discussed in Section 7–1, electromagnetic fields cannot exist in a perfect conductor, and the boundary conditions must be satisfied at the boundary between the propagation media, usually free space and the perfect conductor. An electromagnetic wave is reflected from the ground plane and adds to the direct wave at the receiving antenna. Figure 9–9 illustrates this situation. To satisfy the boundary condition, the tangential component of the electric field intensity $\vec{E}_t$ must be zero. This criterion is achieved by the creation of the reflected wave in which a 180° phase shift cancels the tangential component of the direct wave.

The radiation patterns and antenna impedance of an antenna above a ground plane are different from the antenna, as if it existed without the ground plane.

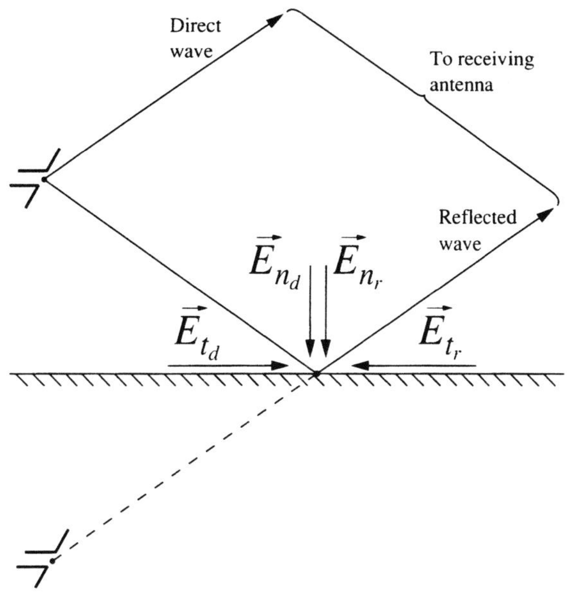

FIGURE 9–9 Effect of ground planes on an antenna.

Image theory is used to determine the characteristics of an antenna above a ground plane. The reflected wave is like a direct wave from an identical antenna located within the ground plane the same distance from the boundary as the real antenna is above the ground plane. This situation is illustrated in Fig. 9–9. The image antenna is similar to an image formed in a mirror at optical frequencies.

An important and commonly used antenna is the λ/4 **monopole antenna** on a ground plane as shown in Fig. 9–10(a). This is an unbalanced antenna since the feed point is between the monopole and ground, and it has vertical polarization. The radiation resistance is 36.5 Ω, and the antenna impedance has a reactive component of $+j21$ Ω. When the monopole is located close to the ground plane, the image antenna forms a dipole as shown in Fig. 9–10(b). The E plane radiation pattern is that of a λ/2 dipole with only $\frac{1}{2}$ of the pattern above the ground plane as illustrated in Fig. 9–10(b).

The ground plane can be achieved either by a grounded metal disc or by radial wires as shown in Fig. 9–11. The roof of a vehicle such as a car or truck can form a ground plane for a λ/4 monopole. The radiation pattern tilts upward as the ground plane becomes smaller and less of a perfect conductor. The maximum gain for a monopole above a semi-infinite perfect ground plane is directly along the ground plane as shown in Fig. 9–11(b). Figure 9–11(c) illustrates the E plane radiation pattern for a λ/4 monopole antenna over a finite-size, nonideal ground plane.

Some vertical monopole antennas are longer than λ/4. When the length exceeds 5λ/8, the radiation pattern forms multiple lobes, and the main lobe is

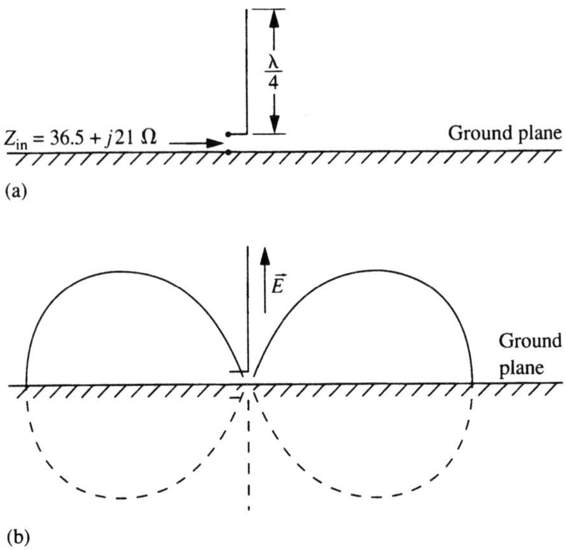

FIGURE 9–10 Quarter-wave monopole antenna above a ground plane.

at an angle above the ground plane. At one wavelength height, the main lobe is above a 45° angle to the ground plane. One of the reasons for making the antenna longer is to increase the radiation resistance, thereby increasing the power radiated for the same antenna current. Large antenna currents have a disadvantage because the power lost in the ohmic resistance of the antenna wires and transmission lines feeding the antenna increases. Most commercial AM broadcast transmitting antennas are vertical monopoles whose heights are $\lambda/2$ to $5\lambda/8$.

9–11 ■ HORNS

The open end of a waveguide can transmit or receive electromagnetic radiation as an antenna. However, the sudden change from the impedance of the waveguide to the impedance of free space reflects energy back to the source due to the impedance mismatch. This problem can be overcome by flaring the four walls of the waveguide to provide a gradual transition from the waveguide impedance to free-space impedance. This structure is referred to as a **waveguide horn.** There are three basic types of waveguide horns: (a) the E plane sectional horn, (b) the H plane sectional horn, and (c) the pyramidal horn, as illustrated in Fig. 9–12.

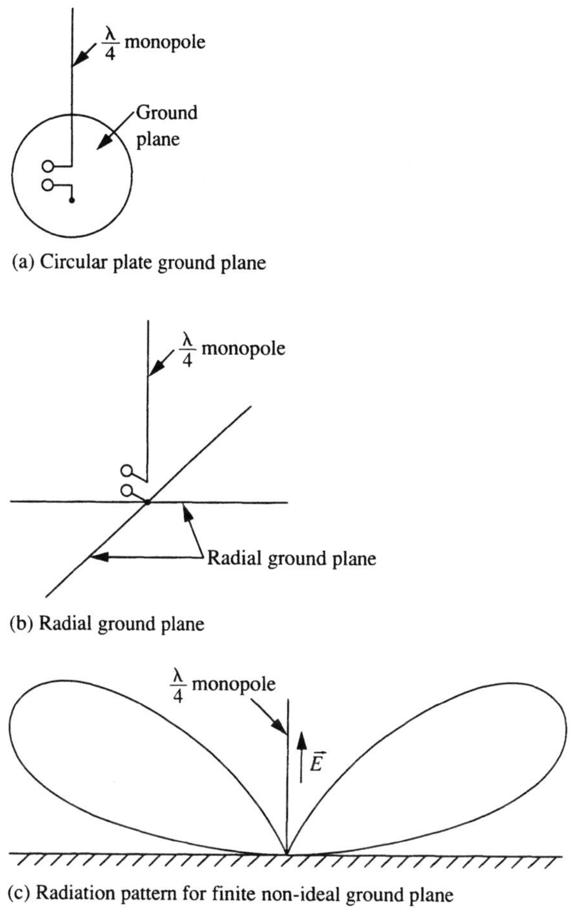

FIGURE 9–11 λ/4 monopole ground planes.

The directivity of a waveguide horn is a function of the frequency, aperture size, and flare length in both the E plane and H plane dimensions. The procedures for these calculations are beyond the scope of this text; however, they can be found in any advanced text on antenna design.

The radiation patterns for a pyramidal horn are shown in Fig. 9–3 in both a polar plot and a rectangular plot. The isotropic gain of a standard-gain horn usually varies from approximately 15 dBi at the lower end of a waveguide band to 17.5 dBi at the upper end of the waveguide band.

The 3-dB beamwidth of the horn is a function of the aperture size. The larger the horn dimension in the flare direction, the narrower the 3-dB beamwidth. Therefore, a sectional horn produces a fan-shaped beam, narrow in one plane

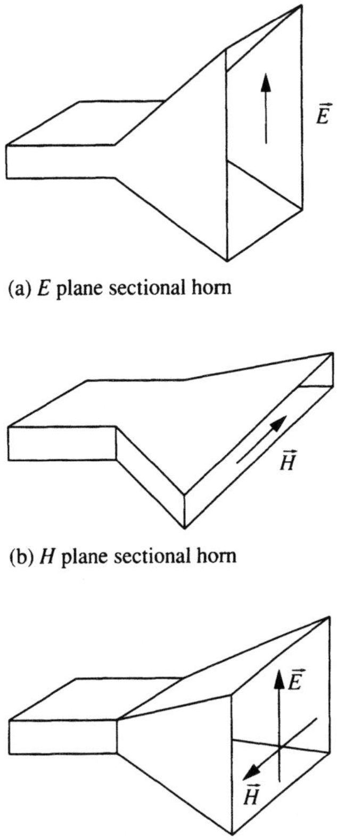

FIGURE 9–12 Waveguide horns. (c) Pyramidal horn

and broad in the other plane. The E plane sectional horn produces a narrow beam in the E plane and a wide beam in the H plane. An H plane sectional horn produces a narrow beam in the H plane and a wide beam in the E plane. A pyramidal horn produces a narrow beam in both planes.

The larger the aperture of a horn, the narrower the beamwidths will be and, therefore, the higher the directivity. Horns have good efficiency, and high isotropic gains can be achieved by constructing horns with a large aperture. This requires a long antenna because the flare must be gradual to provide good impedance matching. Therefore, a large, high-gain horn becomes not only heavy but very expensive to construct. To overcome this problem, a different type of aperture antenna, the parabolic reflector antenna, is used, and its properties will be discussed in the next section.

9–12 ■ PARABOLIC REFLECTOR ANTENNAS

The surface generated by the revolution of a parabola about its axis is a **paraboloid.** The paraboloid has a focal point such that any plane wave electromagnetic radiation incident upon the paraboloid surface will be directed to the focal point. Conversely, if an isotropic radiator is placed at the focal point, it will radiate a spherical wave, a portion of which will be reflected by the paraboloid reflector surface into a plane wave. The plane wave exists over an area equal to the projection of the parabolic reflector, which is a circle with a diameter D, as illustrated in Fig. 9–13(a).

The **parabolic reflector antenna** is shown in Fig. 9–13(b). It consists of a parabolic reflector whose surface is a paraboloid and a primary antenna located at the focal point. The purpose of the primary antenna is to illuminate the reflector with a uniform electromagnetic wave. If an isotropic antenna were

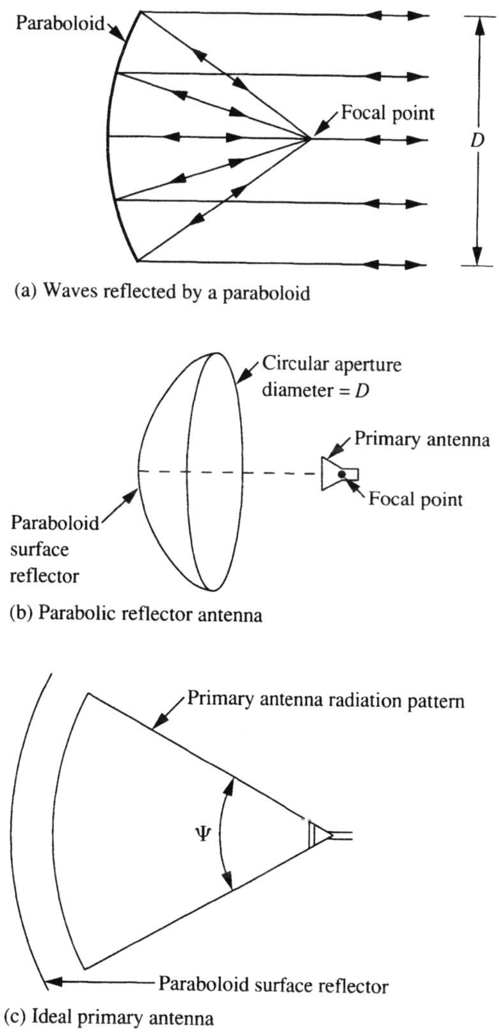

FIGURE 9–13 Parabolic reflector antenna.

used as a primary antenna, it would provide a uniform illumination over the reflector surface. However, it would also radiate and receive radiation from all other directions. This is defined as antenna spillover and is a problem because the antenna could receive undesired radiations from other sources. An ideal primary antenna would have uniform gain over the subtended angle of the reflector ψ and zero gain outside this solid angle. A pyramidal horn antenna is a good primary antenna for the parabolic reflector antenna. It is usually designed such that the gain at the edges of the paraboloid ($\pm\psi/2$) is approximately 10 dB below the peak on-axis gain of the horn. The effect of the nonuniform illumination is taken into account by the **illumination efficiency** η_I. The effective area of the parabolic reflector antenna is

$$A_e = \eta_I A_p \tag{9–34}$$

where A_p is the projected physical aperture of the parabolic reflector and is given by

$$A_p = \frac{\pi D^2}{4} \tag{9–35}$$

The isotropic gain of a parabolic reflector antenna can be determined with (9–25), (9–34), and (9–35). Rearranging (9–25) and substituting (9–34) and (9–35) into (9–25) yields

$$G_i = \frac{4\pi A_e}{\lambda^2} = \frac{4\pi}{\lambda^2}\left(\frac{\eta_I \pi D^2}{4}\right) = \eta_I\left(\frac{\pi D}{\lambda}\right)^2 \tag{9–36}$$

The illumination efficiency η_I usually varies between 0.55 and 0.75.

The 3-dB beamwidth, expressed in radians, for most parabolic reflector antennas is

$$\theta_{3\text{dB}} = 1.2\frac{\lambda}{D}\ \text{rad} \tag{9–37}$$

The 3-dB beamwidth is commonly expressed in degrees as follows:

$$\theta_{3\text{dB}} = 70\frac{\lambda}{D}\ \text{degrees} \tag{9–38}$$

EXAMPLE 9–11

An 11-GHz parabolic reflector antenna has a physical diameter of 15 m. The illumination efficiency is 70%. Determine (a) the effective area, (b) the gain referenced to an isotropic antenna, and (c) the 3-dB beamwidth in degrees.

Solution

The wavelength at 11 GHz is

$$\lambda = \frac{c}{f} = \frac{3 \times 10^8}{11 \times 10^9} = 0.02727\ \text{m} = 2.727\ \text{cm} \tag{9–39}$$

(a) The effective area is determined from (9–35) and (9–36). The physical area is

$$A_p = \frac{\pi D^2}{4} = 176.7 \text{ m}^2 \tag{9–40}$$

The effective area is

$$A_e = \eta_I A_p = 0.70(176.7) = 123.7 \text{ m}^2 \tag{9–41}$$

(b) The isotropic gain is

$$G_i = \frac{4\pi A_e}{\lambda^2} = \frac{4\pi(123.7)}{(2.727 \times 10^{-2})^2} = 2.09 \times 10^6 \tag{9–42}$$

The isotropic gain is expressed in dBi as

$$G_i = 10 \log(2.09 \times 10^6) = 63.2 \text{ dBi} \tag{9–43}$$

(c) The 3-dB beamwidth in degrees can be determined from (9–38):

$$\theta_{3\text{dB}} = 70\frac{\lambda}{D} = 70\left[\frac{0.02727}{15}\right] = 0.1273 \text{ degrees} \tag{9–44}$$

PROBLEMS

9–1. The power density 5 km from a transmitting antenna is $\mathcal{P} = 3.5\ \mu\text{W/m}^2$. Determine the radiation intensity.

9–2. The power density 60 km from a transmitting antenna is $\mathcal{P} = 0.6\ \mu\text{W/m}^2$. Determine the radiation intensity.

9–3. The radiation intensity from a transmitting antenna is $U = 20$ W/sr. Determine the power density at a receiving antenna located 5 km from the transmitting antenna.

9–4. The radiation intensity from a transmitting antenna is $U = 1.5$ W/sr. Determine the power density at a receiving antenna located 20 km from the transmitting antenna.

9–5. Determine the distance from an 88-MHz half-wavelength dipole to the boundary between the near field and the far field.

9–6. Determine the distance from the 156-MHz half-wavelength dipole to the boundary between the near field and the far field.

9–7. A parabolic reflector antenna with a diameter of 12.0 m operates at 6.4 GHz for space communications. Determine the far field distance for this antenna.

9–8. A parabolic reflector antenna with a diameter of 2 m operates at 3.9 GHz for space communications. Determine the far field distance for this antenna.

9–9. An antenna is transmitting 10 W of power. The maximum power density at a distance of 2 km is 4 mW/m^2. Determine the directivity of the antenna.

9–10. An antenna is transmitting 250 W of power. The maximum power density at a distance of 1 km is 0.5 mW/m^2. Determine the directivity of the antenna.

9–11. An antenna with a directivity of +23 dBi is transmitting a power of 1.5 kW. Determine the maximum power density at a distance of 10 km from the antenna.

9–12. An antenna with a directivity of +6 dBi is transmitting a power of 0.5 kW. Determine the maximum power density at a distance of 5 km from the antenna.

9–13. An antenna has an antenna efficiency η_a of 92%. If the directivity is +46 dBi, determine the antenna gain in dBi.

9–14. An antenna has an antenna efficiency η_a of 94%. If the directivity is +13 dBi, determine the antenna gain in dBi.

9–15. Determine the effective area of a 2.3-GHz antenna whose isotropic gain G_i is +43 dBi.

9–16. Determine the effective area of a 1-GHz antenna whose isotropic gain G_i is +7 dBi.

9–17. An antenna has a current of 5 A flowing into the antenna, and it is transmitting 2.5 kW of power. Determine the radiation resistance of the antenna.

9–18. An antenna has a current of 2.5 A flowing into the antenna, and it is transmitting 0.8 kW of power. Determine the radiation resistance of the antenna.

9–19. A 6.4-GHz parabolic reflector antenna has a physical diameter of 15 m. The illumination efficiency is 65%. Determine the following:
(a) the effective area
(b) the gain referenced to an isotropic antenna
(c) the 3-dB beamwidth in degrees

9–20. A 4-GHz parabolic reflector antenna has a physical diameter of 1.5 m. The illumination efficiency is 70%. Determine the following:
(a) the effective area
(b) the gain referenced to an isotropic antenna
(c) the 3-dB beamwidth in degrees

9–21. Determine the effective area of a $\lambda/2$ dipole at 3.9 MHz.

9–22. Determine the effective area of a $\lambda/2$ dipole at 54 MHz.

10

Free-Space Propagation of Electromagnetic Waves

OBJECTIVES

After completing this chapter, the reader should be able to:

1. Describe the **reflection** and **refraction** of an electromagnetic wave that encounters a boundary between two dielectric media.
2. Discuss **Snell's law.**
3. Determine the angle of transmission within a medium for a specific angle of incidence using Snell's law.
4. Define the **index of refraction.**
5. Define the **critical angle of incidence** at a boundary between two media.
6. Define the **Friis transmission formula.**
7. Analyze line-of-sight communication systems and determine the signal-to-noise ratio using the Friis transmission formula.
8. Define **effective isotropic radiated power** (**EIRP**).
9. Define and determine the **free-space path loss** L_p.
10. Describe and determine **receiver sensitivity.**
11. Define the **noise factor** and **noise figure** of a device.
12. Define the **effective noise temperature** of a device and determine its value from the noise figure.
13. Define **system noise temperature.**
14. Determine the minimum required received power at a receiver as a function of the system noise and required **signal-to-noise ratio.**
15. Discuss **ground wave propagation** and explain the difference between the **surface wave** and the **space wave.**
16. Describe the effect of **direct** and **indirect space waves** on the **line-of-sight** propagation paths.
17. Define the **ionosphere** and describe its role in **sky wave** propagation paths.
18. Discuss the various layers of the ionosphere.
19. Define the **maximum usable frequency** (**MUF**) and the **lowest usable frequency** (**LUF**) for a propagation path "bounced" off the ionosphere.

10–1 ■ INTRODUCTION

The transmission of electromagnetic energy can be accomplished by transmission lines, as discussed in Chapter 7, by waveguides, as discussed in Chapter 8, or by propagation through free space, to be presented in this chapter. Antennas, which provide the transition from transmission lines or waveguides to free space, were discussed in Chapter 9. The concepts of electric and magnetic fields were introduced in Chapter 6 along with Maxwell's equations. The concept of an electromagnetic plane wave propagating in a lossless medium, such as free space, was introduced, and the Poynting vector was defined.

This chapter will consider the propagation of plane waves through media that can be defined by their permittivity ϵ, permeability μ, and conductivity σ. The electric and magnetic fields are transverse to the direction of propagation. The first part of the chapter will examine the behavior of the plane wave propagation at the boundary between two media. At the boundary, some of the electromagnetic energy will be reflected, while the remaining portion will be transmitted into the second medium. The directions of propagation for both the reflected and the transmitted wave are altered at the boundary.

The second part of this chapter will address the Friis transmission formula. This formula is most important in establishing a relationship between transmitted power, received power, antenna gain, and distance between the transmitter and receiver. Also, definitions of terms commonly employed in communications relating to free-space propagation will be introduced.

The final portion of this chapter will present propagation of electromagnetic signals by both the ground wave and the sky wave. The sky wave signals are electromagnetic signals that are "bounced" off the ionosphere. The ground wave includes the surface wave, the direct line-of-sight wave, and the ground-reflected line-of-sight wave.

10–2 ■ REFLECTION AND REFRACTION

When an electromagnetic wave encounters a boundary between two media, some of the energy is reflected by the boundary while the remaining portion is transmitted into the second medium. This situation is similar to the transmission line case where the traveling waves on the transmission line arrive at an impedance mismatch: A portion of the wave is reflected by the mismatch, and the remaining portion is transmitted past the impedance mismatch.

The direction of propagation for the plane wave can approach the boundary at any angle. The direction of propagation of the reflected plane wave is a function of the angle of the incident wave θ_i and the smoothness of the surface of the boundary between the two media in terms of the wavelength of the electromagnetic propagation.

The reflection from a smooth surface is illustrated in Fig. 10–1. An electromagnetic wave $\mathcal{P}_i$ is propagating in free space (medium 1) and encounters a

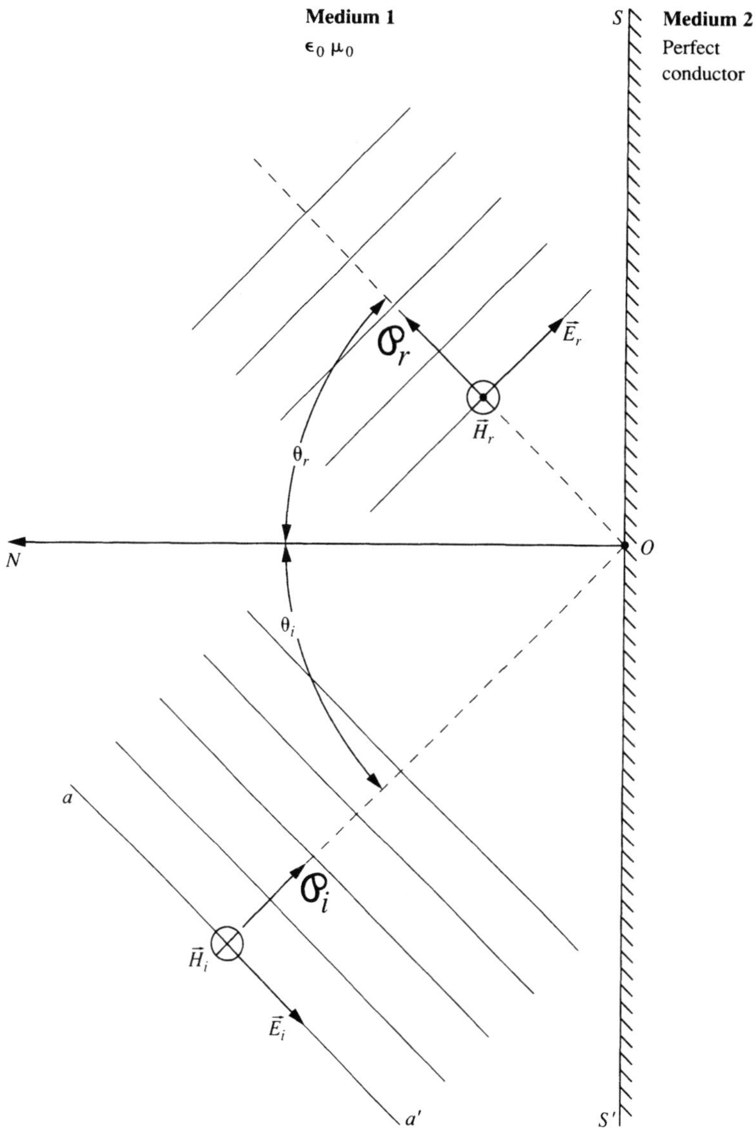

FIGURE 10–1 Reflection at boundary (specular surface).

perfect conductor boundary (medium 2). The plane wave, which contains the electric and magnetic fields, is represented by the line aa'. The smooth surface acts like a mirror, and the angle of reflection θ_r will equal the angle of incidence θ_i. If medium 2 had been a dielectric, some of the energy in $\mathcal{P}_i$ would have been transmitted into medium 2, and the magnitude of $\mathcal{P}_r$ would have been reduced. The angle of reflection θ_r would remain the same for a smooth boundary surface.

A more convenient method to describe the effects of a plane wave traveling between two media in cases where the physical dimensions of the media are large compared to a wavelength is to employ geometric optics. Rays perpendicular to the wavefronts are used to represent the plane wave propagation of electromagnetic energy. This technique is used extensively at optical frequencies to describe the propagation through different media. The angles of incidence, reflection, and transmission are measured from the normal (perpendicular) to the boundary surface and the direction of the ray or Poynting vector. This concept is illustrated in Figs. 10–1 and 10–2. In Fig. 10–1, line *ON* is the normal to the surface *SS'*, and the angles θ_i and θ_r are measured between the Poynting vectors $\mathcal{P}_i$ and $\mathcal{P}_r$ and this line.

Figure 10–2 uses geometric optics to describe the angle of propagation for the transmitted wave θ_t in medium 2 as a function of the angle of incidence θ_i in medium 1 and the characteristics of both media. The permittivity and permeability of medium 1 are ϵ_1 and μ_1; and for medium 2, ϵ_2 and μ_2. The velocity of propagation in media 1 and 2 can be determined from (1–27):

$$\nu_1 = \frac{1}{\sqrt{\epsilon_1\mu_1}} \quad \text{and} \quad \nu_2 = \frac{1}{\sqrt{\epsilon_2\mu_2}} \tag{10–1}$$

The wave traveling in medium 1 is incident upon the boundary between the two media at point *O*. The angle of incidence is θ_i. Referring to triangle $XX'Y$ in medium 1, the angle XYX' is equal to the angle of incidence θ_i. The

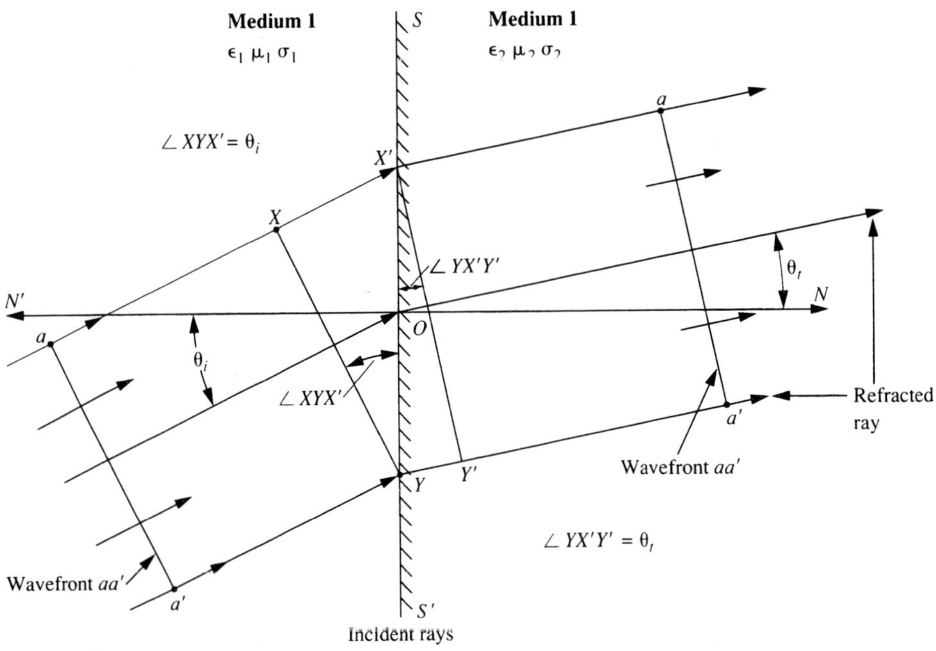

FIGURE 10–2 Refracted waves using geometric optics.

angle of transmission θ_t in medium 2 is equal to the angle $YX'Y'$ in triangle $YX'Y'$. Since the sine of angle XYX' is XX'/YX' and the sine of angle $YX'Y'$ is YY'/YX', then

$$\frac{\sin\theta_i}{\sin\theta_t} = \frac{XX'/YX'}{YY'/YX'} = \frac{XX'}{YY'} \tag{10–2}$$

The phase front of the plane wave aa' travels through a distance XX' in medium 1 in time t. During the same time t, the wave aa' travels through distance YY' in medium 2. The fact that XX' is greater than YY' indicates that the velocity of propagation in medium 2 is slower than in medium 1. Since distance is velocity times time, then $XX' = \nu_1 t$ and $YY' = \nu_2 t$. Substituting into (10–2) yields

$$\frac{\sin\theta_i}{\sin\theta_t} = \frac{XX'}{YY'} = \frac{\nu_1 t}{\nu_2 t} = \frac{\nu_1}{\nu_2} \tag{10–3}$$

Substituting (10–1) into (10–3), we have

$$\frac{\sin\theta_i}{\sin\theta_t} = \frac{\nu_1}{\nu_2} = \frac{\dfrac{1}{\sqrt{\epsilon_1\mu_1}}}{\dfrac{1}{\sqrt{\epsilon_2\mu_2}}} = \sqrt{\frac{\epsilon_2\mu_2}{\epsilon_1\mu_1}} \tag{10–4}$$

Equation (10–4) is one form of **Snell's law.**

The **index of refraction** n in a medium is defined as the ratio of the velocity of propagation in a vacuum to the velocity of propagation in the medium. Substituting (10–1), the index of refraction can be related to the permittivity and permeability of the material:

$$n_1 = \frac{c}{\nu_1} = \frac{\dfrac{1}{\sqrt{\epsilon_0\mu_0}}}{\dfrac{1}{\sqrt{\epsilon_1\mu_1}}} = \frac{\sqrt{\epsilon_1\mu_1}}{\sqrt{\epsilon_0\mu_0}} = \sqrt{\frac{\epsilon_{r_1}\epsilon_0\mu_{r_1}\mu_0}{\epsilon_0\mu_0}} = \sqrt{\epsilon_{r_1}\mu_{r_1}} \tag{10–5}$$

where

n_1 = index of refraction in medium 1

ϵ_{r_1} = dielectric constant of medium 1

μ_{r_1} = relative permeability of medium 1

Since $\mu_{r_1} = 1$ in all media in which propagation will be considered in this textbook, (10–5) becomes

$$n_1 = \sqrt{\epsilon_{r_1}} \tag{10–6}$$

Thus, Snell's law [Eq. (10–4)] can be expressed as

$$\frac{\sin\theta_i}{\sin\theta_t} = \frac{n_2}{n_1} = \sqrt{\frac{\epsilon_{r_2}}{\epsilon_{r_1}}} \tag{10–7}$$

Referring to (10–7), the angle of transmission θ_t is less than the angle of incidence if the dielectric constant in medium 2 is greater than the dielectric constant in medium 1; or $\theta_t < \theta_i$ if $\epsilon_{r_2} > \epsilon_{r_1}$, and $\theta_t > \theta_i$ if $\epsilon_{r_2} < \epsilon_{r_1}$.

Since the angle of transmission θ_t can be greater than the angle of incidence θ_i, then θ_t can be 90°. In this case, no energy is propagated into medium 2, and all energy either is reflected or propagates only on the surface of the boundary between the two media. If $\epsilon_{r_1} > \epsilon_{r_2}$, there is a critical angle of incidence θ_{i_c} for which the angle of transmission θ_t would be 90°. Substituting $\sin\theta_t = \sin 90° = 1$ into (10–7) yields

$$\theta_{i_c} = \sin^{-1}\sqrt{\frac{\epsilon_{r_2}}{\epsilon_{r_1}}} = \sin^{-1}\frac{n_2}{n_1} \qquad \textbf{(10–8)}$$

The concept of the critical angle in which a ray is incident upon a boundary between two dielectric media at an angle equal to or greater than the critical angle is the essential basis that allows the concept of fiber optics to propagate optical wavelength electromagnetic signals (light waves). This concept will be discussed in much greater detail in Chapter 11.

EXAMPLE 10–1

Polystyrene has a dielectric constant of 2.6 and a permeability $\mu_r = 1$. Determine the index of refraction.

Solution

Substituting the dielectric constant into (10–6), we find that

$$n_1 = \sqrt{\epsilon_{r_1}} = \sqrt{2.6} = 1.6125 \qquad \textbf{(10–9)}$$

EXAMPLE 10–2

The silicon glass used in optical fiber transmission lines has an index of refraction n of 1.46. Determine the dielectric constant of this medium and the velocity of propagation of the electromagnetic wave traveling in the optical fiber.

Solution

Rearranging (10–6) yields

$$\epsilon_r = (\sqrt{\epsilon_r})^2 = n^2 = (1.46)^2 = 2.13 \qquad \textbf{(10–10)}$$

The velocity of propagation can be determined from (1–31):

$$v = \frac{c}{\sqrt{\epsilon_r}} = \frac{c}{n} = \frac{3 \times 10^8}{1.46} = 2.055 \times 10^8 \text{ m/s} \qquad \textbf{(10–11)}$$

EXAMPLE 10–3

A plane wave is incident upon the boundary between two media at an angle of incidence θ_i of 45° (see Fig. 10–2). If the first medium is free space while the second has a dielectric constant of 2.6, determine the angle of transmission θ_t in medium 2.

Solution

From Snell's law as given by (10–7), we have

$$\frac{\sin\theta_i}{\sin\theta_t} = \sqrt{\frac{\epsilon_{r_2}}{\epsilon_{r_1}}} = \sqrt{\frac{2.6}{1}} = 1.612 \qquad \textbf{(10–12a)}$$

Solving for $\sin\theta_t$ yields

$$\sin\theta_t = \frac{\sin\theta_i}{1.612} = \frac{\sin 45°}{1.612} = \frac{0.7071}{1.612} = 0.4387 \qquad \textbf{(10–12b)}$$

and

$$\theta_t = \sin^{-1}(0.4387) = 26.02° \qquad \textbf{(10–12c)}$$

EXAMPLE 10–4

In Example 10–3, if medium 1 has a dielectric constant of 2.6 and medium 2 is free space, determine the angle of transmission θ_t in medium 2 if the angle of incidence θ_i is 20°.

Solution

From Snell's law as given by (10–7), we have

$$\frac{\sin\theta_i}{\sin\theta_t} = \sqrt{\frac{\epsilon_{r_2}}{\epsilon_{r_1}}} = \sqrt{\frac{1}{2.6}} = 0.6202 \qquad \textbf{(10–13a)}$$

Solving for $\sin\theta_t$ yields

$$\sin\theta_t = \frac{\sin\theta_i}{0.6202} = \frac{\sin 20°}{0.6202} = \frac{0.3420}{0.6202} = 0.5515 \qquad \textbf{(10–13b)}$$

and

$$\theta_t = \sin^{-1}(0.5515) = 33.47° \qquad \textbf{(10–13c)}$$

EXAMPLE 10–5

Determine the critical incidence angle in medium 1 of Example 10–4.

Solution

The critical angle of incidence can be determined from (10–8) as

$$\theta_{i_c} = \sin^{-1}\sqrt{\frac{\epsilon_{r_2}}{\epsilon_{r_1}}} = \sin^{-1}\sqrt{\frac{1.0}{2.6}} = 38.33° \qquad \textbf{(10–14)}$$

Therefore, if the angle of incidence of the electromagnetic wave in medium 1 is 38.33°, the wave will propagate along the boundary between the two media. If it is greater than 38.33°, the wave will not propagate in medium 2 but will be completely reflected back into medium 1.

10–3 ■ FRIIS TRANSMISSION FORMULA

In this section, a development of the Friis transmission formula will be presented. This formula is most important in establishing a relationship between transmitted power, received power, antenna gain, and distance between transmitter and receiver. Its use permits a trade-off between various system link parameters in establishing proper operating levels for a complete communication system.

The Friis transmission formula applies to a system in which there is a transmitting antenna, a receiving antenna, and a direct electromagnetic path without obstacles between the two antennas, as illustrated in Fig. 10–3. It does not apply to the various types of indirect propagations such as sky wave "bounce" off the ionosphere. These systems will be discussed briefly in later sections of this chapter. Such indirect propagation techniques have depended on empirical data to determine proper operating conditions, and such results are not always reliable.

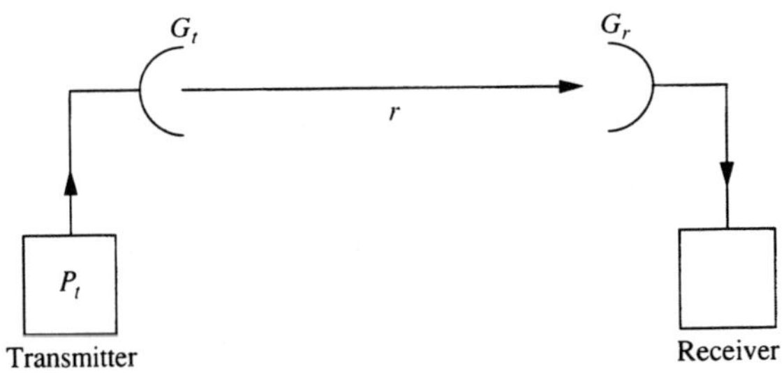

FIGURE 10–3 Direct ray model used in developing the Friis transmission formula.

The development of the several steps leading to the Friis formula will contribute to its understanding. Assume that an isotropic point radiator is transmitting a power P_t in watts. Consider an imaginary spherical surface of radius r surrounding the radiator, and assume that the transmitter is at the center of the sphere. Let $\mathcal{P}$ represent the power density on the surface of the sphere at any point. Since the power radiates equally well in all directions, it divides uniformly across the imaginary spherical surface and is

$$\mathcal{P} = \frac{P_t}{4\pi r^2} \tag{10–15}$$

An actual transmitting antenna having a gain of G_t with respect to an isotropic antenna would produce a power density $\mathcal{P}'$ in the direction of maximum radiation intensity. This power density is given by

$$\mathcal{P}' = G_t\,\mathcal{P} = \frac{G_t P_t}{4\pi r^2} \tag{10–16}$$

It is the power density at the receiving antenna as shown in Fig. 10–3 from a transmitter whose power is P_t with a transmitting antenna gain of G_t.

The product of the actual transmitter power and the transmitting antenna gain is defined as the **effective isotropic radiated power** or **EIRP.** It is the power that a transmitter would have to radiate if the transmitting antenna were an isotropic radiator in order to achieve the same power density at the receiving antenna; i.e.,

$$\text{EIRP} = P_t G_t \tag{10–17}$$

The receiving antenna has a capture area which was discussed in Section 9–6. The receiving antenna will "capture" a power P_r given by

$$P_r = A_e \mathcal{P}' = \frac{A_e G_t P_t}{4\pi r^2} \tag{10–18}$$

where the effective or capture area A_e was defined by (9–25). Substituting (9–25) into (10–18) yields

$$P_r = A_e \mathcal{P}' = \left[\frac{\lambda^2}{4\pi} G_r\right]\left[\frac{G_t P_t}{4\pi r^2}\right] = \frac{\lambda^2 G_t G_r P_t}{(4\pi)^2 r^2} \tag{10–19}$$

where G_r is the gain of the receiving antenna. The result of (10–19) is the **Friis transmission formula** for one-way direct ray propagation from transmitter to receiver. It is also referred to as the **one-way link equation.**

Several deductions can be made from the Friis transmission formula. The power at the receiver is directly proportional to the product of the transmitted power, the transmitting antenna power gain, and the receiver antenna power gain. Within limits, these three parameters may be adjusted without affecting performance, provided that the product remains the same. For example, a transmitter with a power output of 2 W having an antenna with a gain of 100 would produce the same received power (in the direction of maximum radiation intensity) as a transmitter with a power output of 50 W and a gain of 4. As a second example, if a receiving antenna were replaced by one having half the previous gain, the original received power could be restored by either doubling the transmitter output power or replacing the transmitting antenna with one having twice the gain.

The denominator of (10–19) indicates that the received power varies inversely with the square of the distance r. Thus, each time the distance from the transmitted antenna is doubled, the received power is reduced by one-fourth.

The decibel variation as a result of the $1/r^2$ law is interesting. It can be readily shown that if the distance between the two antennas is doubled, the received power is reduced by about 6 dB. If the distance between the two antennas is increased by a factor of 10, the received power is reduced by exactly 20 dB.

The variation of received power with respect to wavelength λ is more subtle than the other quantities and warrants a special discussion before erroneous impressions are formed. If a direct inspection of (10–19) is made without additional insight, one might deduce that the received power is directly proportional to the square of wavelength, which would suggest that the received power would increase significantly at lower frequencies. This deduction would be correct *if* all the other parameters in the equation could remain constant as the wavelength increased. The fallacy in the argument is that G_t and G_r relate to the gains of given antennas at a specific operating frequency. As wavelength increases and frequency decreases, it is necessary that antennas specifically designed for the new frequencies be employed to provide the same values of antenna gains G_t and G_r and optimum conditions for transmission. The gain of a fixed-aperture antenna, such as a parabolic reflector antenna, from (9–36) is

$$G_i = \eta_I\left(\frac{\pi D}{\lambda}\right)^2 \tag{10–20}$$

As can be seen by studying (10–20), the gain is directly proportional to the square of the ratio of the physical diameter to the wavelength. Therefore, the sizes of both the transmitting and receiving antennas increase as the square of the wavelength to maintain the same gain for each antenna.

The Friis transmission equation (10–19) can be rearranged as

$$\frac{P_t}{P_r} = \left(\frac{1}{G_tG_r}\right)\left[\left(\frac{4\pi r}{\lambda}\right)^2\right] \tag{10–21}$$

The ratio of the transmitted power P_t to the received power P_r between isotropic transmitting and receiving antennas located in free space is defined as the **free-space path loss** L_p. Letting $G_t = 1$ and $G_r = 1$ in (10–21), we obtain

$$L_p = \frac{P_t}{P_r} = \left(\frac{4\pi r}{\lambda}\right)^2 = \left(\frac{4\pi fr}{c}\right)^2 \tag{10–22}$$

where the last term was obtained by substituting $\lambda = c/f$.

The free-space path loss increases as the square of the distance between the transmitting and receiving antennas. Also, the free-space path loss increases as the square of the frequency; however, this can be misleading. To obtain a better understanding of the Friis transmission equation and its implications, we will consider three different antenna situations.

Case I: Isotropic Transmitting and Receiving Antennas

There is a trade-off between gain and directivity. That is, as the gain of an antenna is increased, the beamwidth of the antenna decreases; and therefore more stringent pointing of the antenna is required. There are communication situations in which no exact orientation of the antennas can be assumed, and the gain must be sacrificed to allow uniform propagation in all directions. An example of this situation is communication between two aircraft, either of which can assume any random position relative to the other.

From Eqs. (10–21) and (10–22), with $G_t = G_r = 1$, the loss between transmitter and receiver increases as the square of the frequency. This indicates that less power is required at lower frequencies, with the only limitation being the physical size of the isotropic antennas.

Case II: One Antenna Isotropic, the Other with Fixed Physical Aperture Size

This is a fairly common situation where one antenna can have a narrow beamwidth, since it is at a fixed location, and the position of the other antenna is known. However, the orientation of the other antenna can assume any position. An example is a ground-based antenna tracking an aircraft or satellite.

Substituting $G_r = 1$ and (9–36) into (10–19) yields

$$P_r = \frac{\lambda^2 G_t G_r P_t}{(4\pi)^2 r^2} = \frac{\lambda^2\left[\eta_I\left(\frac{\pi D}{\lambda}\right)^2\right](1)P_t}{(4\pi)^2 r^2} = \frac{\eta_I D^2 P_t}{16 r^2} \tag{10–23}$$

Referring to (10–23), the received power P_r for Case II is independent of frequency and is a function only of the diameter D of the fixed-physical-aperture antenna, the distance between the transmitter and receiver r, and the transmitter power P_t.

Case II can also exist for the situation where the transmitting antenna must be isotropic while the receiving antenna has a fixed physical aperture. Let $G_t = 1$ and rearrange (9–25) to relate the receiving antenna gain G_r to the effective area of the antenna. Substituting these into (10–19) results in

$$P_r = \frac{\lambda^2 G_t G_r P_t}{(4\pi)^2 r^2} = \frac{\lambda^2(1)\left[\frac{4\pi A_e}{\lambda^2}\right]P_t}{(4\pi)^2 r^2} = \frac{A_e P_t}{4\pi r^2} \tag{10–24}$$

Since the effective area A_e of a fixed-aperture-area antenna is equal to the illumination efficiency η_I times the physical area A_p, Equation (10–24) becomes

$$P_r = \frac{A_e P_t}{4\pi r^2} = \frac{\eta_I A_p P_t}{4\pi r^2} = \frac{\eta_I D^2 P_t}{16 r^2} \tag{10–25}$$

Again, Equation (10–25), which is the same as (10–23), illustrates that for Case II, the received power is independent of frequency.

Case III: Both Antennas with Fixed Aperture Area

Case III is one in which both transmitting and receiving locations are fixed in both orientation and location. An example is a microwave radio communication system used for telephone communication and satellite communication systems.

Substituting (9–36) into (10–19) yields

$$P_r = \frac{\lambda^2 G_t G_r P_t}{(4\pi r)^2} = \frac{\lambda^2\left[\eta_I\left(\frac{\pi D}{\lambda}\right)^2\right]^2 P_t}{(4\pi r)^2} = \frac{\eta_I^2 \pi^2 D^4 P_t}{16\lambda^2 r^2} \tag{10–26}$$

The received power increases as the square of the decrease in wavelength for Case III. Therefore, the higher the frequency, the less transmitter power is required to maintain the same received power level. As a result, satellite communication systems are designed for higher operating frequencies.

There are several problems associated with employing higher frequencies in the design of Case III communication systems. First, the beamwidth becomes narrower as the wavelength is decreased, and the accuracy required in pointing is increased. Second, losses in the atmosphere due to water vapor absorption, oxygen absorption, and rain attenuation can present sufficient problems in certain microwave-frequency regions. The attenuation due to water vapor and oxygen in dB/km for the microwave frequency region from 10 to 300 GHz is illustrated in Fig. 10–4.

EXAMPLE 10–6

The transmitter power is 25 kW, and the transmitting antenna gain is 23 dBi. Determine the EIRP.

Solution

The antenna gain must be expressed as a factor instead of in decibels referenced to an isotropic antenna whose gain is 1 or 0 dBi. Therefore, the antenna gain is

$$G_t = 10^{23/10} = 199.53 \qquad \textbf{(10–27)}$$

The EIRP is determined from (10–17):

$$\text{EIRP} = P_t G_t = (25 \times 10^3)(199.53) = 4.988 \text{ MW} \qquad \textbf{(10–28)}$$

EXAMPLE 10–7

A satellite communication system is to be used to transmit a video signal from the satellite to an earth station. The frequency is 4.0 GHz, the transmitter power output is 20 W, the satellite transmitting antenna has a gain of 17 dBi, and the earth station has a receiving gain of 50 dBi. If the distance from the satellite to the earth station is 36,000 km, determine the power received at the earth station using the Friis transmission formula.

Solution

The transmitting and receiving antenna gains must be expressed as factors instead of in decibels relative to an isotropic antenna. Therefore,

$$G_t = 10^{17/10} = 50.12 \qquad \textbf{(10–29)}$$

$$G_r = 10^{50/10} = 1 \times 10^5 \qquad \textbf{(10–30)}$$

The wavelength is

$$\lambda = \frac{c}{f} = \frac{3 \times 10^8}{4.0 \times 10^9} = 7.5 \times 10^{-2} = 7.5 \text{ cm} \qquad \textbf{(10–31)}$$

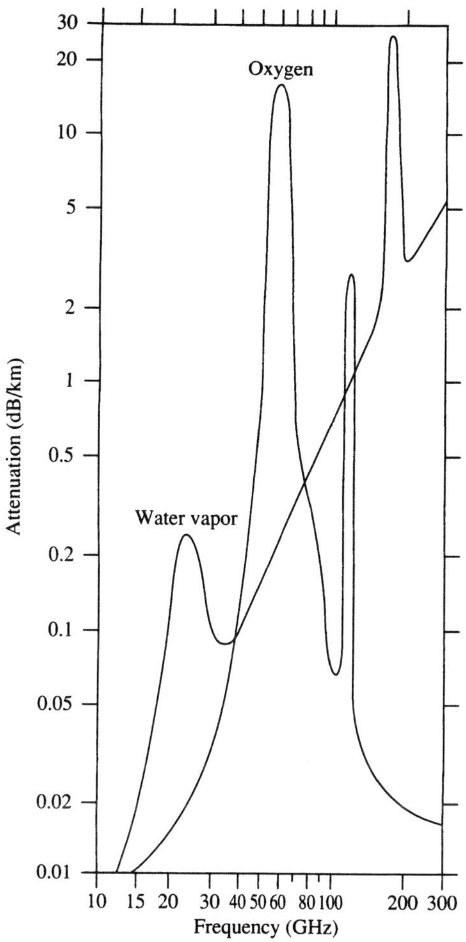

FIGURE 10–4 Atmospheric attenuation due to water vapor and oxygen. (W. D. Stanley, *Electronic Communications Systems,* Englewood Cliffs, NJ: Prentice-Hall, 1982, p. 522, Fig. 11–11. Used with permission.)

Substituting into the Friis transmission formula yields

$$P_r = \frac{\lambda^2 G_t G_r P_t}{(4\pi)^2 r^2} \tag{10–32a}$$

$$= \frac{(7.5 \times 10^{-2})^2 (50.12)(10^5)(20)}{(4\pi)^2 (3.6 \times 10^7)^2} = 2.755 \text{ pW} \tag{10–32b}$$

EXAMPLE 10–8

Determine the free-space path loss in decibels for the satellite-to-earth-station communication link in Example 10–7.

Solution

The free-space path loss is determined from (10–22):

$$L_p = \left(\frac{4\pi r}{\lambda}\right)^2 = \left(\frac{4\pi(3.6 \times 10^7)}{7.5 \times 10^{-2}}\right)^2 = 3.638 \times 10^{19} \quad (10\text{–}33)$$

The free-space path loss in decibels is

$$L_p = 10 \log(3.638 \times 10^{19}) = 195.61 \text{ dB} \quad (10\text{–}34)$$

10–4 ■ RECEIVER SENSITIVITY

The required level of received power at the receiving site in a communication link is a function of the receiver sensitivity, the system noise temperature, and the required **signal-to-noise ratio, S/N**. The **receiver sensitivity** is the received power required such that the received power level at the input of a receiver is equal to the noise power level in the receiver referenced back to the input. The noise level in a receiver, referenced to the input of the receiver, is

$$N_{\text{rec}} = kT_0BF \quad (10\text{–}35)$$

where

k = Boltzmann's constant = 1.38×10^{-23} W/K · Hz

T_0 = 290 K

B = receiver bandwidth

F = noise factor of receiver

The noise in a communication receiver is thermal noise generated in the active and passive devices throughout the receiver. However, it is more convenient, and the accepted industry practice, to assume an ideal, noise-free receiver with an external noise source at the input which produces the same noise output as the actual receiver. The **noise factor** of any device is defined as

$$F = \frac{\left(\dfrac{S}{N}\right)_{\text{in}}}{\left(\dfrac{S}{N}\right)_{\text{out}}} \quad (10\text{–}36)$$

Noise factor is typically expressed in decibels as a **noise figure** where

$$F_{\text{dB}} = 10 \log F \quad (10\text{–}37)$$

A more useful concept in expressing the internally generated noise is that of **effective noise temperature, T_e**. It is the temperature of the theoretical noise source at the input of the receiver which represents the internally generated receiver noise. The effective noise temperature is related to the noise figure by

$$T_e = T_0(F - 1) \tag{10–38}$$

where

T_0 = reference temperature of 290 K
F = noise figure expressed as a factor, *not* in dB

The overall noise power at the input of a communication receiver comes from three sources:

1. the internally generated receiver noise, T_e
2. externally generated, undesired, noiselike electromagnetic radiation T_{ext} received by the receiving antenna. Atmospheric noise, galactic noise, and man-made noise are typical of these types of undesired noise sources. The value of the externally generated noise is that referenced to the input of the receiver. Any ohmic transmission line losses would have reduced the level of this noise as compared to the level at the input to the antenna.
3. ohmic loss noise T_{ohm}. Any ohmic loss, such as that in the transmission line from the antenna to the receiver, which is at a physical temperature above absolute zero, 0 K, emits thermally generated electromagnetic noise. The ohmic loss in the transmission line also attenuates the desired received signal in addition to generating undesired noise power.

The sums of the three noise sources listed above are added together to obtain the receiver **system noise temperature, T_{sys}**:

$$T_{sys} = T_e + T_{ext} + T_{ohm} \tag{10–39}$$

The system noise level, N_{sys}, is

$$N_{sys} = kT_{sys}B \tag{10–40}$$

The signal-to-noise ratio required is determined by the type of modulation used in the communication link and the accuracy or quality that is required in the communication system. Therefore, the required received power level is

$$P_r = \left(\frac{S}{N}\right)N_{sys} \tag{10–41}$$

where

$\frac{S}{N}$ = signal-to-noise ratio

EXAMPLE 10–9

A 2.3-GHz communication receiver has a noise figure of 6 dB and a bandwidth of 10 MHz. Determine (a) the noise level in the receiver referred to the input, (b) the receiver sensitivity, and (c) the effective noise temperature of the receiver.

Solution

(a) The noise factor for a 6-dB noise figure can be determined from (10–37):

$$F = 10^{F_{dB}/10} = 10^{6/10} = 3.98 \qquad \textbf{(10–42)}$$

The noise level in the receiver referred to the input is determined by (10–35):

$$N_{rec} = kT_0BF = (1.38 \times 10^{-23})(290)(10 \times 10^6)(3.98) \qquad \textbf{(10–43a)}$$
$$= 1.593 \times 10^{-13}\,\text{W} = 159.3\,\text{fW} \qquad \textbf{(10–43b)}$$

(b) The receiver sensitivity is equal to the noise level in the receiver referred to the input, or

$$S = N_{rec} = 159.3\,\text{fW} \qquad \textbf{(10–44)}$$

(c) The effective noise temperature of this receiver is determined from (10–38):

$$T_e = T_0(F - 1) = (290)(3.98 - 1) = 864.20\,\text{K} \qquad \textbf{(10–45)}$$

EXAMPLE 10–10

The 2.3-GHz communication receiver in Example 10–9 is being used in a system where there are 50 K of external noise and 200 K of noise from the transmission line and antenna ohmic losses. The communication system requires a signal-to-noise ratio of +15 dB. Determine the required received signal power level at the input to the communication receiver.

Solution

The system noise temperature T_{sys} is the sum of the effective noise temperature of the receiver T_e, the external noise T_{ext}, and the ohmic loss noise T_{ohm}; and it is determined from (10–39):

$$T_{sys} = T_e + T_{ext} + T_{ohm} = 864.20 + 50 + 200 = 1114.20\,\text{K} \qquad \textbf{(10–46)}$$

The system noise level referred to the input is determined from (10–40):

$$N_{sys} = kT_{sys}B = (1.38 \times 10^{-23})(1114.20)(10 \times 10^6) \qquad \textbf{(10–47a)}$$
$$= 153.8\,\text{fW} \qquad \textbf{(10–47b)}$$

The required received power for a signal-to-noise ratio of +15 dB is determined from (10–41). However, the signal-to-noise ratio must be converted from a decibel value to a numeric factor prior to substitution into (10–41). Thus,

$$\frac{S}{N} = 10^{(S/N_{dB})/10} = 10^{15/10} = 10^{1.5} = 31.62 \qquad \textbf{(10–48)}$$

$$P_r = \left(\frac{S}{N}\right)N_{sys} = (31.62)(1.538 \times 10^{-13}) = 4.863\text{ pW} \quad \textbf{(10–49)}$$

10–5 ■ GROUND WAVE PROPAGATION

The **ground wave** is commonly referred to as the electromagnetic propagation that travels from a transmitting antenna to a receiving antenna along or near the surface of the earth without leaving the lower portion of the earth's atmosphere. The ground wave can be divided into a **surface wave,** which travels in contact with the earth, and a **space wave,** which travels in the lower portion of the earth's atmosphere near the surface. The space wave consists of two components: the **direct wave** and the **indirect** or **ground-reflected wave.** Figure 10–5 illustrates the two space waves.

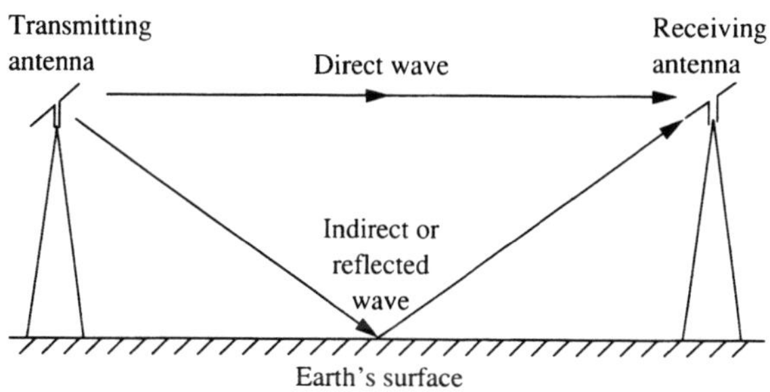

FIGURE 10–5 Space wave portion of the ground wave.

The surface wave is used primarily for vertical polarization propagation below 3 MHz. The horizontally polarized signal is quickly attenuated since the electric field vector is parallel to the earth's surface. The vertical polarized surface wave is the propagation mode for the standard AM broadcast band. It can provide coverage up to about 100 miles in the AM band during the daytime. At 300 kHz, vertically polarized surface waves can provide coverage to about 500 miles over good conductivity earth and to 1000 miles over sea water. The use of the surface wave above 3 MHz is very limited due to its restricted coverage. At 30 MHz, the typical range is 10 miles or less.

The space wave is very poor for communications below 3 MHz. The reflected wave undergoes a 180° phase shift when it is reflected from the earth's surface. For antennas located near the earth's surface and operating at long wavelengths, the distance that the direct and indirect waves travel is essentially the same. Therefore, when arriving at the receiving site, they will be 180° out of phase and will cancel. The vertically polarized signal is reflected better than the horizontally polarized signal. Also, as the frequency is increased, the path difference in terms of a wavelength is increased, and the cancellation between the direct and indirect waves is less significant.

The direct wave **line-of-sight propagation** at frequencies above 30 MHz with horizontally polarized antennas is the preferred mode of propagation. The Friis transmission formula, which was discussed in detail in Section 10–3, can be used to determine the performance of these direct wave line-of-sight propagation links. Other factors that affect the direct wave line-of-sight propagation—such as multipath fading, terrain effects, tropospheric ducting, and atmospheric attenuation—must be taken into account. However, these topics are beyond the scope of this text.

10–6 ■ SKY WAVE PROPAGATION

The sky wave is radiated from a transmitting antenna and is directed into space away from the earth's surface. The antenna also radiates a ground wave toward the surface as discussed in the previous section. As the sky wave travels through the earth's atmosphere, it encounters a region of the atmosphere known as the **ionosphere.** The ionosphere consists of several layers of ionized gases that are located from 40 to 300 km above the earth's surface. The electromagnetic radiation is both refracted and attenuated by the ionosphere, depending on the frequency of the electromagnetic radiation, the radiation angle from the transmitting antenna, and the degree of ionization of the atmospheric gases at that level in the ionosphere. If the electromagnetic radiation is within a certain frequency range, typically 3 to 30 MHz, the signal will be refracted, i.e., bent back, toward the earth's surface and received by a receiving site on the earth. If the frequency is below a certain frequency, typically 3 MHz, it will be totally absorbed by the ionosphere. If the frequency is above a certain frequency, typically 30 MHz, it will pass through the ionosphere and will continue into outer space. It should be noted here that the ionosphere is between the earth's surface and the orbits of the communication satellites. Therefore, satellite communication systems must use frequencies sufficiently high enough to be unaffected by the passage through the ionosphere. These are typically in the microwave frequency region above 3 GHz.

The behavior of a typical frequency that can be refracted back to the earth's surface is illustrated in Fig. 10–6. Six different rays, numbered 1 through 6, are shown as a function of radiation angle. Rays 1, 2, and 3 are radiated at too high an angle and therefore are not bent sufficiently to return to the earth's surface. Ray 4 is at an angle, known as the **critical angle,** such that it will be returned to the surface. The skip distance is the distance along the earth's surface from the transmitting site to the point on the surface where ray 4 returns. Since ray 4 is at the critical angle, this is the closest distance, outside of the area covered by the ground wave, that a receiving station could receive propagation from the transmitter. The range between the ground wave and

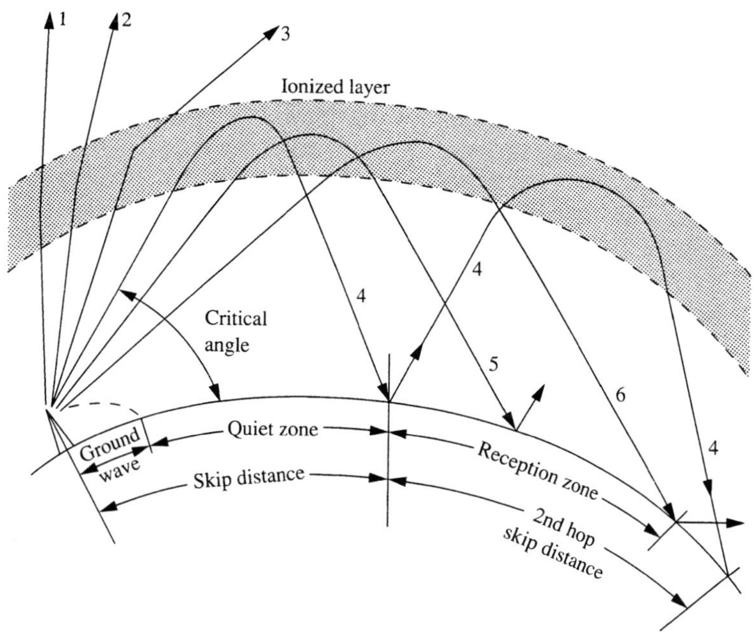

FIGURE 10–6 Ionospheric propagation.

the skip distance is the quiet zone, where no transmissions can be received. If ray 6 represents the lowest radiation angle that the transmitting angle can radiate a sky wave, then the distance from where ray 4 returns and ray 6 returns is the range over which reception of propagation from the transmitter can be received. This region is the reception zone. The maximum range for one hop of this type of propagation is typically 4000 km or 2500 miles.

However, greater ranges can be obtained since the propagation can be reflected from the earth's surface back toward the ionosphere and refracted back to the earth a second time, as illustrated in Fig. 10–6 by ray 4. Multiple hops up to 5 to 7 times have been achieved and result in propagations over very large ranges around the earth's surface.

The ionosphere is a region where the pressures are so low that the constituents of the atmosphere are ionized; i.e., a molecule loses an electron and becomes a positively charged ion. The region therefore consists of both free electrons and ions. With the low pressure, the density of molecules is such that it takes a longer time for recombination to occur. The major source of energy for the ionization process comes from solar ultraviolet radiation, although solar X rays and meteor radiation also play a role. The major ionization process begins right after sunrise, peaks around local noon, and decays after sunset.

During daytime, the ionization is in four distinct layers due to the ionization requirements of the different constituents of the atmosphere. The ionization peaks about at the midrange of each layer and tapers off both above and below the altitude of each layer's maximum ionization. The lowest ionized layer is the D layer which lies between 60 and 92 km (37 to 57 miles) above the surface. The next ionized layer is the E layer which lies between 100 and 115 km (62 to 71 miles) above the surface. The next ionized layer is the F layer which lies between 160 and more than 500 km (100 to over 310 miles) above the surface. Actually, during daytime, the F layer breaks into two distinct layers: the F1 layer and the F2 layer. After the sun sets, the ionization in the D and E layers ceases due to the fairly rapid recombination of electrons and ions, and these layers disappear. The F1 layer, which is lower and weaker than the F2 layer, also disappears at night, and the F2 layer drops in altitude during the night. The F2 layer is slow to recombine and therefore lasts throughout the night, reaching a minimum sometime after midnight.

The use of the ionosphere to "bounce" electromagnetic radiation off the ionosphere and back to the earth is a technique that requires a great deal of information on a large number of factors. These factors include (a) location on the earth of both the transmitting and receiving stations, (b) time of day, (c) season of the year, and (d) degree of solar activity.

The proper choice of the frequency of the electromagnetic radiation is important to the successful completion of a communication link. There is a frequency that is defined as the **maximum usable frequency (MUF).** If a variable-frequency transmitter is used to radiate an electromagnetic wave straight up (at a 90° angle), the highest frequency that is reflected back to the earth for that layer in the ionosphere is the **vertical incidence critical frequency.** The MUF is the vertical incidence critical frequency for that layer. The MUF is a variable that changes significantly as a function of the time of day, solar activity, propagation path distance, and propagation path between the transmitting and receiving stations. For example, the MUF between the eastern United States and Europe has varied from 7 MHz to 70 MHz because of these variables. Predictions for the MUF as a function of the propagation path, time of year, and time of day are published periodically by several agencies worldwide.

The degree of refraction in the ionosphere is a function of the frequency. Therefore, since the transmitting antenna typically radiates at a constant radiation angle relative to the surface, the skip distance is a function of frequency. The shorter the distance between the transmitter and receiver, the lower the MUF will be for that propagation path.

A second important frequency is the **lowest usable frequency (LUF).** If the frequency of the electromagnetic radiation is lowered from the MUF, the losses due to absorption in the ionosphere will increase as the frequency decreases. The frequency at which these losses cause the signal strength to fall below the background noise is the LUF. Therefore, it is desirable to operate as closely to the MUF as the available frequency allocation will allow. Since each type of communication system is restricted to operate in specified frequency-allocation bands, sometimes the frequency range from the LUF to the MUF is so narrow that there will be no allocated frequency band for successful communications.

PROBLEMS

10–1. Polyethylene has a dielectric constant of 2.3 and a permeability $\mu_r = 1$. Determine the index of refraction.

10–2. Teflon has a dielectric constant of 2.1 and a permeability $\mu_r = 1$. Determine the index of refraction.

10–3. A material has an index of refraction η of 1.8. Determine the dielectric constant of this medium and the velocity of propagation of the electromagnetic wave traveling in the optical fiber.

10–4. A material has an index of refraction η of 3.1. Determine the dielectric constant of this medium and the velocity of propagation of the electromagnetic wave traveling in the optical fiber.

10–5. A plane wave is incident upon the boundary between two media and is at an angle of incidence θ_i of 30° (see Fig. 10–2). If the first medium is free space while the second has a dielectric constant of 2.1, determine the angle of transmission θ_t in medium 2.

10–6. A plane wave is incident upon the boundary between two media and is at an angle of incidence θ_i of 60° (see Fig. 10–2). If the first medium is free space while the second has a dielectric constant of 9.9, determine the angle of transmission θ_t in medium 2.

10–7. In Problem 10–5, if medium 1 has a dielectric constant of 2.1 and medium 2 is free space, determine the angle of transmission θ_t in medium 2 if the angle of incidence θ_i is 30°.

10–8. In Problem 10–6, if medium 1 has a dielectric constant of 9.9 and medium 2 is free space, determine the angle of transmission θ_t in medium 2 if the angle of incidence θ_i is 15°.

10–9. Determine the critical angle of incidence in Problem 10–7.

10–10. Determine the critical angle of incidence in Problem 10–8.

10–11. The transmitter power is 5 kW, and the transmitting antenna gain is 33 dBi. Determine the EIRP.

10–12. The transmitter power is 1 kW, and the transmitting antenna gain is 17 dBi. Determine the EIRP.

10–13. A satellite communication system is to be used to transmit a video signal from the satellite to an earth station. The frequency is 12.0 GHz, the transmitter power output is 10 W, the satellite transmitting antenna has a gain of 13 dBi, and the earth station has a receiving gain of 45 dBi. If the distance from the satellite to the earth station is 42,000 km, use the Friis transmission formula to determine the power received at the earth station.

10–14. A satellite communication system is to be used to transmit a video signal from the satellite to an earth station. The frequency is 3.8 GHz, the transmitter power output is 25 W, the satellite transmitting antenna has a gain of 15 dBi, and the earth station has a receiving gain of 43 dBi. If the distance from the satellite to the earth station is 40,000 km, use the Friis transmission formula to determine the power received at the earth station.

10–15. Determine the free-space path loss in decibels for the satellite-to-earth-station communication link in Problem 10–13.

10–16. Determine the free-space path loss in decibels for the satellite-to-earth-station communication link in Problem 10–14.

10–17. A 4.0-GHz communication receiver has a noise figure of 3.5 dB and a bandwidth of 6 MHz. Determine the following:
(**a**) noise level in the receiver
(**b**) receiver sensitivity
(**c**) effective noise temperature of the receiver

10–18. A 12.0-GHz communication receiver has a noise figure of 5 dB and a bandwidth of 2.5 MHz. Determine the following:
(**a**) noise level in the receiver
(**b**) receiver sensitivity
(**c**) effective noise temperature of the receiver

10–19. The 4.0-GHz communication receiver in Problem 10–17 is being used in a system where there are 75 K of external noise and 150 K of noise from the transmission line and antenna ohmic losses. The communication system requires a signal-to-noise ratio of +12 dB. Determine the required received signal power level at the input to the communication receiver.

10–20. The 12.0-GHz communication receiver in Problem 10–18 is being used in a system where there are 100 K of external noise and 250 K of noise from the transmission line and antenna ohmic losses. The communication system requires a signal-to-noise ratio of +6 dB. Determine the required received signal power level at the input to the communication receiver.

11

Fiber Optics Technology

OBJECTIVES

After completing this chapter, the reader should be able to:

1. Describe briefly the historical background of the development of **fiber optics** technology.
2. Discuss the advantages and disadvantages of fiber optic communication systems.
3. Describe the electromagnetic spectrum from the audio-frequency region through the ultraviolet-frequency region.
4. Discuss the spectral region used for light-wave communications and its relationship to the radio-frequency and microwave-frequency regions of the electromagnetic spectrum.
5. Determine the wavelengths and frequencies used in the light wave communication–frequency region of the electromagnetic spectrum.
6. Describe the construction of a fiber optic transmission line and explain the **core** and **cladding** and their relative **indices of refraction.**
7. Describe how a light wave propagates through a fiber optic transmission line.
8. Describe the difference between **step-index** and **graded-index** fiber optic cables.
9. Describe the difference between **single-mode** and **multimode** fiber optic transmission lines.
10. Define and determine the **acceptance angle, numerical aperture,** and **relative refractive index** for a fiber optic transmission line.
11. Explain **dispersion** in a fiber optic transmission line.
12. Discuss the differences between **modal dispersion, chromatic dispersion, material dispersion,** and **waveguide dispersion.**
13. Describe the losses in fiber optic systems, including **absorption** and **scattering losses, macroscopic** and **microscopic bending losses,** and **coupling losses.**
14. Describe **light-emitting diode** and **laser diode** optical sources and their relative advantages and disadvantages.
15. Explain the difference between **spontaneous** and **stimulated emissions.**
16. Describe **PIN photodiode** and **avalanche photodiode** optical receiving devices.
17. Perform **optical link design** and **analysis.**

11–1 ■ INTRODUCTION

The field of fiber optic technology has made significant advances during the past 25 years in its capability as a transmission medium for the communication of information. At the time of this writing, fiber optic transmission lines are responsible for the majority of global telecommunications. Fiber optic cables are superior to copper wire cables, microwave relay systems, and satellite communication systems for the transmission of high-data-rate communications between two specific points. For example, in 1988, a 6500-km fiber optic transatlantic underseas cable was completed and put into operation. This cable is capable of carrying 80,000 simultaneous telephone conversations.

James Clerk Maxwell in 1873 published his equations and theories that proved that light was an electromagnetic wave the same as radio waves. Alexander Graham Bell in 1880 demonstrated voice transmission using light waves. He modulated a light beam with voice and transmitted it over a 200-m distance to an optical receiver, where it was converted back to sound waves. John Tyndall in 1870 demonstrated that light could be guided and bent around curves by a medium. He used a water stream to guide light over a curved path.

Researchers realized that light wave communications offered significant advantages due to its inherent capability for large bandwidth and small size; thus, significant effort was devoted to the development of a low-loss medium for the guiding of light waves along with the efficient generation and detection of light frequencies. Both communications and medical applications provided the emphasis for this development. In the 1950s, Harry Hopkins and Narinder Kapany invented the first flexible fiberscope to explore inside the human body for medical diagnostic purposes. Kapany developed in 1953 the first practical glass-coated fiber and invented the term **fiber optics.**

The early fiber optics were very lossy due to impurities in the material. Experiments to use fiber optics for communication purposes were started in England in 1964 by Kao and Hockman. Their optical waveguide had losses of approximately 1000 dB/km. Communication system designers believed that losses less than 20 dB/km were necessary for any practical system design. By 1970, Corning Glass had developed the first fibers with losses less than 20 dB/km. At the time of this writing, fiber technology has advanced to the point that typical fiber optic losses are between 0.2 and 2 dB/km, depending on the type of fiber optic cable and the frequency of operation.

There are significant advantages to fiber optic communication systems, including a much greater bandwidth capability than with RF and microwave systems. The greater bandwidths provide high-data-rate communication capability where required, such as the transatlantic cable capable of 80,000 simultaneous telephone conversations. The development in 1981 of low-loss, single-mode fiber optics and existing semiconductor laser sources made the large data rates possible.

Other advantages of fiber optics include its freedom from electrical interference, lack of crosstalk between adjacent fibers in the same cable, security, small size, and low cost per channel. Military applications of fiber optics have increased due to its immunity to electromagnetic pulses (EMP) and to its significant weight reduction in ship cabling systems.

There are few disadvantages, and most are the result of the newness of this technology. The fiber optic is not electrically conductive. Therefore, if electrical power is needed to operate repeaters, additional conductive wiring must be added to the fiber optic cable to carry the electrical power for the repeaters. Mechanical tolerances for assembly of the fiber optics, especially in splicing, are more complicated. Improper splicing and coupling to sources and detectors result in higher coupling losses.

This chapter is devoted primarily to the fiber optic transmission line. First there is a discussion of the optical-frequency spectrum and its relationship to the RF and microwave frequencies within the overall electromagnetic spectrum. The next section is a presentation of the fiber optic transmission line. Following that, a brief description of optical sources and optical detectors is presented. The final section is concerned with the design and link analysis of a fiber optic communication link.

11–2 ■ OPTICAL SPECTRUM

The electromagnetic spectrum from a frequency of 10^3 Hz to 10^{17} Hz is illustrated in Fig. 11–1. Its 14 decades of frequency span the range from the audio frequencies through the radio frequencies, microwave frequencies, infrared, visible light, and ultraviolet frequencies. The wavelengths of these electromagnetic radiation frequencies are also shown in Fig. 11–1.

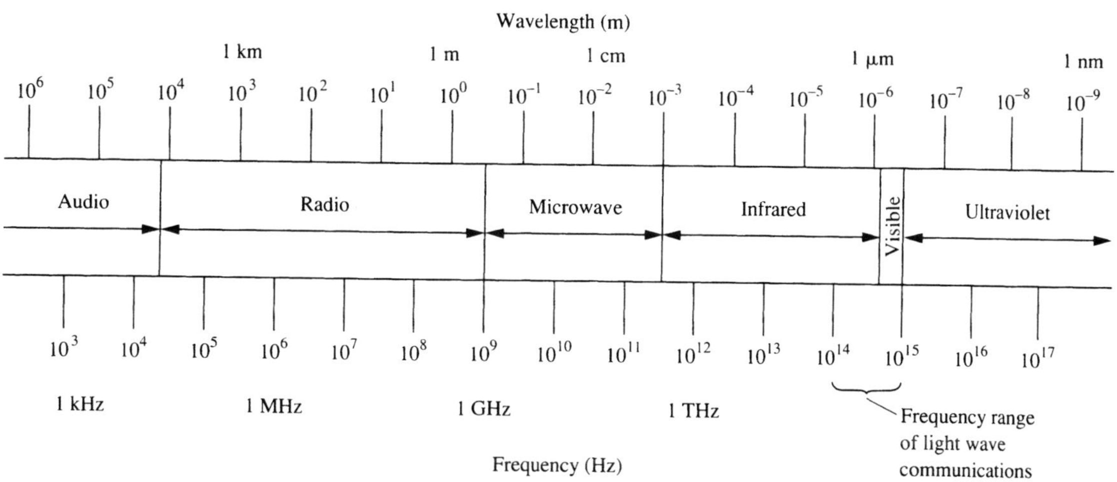

FIGURE 11–1 Electromagnetic spectrum.

As was discussed in previous chapters, the wavelength of an electromagnetic wave is related to the frequency by

$$f = \frac{\nu}{\lambda} \tag{11–1}$$

where

f = frequency of the electromagnetic radiation

ν = velocity of the electromagnetic radiation in the medium

λ = wavelength

Typically, the electromagnetic radiation is in free space, and therefore the velocity is that of free space, or 3×10^8 m/s. This velocity is used in relating the wavelength to frequency in Fig. 11–1. In the frequency range of interest for light wave communications, i.e., 10^{14} to 10^{15} Hz, wavelength rather than frequency is normally used to define the electromagnetic radiation. The wavelength is usually expressed in micrometers (μm), where 1 μm = 10^{-6}. (*Note:* A micrometer is sometimes expressed as a micron.)

The fiber optic optical spectrum is shown in Fig. 11–2. The visible light–frequency range is from 4×10^{14} to 7.5×10^{14} Hz, or 0.75 μm to 0.4 μm. The normal frequency range for light wave communications is from 1.88×10^{14} to 4.29×10^{14} Hz, or 1.6 μm to 0.7 μm. Thus, the frequency range for light wave communication is principally in the near infrared region and is only slightly into the red portion of the visible light region. The reason for this, which will be discussed in the next section, is a function of the region for minimum losses in the fiber optic transmission line.

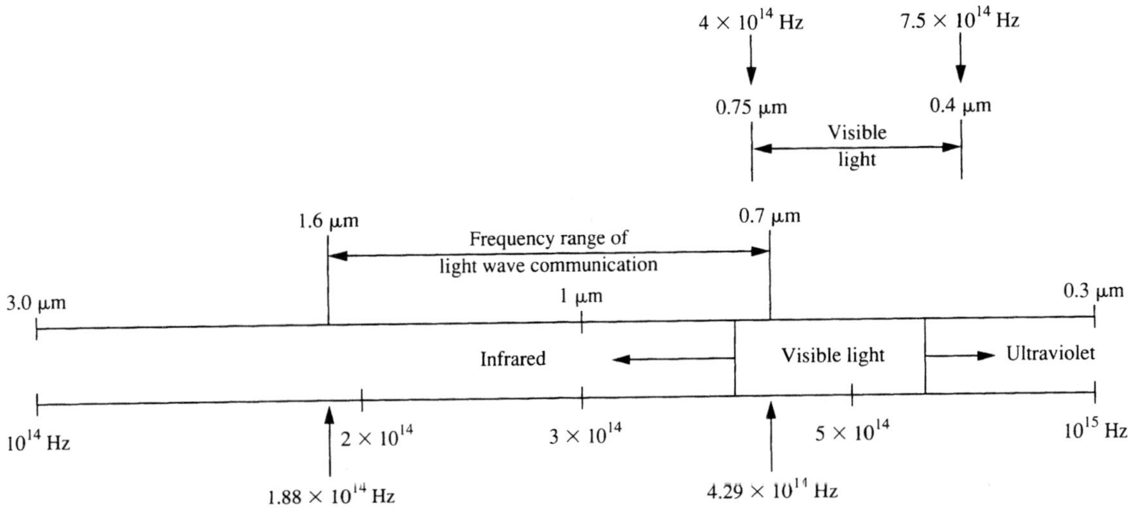

FIGURE 11–2 Fiber optic optical spectrum.

EXAMPLE 11-1

A semiconductor laser optical transmitter has a wavelength in free space of 0.84 μm. Determine the frequency of this electromagnetic radiation.

Solution

The wavelength and frequency are related by (11–1):

$$\lambda = \frac{\nu}{f} = \frac{3 \times 10^8}{0.84 \times 10^{-6}} = 3.57 \times 10^{14}\,\text{Hz} \qquad \textbf{(11–2)}$$

11-3 ■ FIBER OPTIC TRANSMISSION LINES

The basic construction of a fiber optic transmission line is illustrated in Fig. 11–3. The transmission line consists of a glass or plastic cylinder called the **core,** which is surrounded by a second concentric cylinder of either glass or plastic material called the **cladding.** The **index of refraction** n_1 of the core must be greater than the index of refraction of the cladding n_2. It will be shown later by Snell's law that if the light is coupled into the core at the proper angle, it will travel through the optical fiber within the core and not enter the cladding.

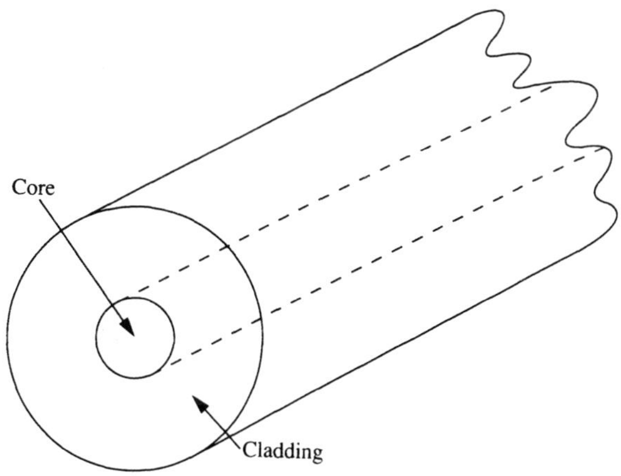

FIGURE 11–3 Basic fiber optic transmission line.

The diameter of the core varies anywhere from 8 to 200 μm, while the overall diameter of the cladding, including the core, varies from 125 to 240 μm. As a reference point, the typical diameter of a human hair is 100 μm. Therefore, the core of the smallest-diameter fiber is only 8% of that of a human hair. The core and cladding diameters have now been standardized throughout the world to enable compatibility between fiber optic components.

The index of refraction of the core, which must be greater than that of the cladding, can be designed to be a constant value throughout the core, or it can vary from a value equal to the cladding at the outer edge of the core to a maximum value at the center. Figure 11–4 illustrates these two types of cores. The core with a constant index of refraction is known as a **step-index** fiber, and the core with a variable index of refraction is known as a **graded-index** fiber.

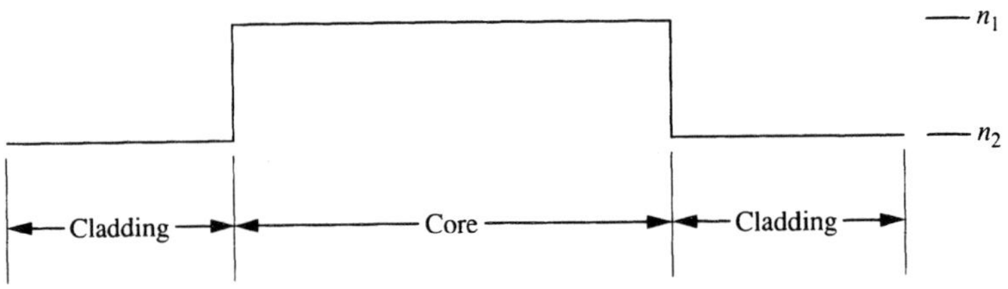

(a) Step-index fiber

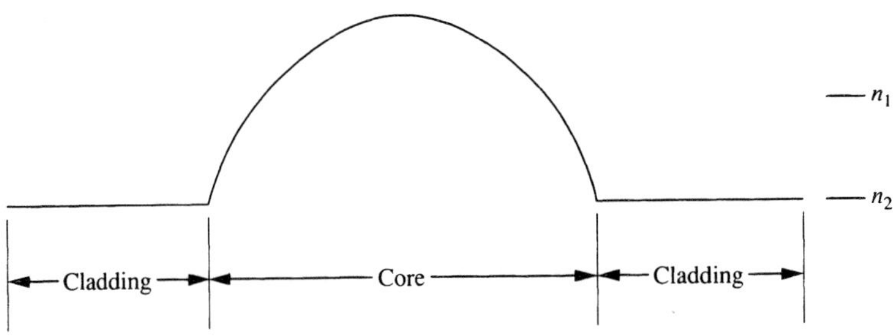

(b) Graded-index fiber

FIGURE 11–4 Index of refraction profiles in fiber optics.

A second form of classification of fiber optic transmission lines is based on how many propagation modes can be supported at the frequency of the light wave. If the diameter of the core is small enough, typically 8 to 10 μm, only one propagation mode can exist in the core. This type of fiber optic transmission line is called **single-mode.** If the core is large, 50 to 200 μm, many propagation modes can exist in the core, and the fiber optic transmission line is called **multimode.** Single-mode fiber optic systems are used for high-data-rate, long-range telecommunication networks, while multimode fiber optic systems are used for lower-data-rate, short-range systems such as computer local area networks (LAN). Single-mode fiber optic systems require laser sources and are much more expensive than multimode fiber optic systems because of the much smaller-diameter core.

The propagation of the light wave electromagnetic radiation through the fiber optic transmission lines can be explained using Snell's law, which was presented in Section 10–2. Optical rays, which point in the direction of power flow (i.e., they point in the direction of the Poynting vector), are used at optical frequencies to indicate the flow of light wave electromagnetic propagation.

Figure 11–5 presents two conditions for refraction of an optical ray incident upon a boundary between two dielectric media. Let n_1 and n_2 be the indices of refraction in medium 1 and medium 2, respectively. Angle θ_1 is the angle between the incident ray in medium 1 and the normal (perpendicular) to the boundary. Angle θ_2 is the angle between the refracted ray in medium 2 and the normal to the boundary. Snell's law states that

$$\frac{\sin \theta_1}{\sin \theta_2} = \frac{n_2}{n_1} \tag{11-3}$$

Figure 11–5(a) is for the case where the index of refraction of the second medium n_2 is greater than that of medium 1; i.e., $n_2 > n_1$. In this case, θ_1 is

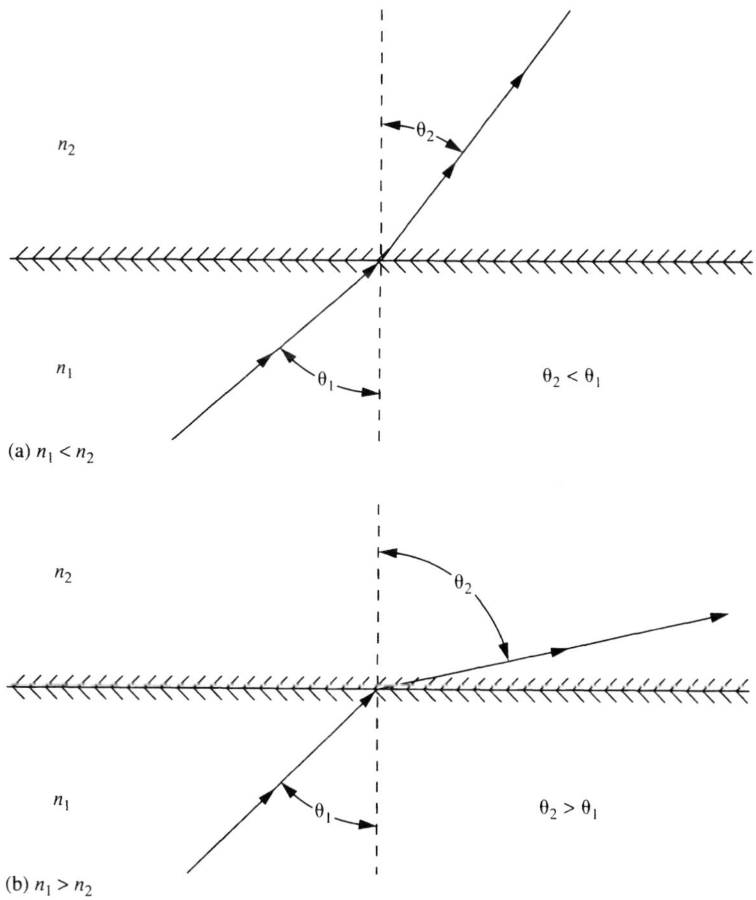

FIGURE 11–5 Snell's law.

greater than θ_2, and the refracted ray is bent away from the boundary and toward the normal to the boundary. Figure 11–5(b) is for the opposite case where $n_2 < n_1$. In this case, the refracted wave in medium 2 is bent toward the boundary and away from the normal. Therefore, in the second case, where the index of refraction in medium 1 is greater than the index of refraction in medium 2, there will be an angle θ_1 such that the angle in medium 2 is 90°. This angle θ_1 is defined as the critical angle:

$$\theta_{1_c} = \sin^{-1}\left(\frac{n_2}{n_1}\right) \tag{11–4}$$

Therefore, if the incident angle θ_1 is equal to or greater than the critical angle, there will be no ray propagated into medium 2, and the entire ray will be reflected back into medium 1. The reflected angle will be equal to the incident angle.

Propagation through the fiber optic core is illustrated in Fig. 11–6. If the ray is propagated into the core such that it is incident upon the boundary between the core and cladding at an angle less than the critical angle, some of the energy will be refracted into the cladding, and some will be reflected back into the core. Each time the ray is incident upon the boundary between

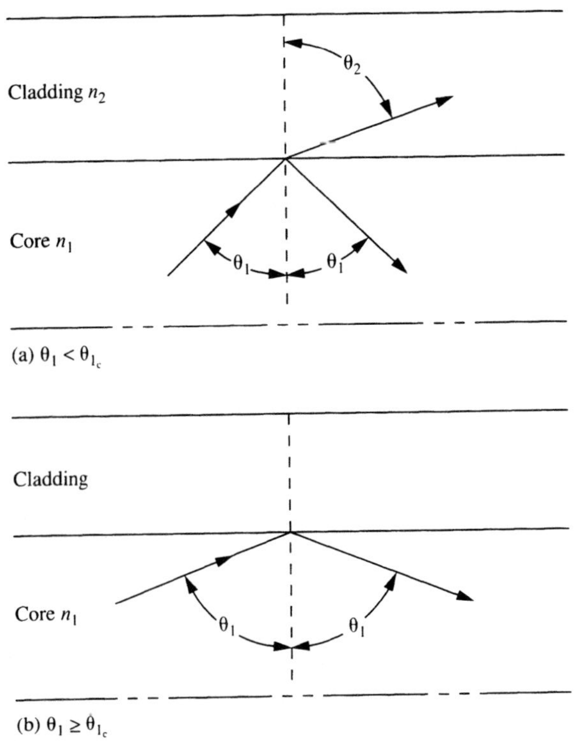

FIGURE 11–6 Propagation in fiber optic core.

the core and cladding, some of the energy will be lost into the cladding, as shown in Fig. 11–6(a). If the incident angle θ_1 is equal to or greater than the critical angle, the optical ray propagating through the core will be completely reflected at the boundary between the core and cladding, and no energy will be lost into the cladding.

The optical transmitting source, which is either a light-emitting diode or a semiconductor laser, must focus the light wave rays onto the core within the fiber optic cable such that the rays are incident upon the boundary between the core and cladding at the critical angle or greater. This situation is illustrated in Fig. 11–7. The optical source usually is in air where the index of refraction n_0 is equal to 1.0. When this ray is refracted into the core, it must be at an angle to the normal at the interface between air and the core of $90° - \theta_{1_c}$ or less to ensure total internal reflection which keeps the light wave propagating within the core. To achieve this, the angle between the ray in air and the normal to the interface θ_a must be equal to or less than θ_a. This angle has been defined as the half-acceptance angle. Therefore, the **acceptance angle,** which is specified for most fiber optic cables, is equal to $2\theta_a$. Another commonly used definition used to specify this angle is the **numerical aperture** or NA which is defined as

$$NA = \sin\theta_a \tag{11–5}$$

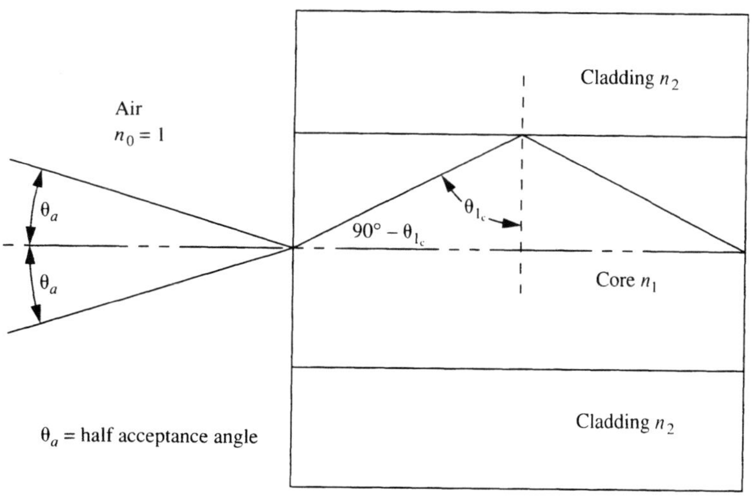

FIGURE 11–7 Acceptance angle.

The numerical aperture and the half-acceptance angle can be related to the indices of refraction of the core and cladding n_1 and n_2 by the following equations:

$$NA = (n_1^2 - n_2^2)^{1/2} \tag{11–6}$$

and

$$\theta_a = \sin^{-1}(NA) = \sin^{-1}[(n_1^2 - n_2^2)^{1/2}] \quad \textbf{(11–7)}$$

Another term used in the fiber optic technology field is the **relative refractive index Δ**, defined as

$$\Delta = \frac{n_1^2 - n_2^2}{2n_1^2} \quad \textbf{(11–8)}$$

Since n_1 and n_2 are approximately equal to each other, the relative refractive index can be approximated by

$$\Delta \approx \frac{n_1 - n_2}{n_1} \quad \textbf{(11–9)}$$

Typical values for Δ range from 0.01 to 0.03.

EXAMPLE 11–2

The core of a fiber optic cable has an index of refraction of 1.45 while the cladding has an index of refraction of 1.40. Determine (a) the critical angle within the core, (b) the half-acceptance angle, (c) the numerical aperture, and (d) the relative refractive index.

Solution

(a) The critical angle can be determined from (11–4):

$$\theta_{1_c} = \sin^{-1}\frac{n_2}{n_1} = \sin^{-1}\frac{1.40}{1.45} = 74.91° \quad \textbf{(11–10)}$$

(b) The half-acceptance angle can be determined from (11–7):

$$\theta_a = \sin^{-1}[(n_1^2 - n_2^2)^{1/2}] = \sin^{-1}[((1.45)^2 - (1.40)^2)^{1/2}] \quad \textbf{(11–11a)}$$

$$= \sin^{-1}(0.3775) = 22.18° \quad \textbf{(11–11b)}$$

(c) The numerical aperture can be determined from (11–6):

$$NA = (n_1^2 - n_2^2)^{1/2} = [(1.45)^2 - (1.40)^2]^{1/2} \quad \textbf{(11–12a)}$$

$$= 0.3775 \quad \textbf{(11–12b)}$$

(d) The relative refractive index can be determined from (11–8) or (11–9). Comparing both the exact and approximation formulas, the exact formula gives

$$\Delta = \frac{n_1^2 - n_2^2}{2n_1^2} = \frac{[(1.45)^2 - (1.40)^2]}{2(1.45)^2} = 0.03389 \quad \textbf{(11–13)}$$

while the approximate formula gives

$$\Delta = \frac{n_1 - n_2}{n_1} = \frac{(1.45 - 1.40)}{1.45} = 0.03448 \quad \textbf{(11–14)}$$

11–4 ■ DISPERSION IN FIBER OPTIC SYSTEMS

A pulse will undergo distortion as it travels through a fiber optic transmission line. The pulse is spread out in time with longer rise and fall times and increased pulse width. This characteristic of the transmission line is known as **dispersion.**

The effects of dispersion in a fiber optic transmission line are illustrated in Fig. 11–8. A pulse train at the source end representing a binary value of 101 is shown in Fig. 11–8(a). When this pulse train arrives at the receiving end, each pulse has been spread out in time due to dispersion, and the two pulses now overlap. They are therefore interpreted as binary signal 111, as illustrated in Fig. 11–8(b).

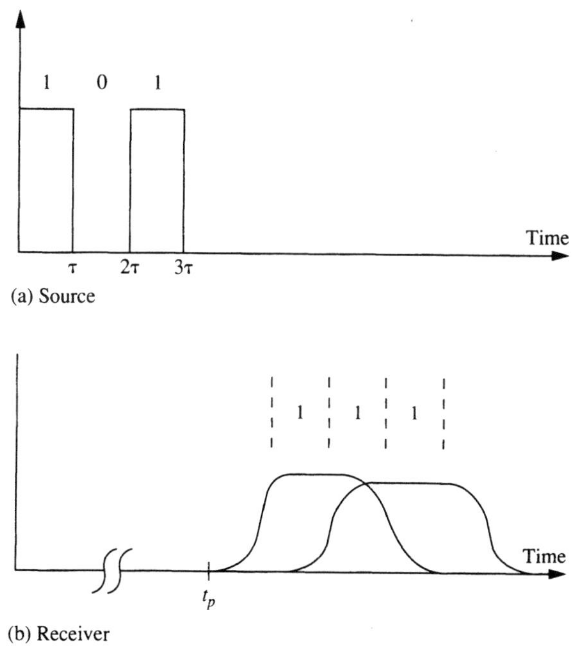

FIGURE 11–8 Effects of dispersion in fiber optic transmission lines on pulse transmission.

Dispersion is the result of two fundamental characteristics. First, in multimode fiber optics, the pulse energy follows more than one path. The shortest path is the zero mode which is down the centerline of the fiber. A much longer path is the critical-angle path which bounces off the boundary between the core and cladding as it moves from the source to the receiver. The length of the critical-angle path is longer by the reciprocal of the cosine of the critical angle. The portion of the pulse energy that travels by the longer paths arrives at a later time than the energy that travels down the centerline of the fiber, resulting in a spreading of the pulse. This type of dispersion is referred to as **modal dispersion.**

The second source of dispersion is known as **chromatic dispersion.** It is a result of the fact that the velocity of propagation of the signal within the fiber is a function of the wavelength or frequency of the light wave signal. The frequency of a light wave source is not a single monochromatic frequency, but it is a band of frequencies that have a finite bandwidth or, as commonly referred to in the light wave region, a finite linewidth. Light-emitting diodes (LEDs) have a fairly broad linewidth, while lasers have a much narrower linewidth.

There are two sources of chromatic dispersion. In the first, **material dispersion,** the index of refraction of the material varies with wavelength. The index of refraction is the ratio of the velocity of propagation in free space to the velocity of propagation in the material. Therefore, the longer wavelength components of the light wave pulse signal take a different time to travel from the source to the receiver than the shorter wavelength components of the pulse signal, thereby causing a pulse spreading.

The second form of chromatic dispersion is **waveguide dispersion.** A fiber optic transmission line is a circular waveguide. The electric and magnetic fields within this waveguide can be solved by Maxwell's equations, especially for single-mode, step-index fibers. As was discussed in Section 8–7, expressions for the phase and group velocities can be obtained and are a function of frequency or wavelength. Since the group velocity varies with frequency within the waveguide, there will be a resulting distortion due to pulse spreading of the light wave pulse signal. Waveguide dispersion is the major source of dispersion in single-mode, step-index fiber optic transmission lines where the light source is a laser (lasers are required for single-mode fiber optic systems).

Dispersion is the major limitation on the data rate that can be sent over a fiber optic transmission line. It determines the minimum spacing, as illustrated in Fig. 11–8, for pulses such that the pulse spread does not result in overlap. The total dispersion is the root sum square for the three types of dispersion. Thus, the effect of each source is squared and the result added to the others, and the square root of the sum is then taken to obtain the total dispersion. If any one source of dispersion is ten times greater than the other two sources, then it is essentially the total source of dispersion.

The dispersion in a fiber optic transmission line is a function of length and can be expressed as an amount of pulse spreading in nanoseconds per kilometer (ns/km). However, dispersion is typically specified in data sheets in either a bandwidth format or a time-delay format. The bandwidth format is in MHz times km units. The approximate baseband bandwidth BW required to determine the presence or absence of a pulse is

$$\mathrm{BW} = \frac{1}{2 \times \text{pulse width}} \qquad \textbf{(11–15)}$$

If the pulse spreads one pulse width, there will be errors. Therefore, the dispersion can be related to the maximum bandwidth-distance product. For example, if the specification is 1000 MHz·km, the fiber can be used to transmit a bandwidth of 100 MHz over a 10-km path or a bandwidth of 1000 MHz over a 1-km path.

Typical values of bandwidth-distance product vary from 0.5 MHz·km to greater than 4000 MHz·km. The 4000 MHz·km fiber is capable of transmitting 565 million bits per second over a distance of 80 km at a wavelength of 1.55 micrometers.

A second method of specifying dispersion in a fiber optic transmission line is in terms of pulse delay time per linewidth of the source per distance. This value is normally expressed in picoseconds per nanometer·kilometer (ps/nm·km). As an example of this specification, a single-mode glass fiber has a dispersive specification of 50 ps/nm·km. The source has a linewidth of 1.5 nm. If the fiber optic transmission line is 10 km in length, the pulse spreading or dispersion is 0.75 ns.

11–5 ■ LOSSES IN FIBER OPTIC TRANSMISSION LINES

The propagation of electromagnetic energy along a lossy transmission line results in the attenuation of the amplitude of the electric and magnetic fields as a function of distance along the line. This attenuation is expressed as an attenuation constant α either in nepers per unit length or in decibels per unit length. Propagation of light wave electromagnetic energy through a fiber optic transmission line is also attenuated as a function of the transmission line length due to losses in the fiber.

The attenuation in a fiber optic transmission line is due to a combination of both **absorption** of the electromagnetic energy within the fiber optic material and **scattering** of the electromagnetic energy. There are also losses due to bending of the fiber. Both **macroscopic bending** and **microscopic bending** create losses in the fiber. Another loss, which is not within the fiber material, is **coupling losses** in the coupling between the source and the fiber optic cable and between the fiber optic cable and the detector or receiver. In addition, there are coupling losses due to misalignment of two fibers that have been spliced together.

One form of absorption is **material absorption,** which is a basic property of the material and is the minimum attainable attenuation. This loss can be changed only by developing lower-loss materials. The second source of absorption is **ionic absorption** due to the presence of ions such as metallic ions and OH^- ions. These are a result of impurities in the silica glass fiber during manufacturing. The OH^- ions are a result of humidity, i.e., water vapor impurities, within the silica glass. Resonant absorption peaks occur for the OH^- ion at 0.95, 1.25, and 1.39 μm. Figure 11–9 illustrates these absorption peaks superimposed on an attenuation curve of a typical fiber over the wavelength range from 0.7 to 1.6 μm. Improved manufacturing procedures are required to improve the ionic absorption.

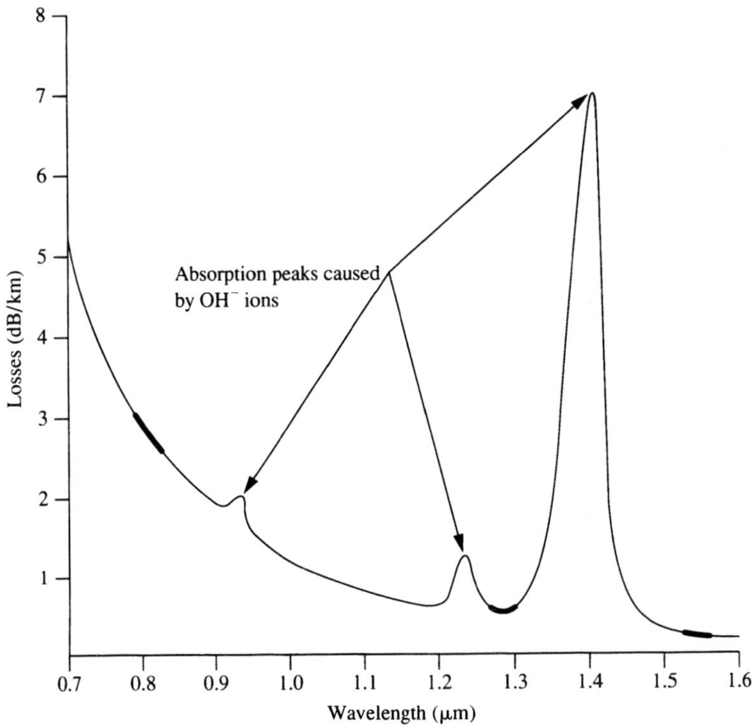

FIGURE 11–9 OH^- ionic absorption peaks.

Scattering losses are due primarily to Rayleigh scattering within the fiber. Rayleigh scattering is a phenomenon in which the light waves are interacting with inhomogeneities within the silica glass that are much smaller than the wavelength of the light wave. These inhomogeneities are due to minute changes in the index of refraction of the material, changes in the composition of the material, and changes in the density of the material. The amount of Rayleigh scattering is inversely proportional to the fourth power of the wavelength ($1/\lambda^4$). Figure 11–10 shows both the material absorption loss and the Rayleigh scattering loss for silica glass. Examination of this figure reveals that there is a significant minimum in the attenuation at 1.55 μm. This region is one of three wavelength regions in which fiber optic systems operate.

Bending losses are illustrated in Fig. 11–11. The microscopic bending losses are due to small bumps or irregularities in the boundary between the core and the cladding. When the light wave ray is incident upon these bumps, the angle is less than the critical angle, and some of the energy is lost into the cladding, as shown in Fig. 11–11(a). In macroscopic bending, illustrated in Fig. 11–11(b),

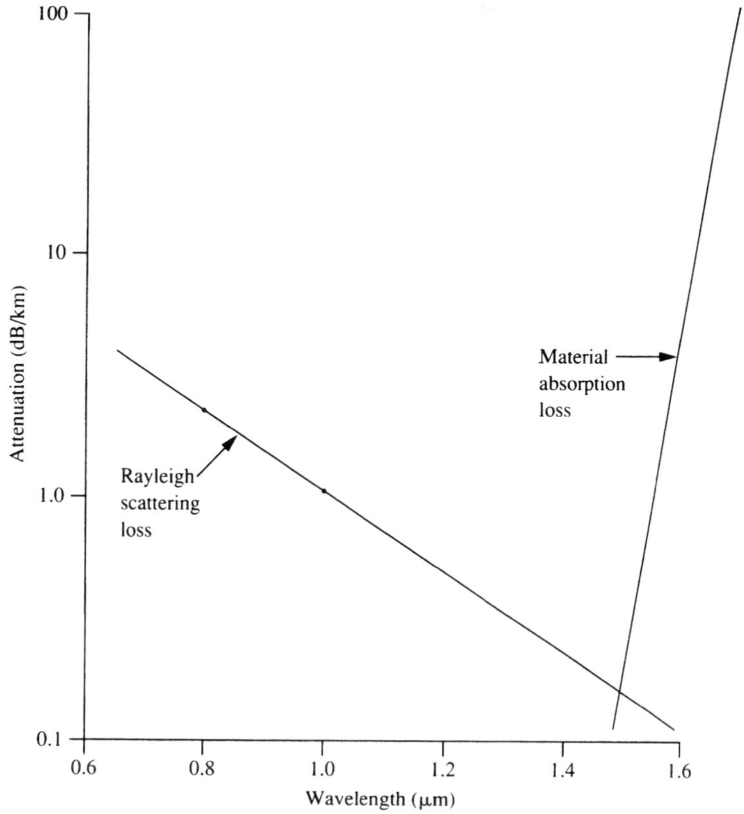

FIGURE 11–10 Absorption and scattering losses in a typical silica glass fiber.

the entire fiber optic cable is bent around a sharp radius. As shown in the figure, when the ray that is bouncing back and forth down through the core reaches the bend, the angle is less than the critical angle due to the bend, and some of the energy is lost into the cladding.

There are three principal wavelength regions chosen for fiber optic systems. The first region is a short-wavelength region centered at 0.85 μm. This is the wavelength where gallium arsenide (GaAs) light-emitting diodes are economically available. The other two regions are referred to as the long-wavelength regions, the first centered at 1.3 μm and the second centered at 1.55 μm. The 1.3-μm region is at the wavelength for minimum dispersion, while the 1.55-μm region is at the minimum between scattering losses and absorption losses. Recent advances in fiber technology have resulted in fibers that have shifted the minimum dispersion from 1.3 to 1.55 μm. Therefore, the minimum dispersion and minimum losses will occur at the same wavelength.

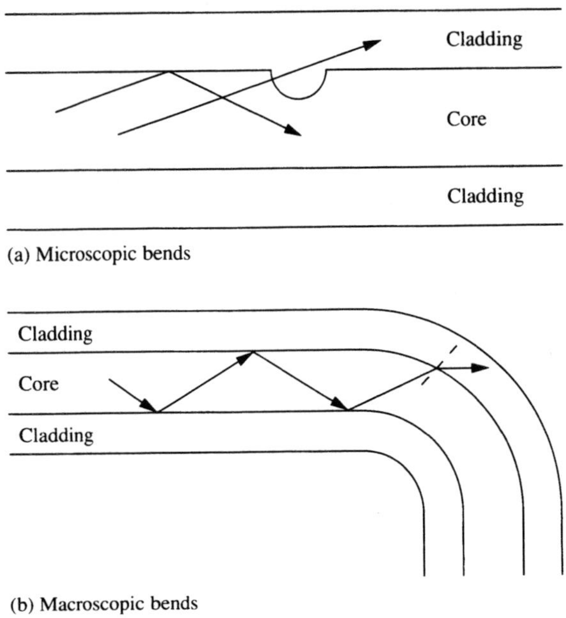

FIGURE 11–11 Bending losses.

11–6 ■ OPTICAL TRANSMITTING DEVICES

Two basic types of semiconductor devices are used as sources of light wave electromagnetic energy. Both of these devices produce radiation in the light wave frequency region when a direct current and a dc voltage are applied to them. Modulation can be achieved by varying the current to these devices. The first device is a **light-emitting diode (LED),** and the second is a **laser diode.**

Light-Emitting Diode (LED)

The LED is a type of forward-biased PN junction diode. When a dc voltage is applied, the electrons and holes will recombine within the junction. During the recombination process, energy is released. By proper selection of materials to fabricate the PN junction, the recombination process will emit electromagnetic energy in the light wave frequency region. This radiation is known as **spontaneous emissions.** The wavelength of the emitted radiation is a function of the gap energy in the semiconductor material used in the PN junction diode. This wavelength is given by

$$\lambda = \frac{hc}{E_g} \tag{11–16}$$

where

h = Planck's constant (6.626×10^{-34} J·s)
c = velocity of light (3×10^8 m/s)
E_g = bandgap energy (in J)

Normally the bandgap energy is given in electron·volts (eV) where 1 eV = 1.6×10^{-19} J. Equation (11–16) can be simplified to

$$\lambda = \frac{1.2424}{E_g} \tag{11–17}$$

where λ is in μm and E_g is in eV.

Radiation from an LED due to sponaneous emission is noncoherent radiation; i.e., the radiation occurs over a broad range of frequencies and is randomly polarized and randomly directed. The **spectral width** of the radiation is defined as the difference between the half-power points of the radiation spectrum. The typical spectrum illustrating the spectral width of a GaAs LED is shown in Fig. 11–12. The spectral width between half-power points is from 0.86 to 0.90 μm or a width of 0.04 μm. The major disadvantages of a wide spectral width are the resulting chromatic dispersion and the limitations on data rate due to the dispersion.

There are two types of LEDs based on how the radiation is emitted by the diode. In one type, called a **surface emitter,** the radiation is radiated from the diode surface over a very large beamwidth approaching a hemisphere. Typically, the half-power beamwidth approaches 120° in both planes. The other type is called an **edge emitter** where the radiation is emitted from the edge of the active region instead of the surface. Edge emitters typically have a fan-shaped beam which results in a more directed radiation. The surface emitters are used only with multimode fibers, while the edge emitters can be used with both single-mode and multimode fibers. The output power of an edge-emitting LED is typically less than that of the surface emitter. However, due to the higher directivity of the edge emitter, both types couple into a fiber at approximately the same power level.

The typical optical output power from an LED is approximately 1 mW. However, due to the difficulty of coupling the power into a fiber, the *power launched* (a specification used for LED fiber optic sources) is usually between 10 and 50 μW. One vendor defines power launched as the optical power exiting one meter of a 0.045-in-diameter optical fiber having an *NA* of 0.67 and an attenuation constant of 0.60 dB/km at 0.94 μm.

The communication of digital data is typically accomplished by intensity modulating the LED, which in turn is accomplished by operating the LED either at a low current for a zero or at a high current for a one. Therefore, the power is low for a zero and high for a one. The speed at which the power shift responds to the change in current is the limiting factor on the data rate in a fiber optic system using an LED optical source. The typical maximum bit rate for an LED source is approximately 50 Mbps.

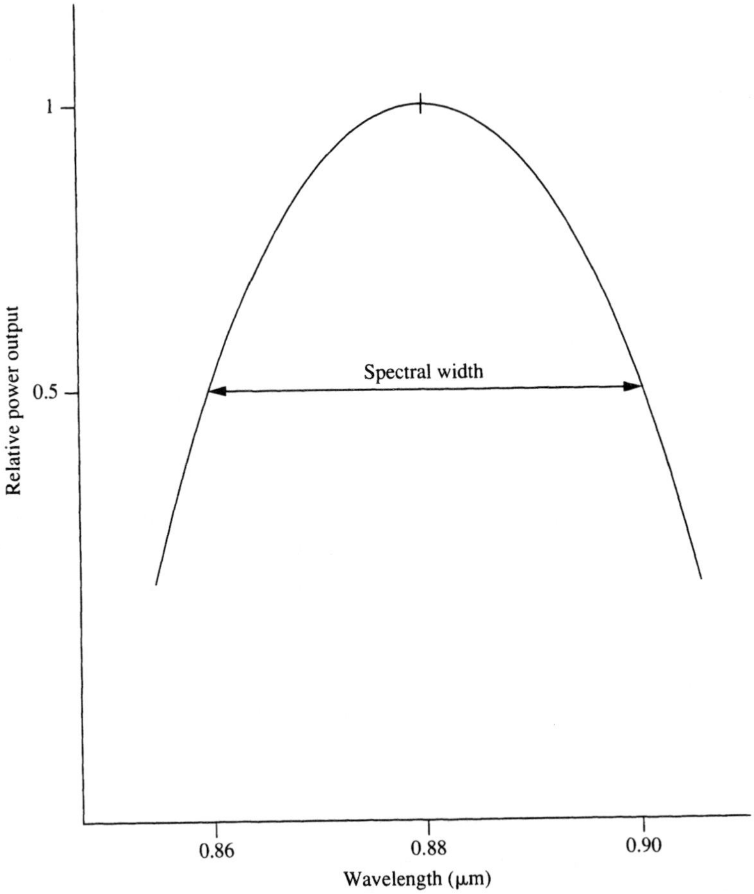

FIGURE 11–12 LED spectrum.

To summarize, the LED optical source is usually used in a multimode fiber system where the data rate is 50 Mbps or less and the data are moved over short distances. The advantages of the LED over the laser diode source are its cost, reliability, and ruggedness; in addition, it is less affected by temperature variations.

Laser Diode

The semiconductor laser diode is very similar to the LED discussed previously in that it is also a forward-biased PN junction. The laser diode is fabricated from the same materials as the LED. However, there are several significant differences. First, the laser uses **stimulated emissions** and **population inversion** within the active semiconductor PN junction. Second, an optical cavity is constructed around the active region to resonate the photons into a coherent, single-frequency optical source.

The term *laser* is an acronym for *light amplification by stimulated emissions of radiation.* Stimulated emission results when a photon is injected into the bandgap region. The injected photon stimulates an electron to drop from the conduction band (excited state) back to the valance band (ground state), releasing a photon of energy at the same wavelength as the photon that stimulated the emission. Population inversion is a condition in which there are a significant number of electrons in the excited state. The optical cavity provides for modes where the photons will bounce back and forth, i.e., resonate, within the cavity and stimulate more and more electrons to emit photons. This action is referred to as *lasing*.

Figure 11–13 is a typical plot of the optical output power as a function of input current for both an LED and a laser diode. The output power as a function

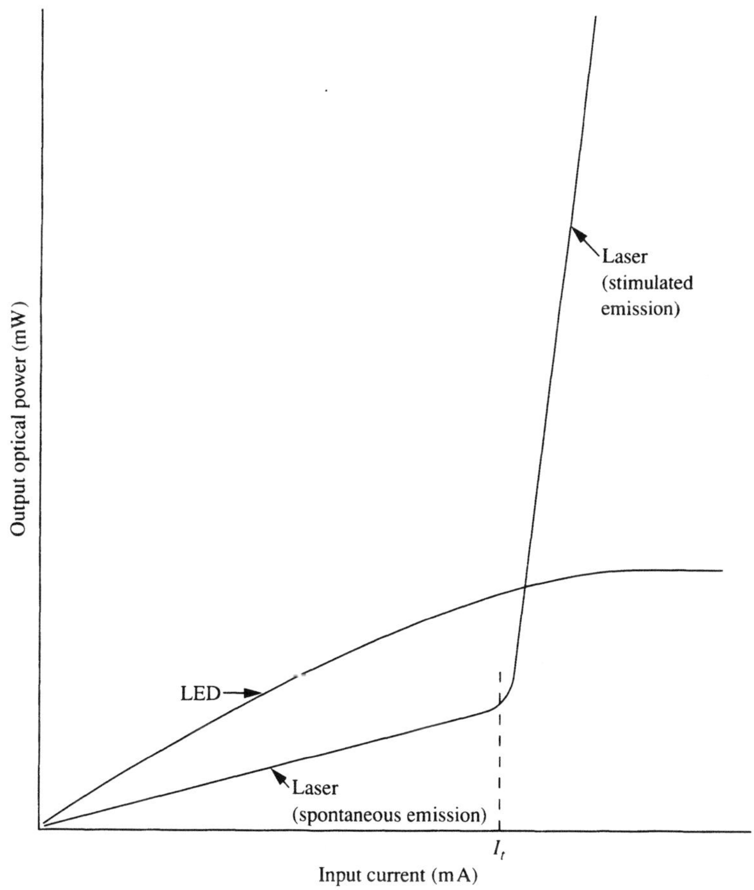

FIGURE 11–13 Power-current curves for LEDs and laser diodes.

of input current in an LED is a function of spontaneous emission only and is a nearly linear function. It becomes slightly nonlinear at high input currents. However, there is a significant difference for the laser. At low input current levels, the laser acts the same as the LED, due also only to spontaneous emissions. Then at a certain current threshold I_t, stimulated emissions take over, and lasing results in a significant increase in output power as a function of input current. The curve for the laser is highly nonlinear due to the pronounced threshold effect. One of the problems with a laser diode source is that the threshold is a function of the laser's physical temperature. The design of the drive circuits for a laser diode must compensate for the variation in threshold current with laser temperature.

Laser diodes offer several significant advantages over LEDs. Since laser diodes can be a single-frequency, coherent source, their spectral width is typically 10^{-5} μm compared to 0.04 μm for the LED source. The spectral width is 4000 times narrower and has a significant reduction in the effects of chromatic dispersion, resulting in a much higher data-rate capability.

The optical output of a laser diode is much more directive than that of an LED. The output beamwidth from a laser is approximately 20° by 30°, allowing a larger percentage of the output power to be coupled into a fiber. For example, an LED can couple only about 1% of its power into a 50-μm, graded-index fiber, while a laser can couple about 50% of its power into the same fiber. Laser diodes can couple from 1 to 10 mW of optical power into a fiber, while LEDs typically couple 10 to 100 μW.

Another important advantage of the laser over the LED is the rate of speed with which the source can be turned on and off. This response time is a limiting factor in the rate at which data can be transmitted over a fiber optic system. The LED sources have response times from 2 to 20 ns, while the laser has response times from 0.01 to 1 ns. The laser sources have a transmitted data rate of up to 1.2 Gbps compared to 50 Mbps for LEDs.

EXAMPLE 11–3

The bandgap energy for gallium arsenide is 1.4 eV. Determine the wavelength of emissions from a GaAs LED optical source.

Solution

The wavelength can be determined from either (11–16) or (11–17). Choosing (11–17) and substituting the bandgap energy, we obtain

$$\lambda = \frac{1.240}{E_g} = \frac{1.240}{1.4} = 0.886\ \mu\text{m} \tag{11–18}$$

EXAMPLE 11–4

Determine the bandgap energy of a material required to construct an LED whose emissions are at a wavelength of 1.55 μm.

Solution

Rearranging (11–17) and solving, we have

$$E_g = \frac{1.240}{\lambda} = \frac{1.240}{1.55} = 0.8 \text{ eV} \tag{11–19}$$

11-7 ■ OPTICAL RECEIVING DEVICES

Optical detectors are used in fiber optic systems as the optical receiving devices. An optical detector performs the reverse function of the optical source. A photon creates an electron-hole pair in the detector which will result in an electrical current flowing. This current is amplified by the receiver and represents the received data.

There are two types of optical detectors used in the wavelength region in fiber optic systems (0.8–1.55 μm): the **PIN photodiode** and the **avalanche photodiode (APD).** The desirable characteristics for optical detectors include good quantum efficiency, spectral response, sensitivity, responsity, low internal noise, and fast response time.

Quantum efficiency in an optical detector is the ratio of the number of electron-hole pairs generated to the number of photons incident upon the detector. Quantum efficiency is a function of the material used in the detector and the wavelength. Silicon is the best material over the frequency range from 0.5 to 0.9 μm. Germanium and indium gallium arsenide phosphide (InGaAsP) are used in the wavelength region from 1.2 to 1.6 μm. Quantum efficiencies between 50% and 80% are typical between 0.5 μm and 1.6 μm.

Spectral response is the ability of the detector to absorb light over the wavelength region of interest. Sensitivity is the level of the received optical signal required to equal the overall noise level in the system; it is reached when the signal-to-noise ratio is unity. Higher signal levels are usually required to achieve the specified bit error rate in the typical digital data communication fiber optic system.

Responsity of the photodetector is the ratio of output current to the input optical power in amperes per watt. Response time for the detector is a function of the bandwidth of the photodetector. The maximum data rate is a function of the bandwidth of the photodetector.

The PIN photodiode consists of three layers of semiconductor material: (a) a highly doped P layer, (b) an intrinsic layer, and (c) a highly doped N layer. The intrinsic layer acts as a depletion region for the PN junction diode. A photon, whose energy equals or exceeds the bandgap energy of the PN junction material, creates an electron-hole pair, with the electron moving to the conduction band. An electrical circuit allows the current to flow through the photodiode. The number of electrons produced on average is 50% to 90% of the number of photons incident into the depletion region. This is the quantum efficiency of the PIN photodetector. PIN photodiodes can operate without an external bias; however, normally they are reversed biased with a voltage between 8 to 10 V. The use of the reverse bias improves the response time of the PIN photodiode.

Avalanche photodiodes (APDs) are constructed differently from PIN photodiodes. They consist of four layers: (a) an N doped region, (b) then a P doped region, followed by (c) a thick intrinsic depletion region, and finally (d) a P^+ doped region. A high-voltage reverse bias is established in the PN junction formed between the first two regions. A photon is incident into the intrinsic region and produces an electron-hole pair. The carriers which now flow through the PN junction produce an avalanche breakdown in the junction, creating a large number of secondary carriers. Therefore, there is a multiplication of carriers within the APD. Due to this multiplication factor, APDs have 5 to 7 dB better sensitivity than PIN photodiodes. APDs require a reverse-bias voltage of approximately 300 V compared to 10 V for a PIN photodiode.

The PIN photodiode is the most common type of detector used in fiber optic systems. It is a simple device that is easy to fabricate and is inexpensive. It uses a low-voltage bias source and has good linearity, fast response time, low environmental temperature sensitivity, low noise, and high reliability. The PIN photodiode is used in most applications except for high-data-rate communication systems that need to have maximum distance between repeaters. For these applications, which usually employ laser optical sources and single-mode step-index fibers, the APD is typically used due to its better sensitivity. APDs are difficult to use since they require a large bias voltage which is temperature sensitive, requiring careful power supply design. APDs are also more expensive.

11–8 ■ OPTICAL LINK DESIGN AND ANALYSIS

The design of an optical fiber optic communication system is very similar to the design of any other type of communication system. The information required is the source power, the losses in the overall system, the sensitivity of the receiving devices, and the required signal-to-noise ratio at the receiving device.

The power level of the optical source and the sensitivity of the optical receiver must be specified in terms of decibels. Since these are absolute power levels, and since decibels are ratios of power, there appears to be a problem. This problem is resolved by expressing the absolute power level as a ratio of the absolute power to a reference power. The three common references are the watt (W), the milliwatt (mW), and the femtowatt (fW), defined as follows:

$$\text{power level in dBW} = 10\log\left(\frac{\text{power in watts}}{1\text{ W}}\right) \tag{11–20}$$

$$\text{power level in dBm} = 10\log\left(\frac{\text{power in milliwatts}}{1\text{ mW}}\right) \tag{11–21}$$

$$\text{power level in dBf} = 10\log\left(\frac{\text{power in femtowatts}}{1\text{ fW}}\right) \tag{11–22}$$

Remember that

$$1\text{ W} = 10^3\text{ mW} = 10^{15}\text{ fW}$$

The term **system gain** refers to the difference in decibels between the transmitting source and the receiver sensitivity. It can be obtained by determining the difference between the source power expressed in dBW, dBm, or dBf; the receiver sensitivity is also expressed in dBW, dBm, or dBf. However, the units must be identical, i.e., both dBW, dBm, or dBf. The system gain must equal or exceed the sum of the system losses, including the required signal-to-noise ratio.

EXAMPLE 11–5

A laser diode optical source has an optical output power of 15 mW. The avalanche photodiode has a sensitivity of 10 μW. Determine the system gain.

Solution

The first step is to express both the source power and the sensitivity in the same units of power. Using milliwatts as the common unit, the sensitivity is 1×10^{-2} mW. Next, converting both to decibels results in

$$\text{source power (dBm)} = 10\log\left(\frac{15\text{ mW}}{1\text{ mW}}\right) = +11.76\text{ dBm} \qquad \textbf{(11–23)}$$

$$\text{sensitivity} = 10\log\left(\frac{1\times 10^{-2}\text{ mW}}{1\text{ mW}}\right) = -20\text{ dBm} \qquad \textbf{(11–24)}$$

The system gain is the difference between the source power in dBm and the sensitivity in dBm, or

$$\text{system gain} = +11.76\text{ dBm} - (-20\text{ dBm}) = 31.76\text{ dB} \qquad \textbf{(11–25)}$$

It is important to note that if the difference between two power levels expressed in dBW, dBm, or dBf is taken, the result is in dB since subtracting decibels corresponds to division, and therefore the result is a ratio.

The losses in a fiber optic system were discussed in Section 11–5. These included coupling losses, fiber attenuation in dB per unit length, and bending losses. The required signal-to-noise ratio at the optical signal, although not an actual loss, can be thought of as a loss since it is the amount by which the signal power at the receiver must exceed, i.e., be greater than, the sensitivity of the receiving device. Therefore, this amount of power cannot be used to overcome other losses. Another factor included in the link budget is an allowance for aging, especially in laser diode sources.

The typical source-to-fiber coupling loss is approximately 6 to 10 dB. The fiber-to-detector coupling loss is usually much less, typically 1 to 2 dB. The fiber attenuation is typically specified in the data sheets and can be anywhere from 10 dB/km to 0.2 dB/km. This value must be multiplied by the length of the fiber system to obtain the total fiber attenuation.

Two different types of link design and analysis are illustrated in the next two examples. First, the available signal-to-noise ratio is calculated for an existing system, and second, the maximum distance between repeaters is calculated to ensure a given signal-to-noise ratio with a specified allowance for aging and bending losses.

EXAMPLE 11–6

An optical source has a power output of 0.5 mW (−3 dBm). The source-to-fiber coupling loss is 6 dB. The fiber-to-receiver detector loss is 2 dB. The fiber attenuation is 3 dB/km, and the fiber optic link is 3 km in length, for a total loss of 9 dB. If the receiver detector sensitivity is 2 μW (−27 dBm), determine the signal-to-noise ratio.

Solution

The system gain, i.e., the difference between the source power and the receiver detector sensitivity, is

$$\text{system gain} = \text{source power} - \text{sensitivity} \qquad \textbf{(11–26a)}$$

$$= (-3\text{ dBm} - (-27\text{ dBm})) = 24\text{ dB} \qquad \textbf{(11–26b)}$$

The sum of the losses is 17 dB, and therefore the signal-to-noise ratio, which is the difference between the system gain and the system losses, is +7 dB.

EXAMPLE 11–7

Determine the maximum distance between repeaters if the optical source power is 10 mW (+10 dBm), the receiver detector sensitivity is 1 μW (−30 dBm), the source-to-fiber coupling loss is 6 dB, and the fiber-to-receiver coupling loss is 2 dB. The required signal-to-noise ratio is +13 dB. An aging allowance of 3 dB is required by the specification. The fiber has an attenuation of 0.2 dB/km. A connector and bending loss allowance of 2 dB is also required.

Solution

The system gain is the difference between the source power and the sensitivity which is +40 dB. The total losses without the fiber attenuation, but including the signal-to-noise ratio and aging and connector/bending allowances, is 26 dB. This leaves 14 dB for the total fiber attenuation loss. If the attenuation per kilometer is 0.2 dB/km, then the total length is given by

$$\text{length} = \frac{14\text{ dB}}{0.2\text{ dB/km}} = 70\text{ km} \qquad \textbf{(11–27)}$$

■ PROBLEMS

11–1. A semiconductor laser optical transmitter has a wavelength in free space of 1.3 μm. Determine the frequency of this electromagnetic radiation.

11–2. A semiconductor laser optical transmitter has a wavelength in free space of 1.5 μm. Determine the frequency of this electromagnetic radiation.

11–3. The core of a fiber optic cable has an index of refraction of 1.49, while the cladding has an index of refraction of 1.39. Determine the following:
(a) critical angle within the core
(b) half-acceptance angle
(c) numerical aperture
(d) relative refractive index

11–4. The core of a fiber optic cable has an index of refraction of 1.48, while the cladding has an index of refraction of 1.46. Determine the following:
(a) critical angle within the core
(b) half-acceptance angle
(c) numerical aperture
(d) relative refractive index

11–5. The bandgap energy for indium arsenide is 0.33 eV. Determine the wavelength of emissions from an InAs LED optical source.

11–6. The bandgap energy for indium gallium arsenide varies from 0.36 to 1.43 eV. Determine the range of wavelengths of emissions available from LED optical sources constructed from this material.

11–7. Determine the bandgap energy of a material required to construct an LED whose emissions are at a wavelength of 1.3 μm.

11–8. Determine the bandgap energy of a material required to construct an LED whose emissions are at a wavelength of 0.9 μm.

11–9. A laser diode optical source has an optical output power of 8 mW. The avalanche photodiode has a sensitivity of 5 μW. Determine the system gain.

11–10. A laser diode optical source has an optical output power of 5 mW. The avalanche photodiode has a sensitivity of 6 μW. Determine the system gain.

11–11. An optical source has a power output of 5 mW. The source-to-fiber coupling loss is 5 dB. The fiber-to-receiver coupling loss is 1.5 dB. The fiber attenuation is 2.5 dB/km, and the length of the fiber is 4 km. If the receiver sensitivity is 0.5 μW, determine the signal-to-noise ratio.

11–12. An optical source has a power output of 3 mW. The source-to-fiber coupling loss is 7 dB. The fiber-to-receiver coupling loss is 2.5 dB. The fiber attenuation is 2.0 dB/km, and the length of the fiber is 6 km. If the receiver sensitivity is 1.5 μW, determine the signal-to-noise ratio.

11–13. Determine the maximum distance between repeaters if the optical source power is 3 mW, the receiver sensitivity is 3 μV, the source-to-fiber coupling loss is 7 dB, and the fiber-to-coupling loss is 1 dB. The required signal-to-noise ratio is +10 dB. Provide for an aging allowance of 4 dB and a future connector and bending loss allowance of 2.5 dB. The fiber has an attenuation of 0.25 dB/km.

11–14. Determine the maximum distance between repeaters if the optical source power is 2 mW, the receiver sensitivity is 2 μV, the source-to-fiber coupling loss is 4 dB, and the fiber-to-coupling loss is 1 dB. The required signal-to-noise ratio is +7 dB. Provide for an aging allowance of 3 dB and a future connector and bending loss allowance of 2.0 dB. The fiber has an attenuation of 0.35 dB/km.

12

Satellite Communications

■ OBJECTIVES

After completing this chapter, the reader should be able to:

1. Describe briefly the historical background of the development of satellite communications.
2. Describe a satellite communication system.
3. Define the difference between the **uplink** and the **downlink.**
4. Describe frequency allocations and types of satellite communication services.
5. Discuss the types of antennas used on communication satellites for uplinks and downlinks.
6. Discuss the types of antennas used by earth stations for the uplinks and the downlinks.
7. Determine the range from any earth station to a particular communication satellite in geosynchronous orbit.
8. Define the G_t/T_{sys} ratio and its application to satellite communication receiving stations.
9. Determine the C/N ratio for both the uplinks and the downlinks in a satellite communication system.

12–1 ■ INTRODUCTION

The need for reliable communications of large quantities of information over long distances provided the motivation that led to the development of satellite communication systems. The use of skywave HF communications and LF ground communications could not handle this need. Communications over cable and radio relay systems were also limited in their capabilities.

The concept of using satellites for communication systems is credited to the science fiction writer Arthur C. Clarke, author of *2001: A Space Odyssey*. He proposed in 1945 that if a satellite could be placed in an equatorial orbit with a radius of 42,242 km, it would have the same rotation rate as that of the earth and would remain fixed over the same location on the earth. Such an orbit is defined as a **geostationary orbit.** This satellite could provide relay communications between any two points on the hemisphere visible from the satellite. If three satellites, spaced 120° apart, were used, and if they could communicate among themselves, then communications between any two points on the earth were possible.

The use of both passive and active satellites to relay communications to various parts of the earth was pursued beginning early in the 1950s. In 1951, the U.S. Navy bounced signals off the earth's natural passive satellite, the moon. The first man-made passive communications satellite, *Echo I*, a metallic reflecting balloon, was launched August 12, 1960, into a 1000-mile-high orbit. It was used to bounce signals between earth stations in New Jersey and California. However, the use of passive satellites was short-lived since the active satellites were vastly superior for communication purposes.

SCORE, the first active communication satellite, was launched by the U.S. Air Force December 18, 1958. It could relay either one voice channel or 70 teletype channels. Due to a battery failure, it lasted only 35 days. The first truly successful active communication satellite was *TELSTAR I,* launched on July 10, 1962, by the Bell System. It used 6 GHz for the uplink and 4 GHz for the downlink, relaying communication between the United States and Europe. It established the 6/4-GHz frequency band for future communication satellites.

The first successful geostationary communication satellite, *SYNCOM II,* was launched on July 26, 1963, by NASA and DOD—only 18 years after Clarke had suggested this concept in his science fiction writings. The first operational commercial geostationary communication satellite, *Intelsat I*, called *EARLY BIRD,* was developed by COMSAT for the INTELSAT Corporation. Launched April 6, 1965, *Intelsat I* had the capability of 480 simultaneous voice channels. During 1969, a series of communication satellites, the *Intelsat III,* were placed over the Atlantic, Pacific, and Indian oceans, thereby providing for worldwide commercial satellite communications. Each *Intelsat III* could carry 2400 simultaneous voice channels. The latest international communication satellites, the *Intelsat 600* series, have the capability of 80,000 simultaneous voice channels.

During the 28 years from 1965 to 1993, there has been a significant increase in communication satellites, especially geosynchronous communication satellites. Almost all of the locations around the earth at geosynchronous orbit either are occupied or will be someday soon. Recently the satellite spacings have been reduced from 4° to 2° to double the number of parking orbit locations available. Also, the use of additional frequency bands, such as the Ku and K microwave-frequency bands, has allowed the deployment of multiple satellites at each location due to frequency diversity.

The first satellite to use the Ku microwave-frequency band was *SBS–1*, launched in 1979. A the end of 1992, there were 33 domestic geosynchronous communication satellites serving the continental United States. As of this writing, seventeen of these operate in the C-band frequency region, ten operate in the Ku-band frequency region, and six operate in both. These 33 satellites provide 400 C-band transponder channels and 170 Ku-band transponder channels.

This chapter will cover the basic design concepts of satellite communication systems. Although the communication links are similar to those in any other type of communication system, there are some special considerations for a satellite communication system. Also, due to standard practice among satellite communication system designers, there are some specific definitions and concepts unique to satellite communication systems. These will be identified in this chapter.

12–2 ■ SPACECRAFT COMMUNICATION SYSTEMS

The most common form of satellite communication system is illustrated in Fig. 12–1. A ground station on the earth's surface transmits information on one frequency, called the **uplink,** to a communication satellite in a geosynchronous orbit. The receiving antenna on the satellite receives the uplink frequency and sends it to a transponder. The purpose of the transponder is to perform a frequency translation from the uplink frequency to a new frequency. This translated signal is then sent to the satellite's transmitter and the transmitting antenna on the satellite. The information is transmitted to a different receiving station on the earth using the **downlink** frequency.

Although Figure 12–1 illustrates information sent in one direction only, the use of different uplink and downlink frequencies provides the capability for each ground station to operate as both a transmitting station and a receiving station, thereby enabling communication in both directions between the two ground stations. Another reason for using separate uplink and downlink frequencies is to prevent the high transmitted power level of the downlink transmitter on the satellite from interfering with the weak uplink signal being received by the satellite.

The first uplink and downlink frequencies employed in communication satellites were in the C-band microwave-frequency region. The uplink frequencies ranged from 5925 to 6425 MHz. The downlink frequencies ranged from 3700 to 4200 MHz, providing a 500-MHz bandwidth. The satellite communication system channels are spaced 20 MHz apart. The even channels have one polariza-

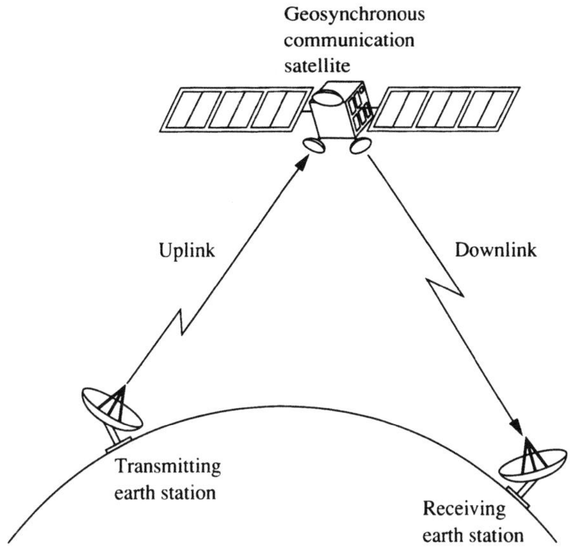

FIGURE 12–1 Basic satellite communication system.

tion, and the odd channels have the orthogonal polarization. Therefore, for each polarization, the channels are spaced 40 MHz apart. The even and odd channels are staggered by 20 MHz. The actual bandwidth used in each channel is 36 MHz with a 4-MHz guard band between each channel. United States satellites have 24 channels, each with a 36-MHz bandwidth. The 24-channel systems are spaced every 20 MHz apart and use 480 MHz of the 500-MHz bandwidth. The remaining 20 MHz is used for satellite control and data. Therefore, polarization diversity (in which the even channels use one polarization and the odd channels use the orthogonal polarization) doubles the communication capability of these systems. The polarizations are either (a) vertical or horizontal polarization if linear polarization is employed or (b) right-handed or left-handed polarization if circular polarization is employed.

The choice of the C-band frequencies was an ideal selection based on external noise and atmospheric loss considerations. The typical geostationary satellite communication system is one where the ground stations know exactly where the satellite is and therefore can use high-gain, large-aperture antennas. The transmitting antennas must also provide a narrow beamwidth such that the earth station does not cause interference to the adjacent communication satellite located only 2° away in the geostationary orbit. The satellite must transmit to many earth stations located in different positions on the earth. Therefore, the transmitting antenna on the satellite must use lower gain and wider beamwidths.

The C-band region has low levels of externally received noise for a receiving antenna looking toward outer space. Figure 12–2 illustrates the external atmo-

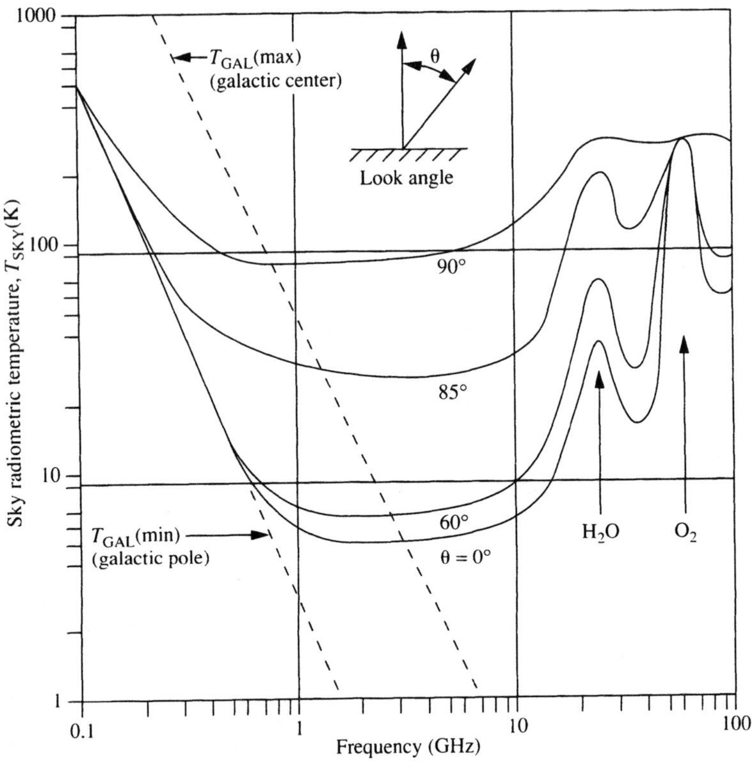

FIGURE 12–2 External atmospheric noise. F. T. Ulaby, R. K. Moore, and A. K. Fung, *Microwave Remote Sensing: Active and Passive,* Vol. 1, *Microwave Remote Sensing Fundamentals and Radiometry* (Reading, MA: Addison-Wesley, 1981) p. 287, Fig. 5.9(b). Used with permission.

spheric noise that would be received by a high-gain, narrow-beamwidth antenna looking toward the sky. The noise as a function of the look angle θ is plotted. The look angle θ is measured from the vertical to the earth's surface. The noise from the galaxy decreases significantly in the microwave-frequency region and is below that of the atmosphere for all frequencies above 3 GHz. The noise level is nominally constant for look angles from straight upward ($\theta = 0°$) to $\theta = 60°$, which is 30° above the earth's surface. The peak at 22 GHz is a result of absorption due to the water vapor in the atmosphere, while the peak at 60 GHz is due to absorption by the oxygen in the atmosphere. Liquid water in the form of rain introduces additional losses and noise. The loss due to rain increases with increasing frequency and is minimum in the C-band region. Therefore, as shown in Fig. 12–2, the C-band region from 3.7 to 6.425 GHz falls in a region of minimum atmospheric noise.

The C-band frequency region does have one significant problem. It is the frequency region assigned to terrestrial microwave radio communication systems. There are tens of thousands of these systems located all over the world, and they carry a large volume of the commercial telephone communications, both voice and data. Since these terrestrial systems existed prior to the development of satellite communication systems, the newer satellite communication systems must not interfere with the terrestrial systems. Therefore, the level of radiated power from the satellite, known as the **effective isotropic radiated power (EIRP),** is limited to a level so that there is no interference. Also, the locations of the earth-based uplink stations were restricted to prevent interference with terrestrial microwave communication systems. It takes significant time and paperwork to obtain approval to transmit from a specific location on the earth's surface.

The Ku-band frequency region was selected in the 1970s for the exclusive use by satellite communication systems, thereby eliminating the problem of interference with terrestrial systems. The downlink frequencies are 11.7 to 12.7 GHz, and the uplink frequencies are 14.0 to 14.5 GHz. The Ku-band frequency region allows increased EIRP levels from the satellite and significantly smaller earth-based antennas for the same gain and beamwidth as the C-band antennas. Also, Ku-band uplinks can operate from any location and can be highly mobile. These capabilities have made possible the use of live television coverage of both news and sports events. Small Ku-band uplink stations can fit into suitcases and travel with news correspondents to any location on the earth.

The only disadvantage to the use of Ku-band is a small increase in atmospheric noise and an increase in rain attenuation. However, the probability of outage due to rain attenuation can be made small by proper design techniques. Also, where outages must be prevented, the use of space diversity can eliminate the problem of rain attenuation. (Space diversity is the use of multiple uplink earth stations at different locations such that the probability of rain at all stations is extremely small.)

The K-band frequency region, 27.5 to 30 GHz for uplinks and 17.7 to 20.2 GHz for downlinks, is the latest frequency allocation for satellite communication services. The National Aeronautics and Space Administration (NASA) launched the *Advanced Communications Technology Satellite (ACTS)* in 1993 to investigate this frequency region for future satellite communication services. There were only a limited number of K-band satellite communication systems operational prior to the launch of *ACTS*.

12–3 ■ FREQUENCY ALLOCATIONS AND TYPES OF SERVICE

The use of any frequency for electromagnetic radiation must be controlled on a worldwide basis to prevent interference and chaos. This control is accomplished by the International Telecommunication Union (ITU) which is an agency of the United Nations. There are at present 166 member nations of the ITU. Each member nation sends delegates to the World Administrative Radio

Conference (WARC), which meets periodically to review and update the International Radio Regulations and the International Table of Frequency Allocations. The last WARC was held in 1992. The frequency region that the WARC covers is from 10 kHz to 300 GHz.

The WARC also decides which type of service can be used at which frequency allocation. Allocations are made both for governmental services, which include the military, and for nongovernmental services. The ITU divides the world into three regions for the purpose of frequency allocations. Region 1 covers Europe, Africa, and the Middle East; region 2 covers North and South America; and region 3 covers the South Pacific, Asia, and the Indian subcontinent.

The WARC delineates three principal types of satellite communication systems as defined in the International Table of Frequency Allocations:

1. Fixed Satellite Service (FSS): A radio communication service between earth stations at given positions when one or more satellites are used; the given position may be a specified fixed point or any fixed point within specified areas.
2. Mobile Satellite Service (MSS): A radio communication service between mobile earth stations and one or more space stations, between space stations used in this service, or between mobile earth stations by means of one or more space stations.
3. Broadcast Satellite Service (BSS): A radio communication service in which signals transmitted or retransmitted by space stations are intended for direct reception by the general public.

TABLE 12–1

Fixed Satellite Service (FSS)	
Downlink	*Uplink*
3.7 to 4.2 GHz	5.85 to 7.075 GHz
4.5 to 4.8 GHz	12.7 to 13.25 GHz
10.7 to 12.2 GHz	14.0 to 14.50 GHz
17.7 to 20.2 GHz	27.5 to 30.00 GHz
Broadcast Satellite Service (BSS)	
Downlink	*Uplink*
2.5 to 2.655 GHz	2.655 to 2.690 GHz
12.2 to 12.7 GHz	17.3 to 17.80 GHz
22.5 to 23.0 GHz	
Mobile Satellite Service (MSS)	
Downlink	*Uplink*
1.530 to 1.559 GHz	1.6265 to 1.6605 GHz
19.2 to 20.2 GHz	29.5 to 30.0 GHz

The FSS service was not intended for use by the general public. However, there are many downlink earth-receiving stations being used by the general public to receive FSS transmissions from geostationary communication satellites. The Direct Broadcast Satellites use the BSS frequency allocations to transmit both video and audio entertainment programs to the general public. The MSS is used to provide voice and data communication services to mobile earth stations such as ships at sea and aircraft. The WARC defines a mobile earth station as one located either on the earth's surface or within the earth's atmosphere.

The frequency allocations for nongovernmental satellite communication services in the ITU region 2 as specified in the 1991 edition of the International Table of Frequency Allocations are listed in Table 12–1.

12–4 ■ SPACECRAFT ANTENNAS

The uplink receiving antenna and the downlink transmitting antenna on a communication satellite can be classified into three types. These antennas are designed to have beamwidths that provide global coverage, hemispheric coverage, or spot coverage. Figure 12–3 illustrates the various types of spacecraft antenna patterns. Figure 12–3(a) shows a global coverage, and Figure 12–3(b) shows two spot beams that straddle the Atlantic Ocean to relay signals from North America to Europe. Figure 12–3(c) illustrates the use of multiple spot beams to cover a larger area, and Figure 12–3(d) shows two hemispheric beams with dual linear polarization which doubles the number of channels through polarization diversity. The effective receiving area or transmitting antenna gain is controlled by the coverage. Global coverage from the geostationary orbit requires a 3-dB beamwidth of 18°. Hemispheric coverage of the continental United States (Conus) requires a 3-dB beamwidth of approximately 8°. Spot beams can be as narrow as 0.5° to 1°. Multiple spot beams can be employed to cover an area such as a single country. A 1° beam covers an area with a diameter of approximately 630 km.

The 3-dB beamwidth of a parabolic antenna as a function of the antenna diameter from (9–38) is

$$\theta_{3\mathrm{dB}} = 70\frac{\lambda}{D} \text{ degrees} \tag{12–1}$$

The isotropic gain of a parabolic reflector antenna as a function of the antenna diameter from (9–36) is

$$G_i = \eta_I\left(\frac{\pi D}{\lambda}\right)^2 \tag{12–2}$$

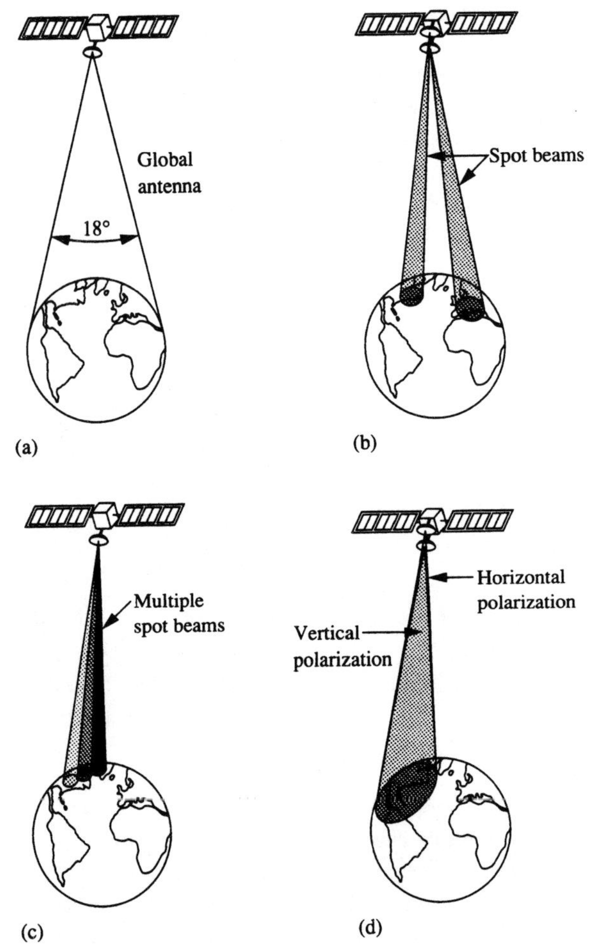

FIGURE 12–3 Spot beaming from satellites: (a) simple global beam, (b) two spot beams, (c) multiple spots, and (d) dual polarized spots.

EXAMPLE 12–1

Determine the diameter and the isotropic gain of uplink and downlink antennas on a communication satellite that provides hemispheric coverage of the continental United States with a 3-dB beamwidth of 8°. The downlink frequency is 3.720 GHz, and the uplink frequency is 5.945 GHz. The illumination efficiency is 70%.

Solution

The diameter of the antenna as a function of the beamwidth can be determined by rearranging (12–1):

$$D = \frac{70\lambda}{\theta_{3dB}} \tag{12–3}$$

where θ_{3dB} is expressed in degrees. The wavelength of the downlink frequency can be determined from

$$\lambda = \frac{c}{f} = \frac{3 \times 10^8}{3.720 \times 10^9} = 8.0645 \text{ cm} \tag{12–4}$$

The wavelength of the uplink frequency is

$$\lambda = \frac{c}{f} = \frac{3 \times 10^8}{5.945 \times 10^9} = 5.0463 \text{ cm} \tag{12–5}$$

The diameter of the downlink antenna is

$$D = \frac{70\lambda}{\theta_{3dB}} = \frac{70(8.0645 \times 10^{-2})}{8} = 0.7056 \text{ m} \tag{12–6}$$

The diameter of the uplink antenna is

$$D = \frac{70\lambda}{\theta_{3dB}} = \frac{70(5.0463 \times 10^{-2})}{8} = 0.4416 \text{ m} \tag{12–7}$$

The isotropic gain of the downlink antenna is

$$G_i = \eta_I \left(\frac{\pi D}{\lambda}\right)^2 = 0.70 \left(\frac{\pi(0.7056)}{(8.0645 \times 10^{-2})}\right)^2 = 528.88 \tag{12–8}$$

The isotropic gain of the uplink antenna is

$$G_i = \eta_I \left(\frac{\pi D}{\lambda}\right)^2 = 0.70 \left(\frac{\pi(0.4416)}{(5.0463 \times 10^{-2})}\right)^2 = 529.07 \tag{12–9}$$

The gains of both the uplink and downlink antennas should be the same since they have the same 3-dB beamwidth. Normally the gains are expressed in decibels. Thus,

$$G_{i_{db}} = 10 \log G_i = 10 \log(529) = 27.24 \text{ dB} \tag{12–10}$$

12–5 ■ EARTH STATION ANTENNAS

The requirements for earth station antennas for both the uplink and the downlink are based on two considerations. First, the antenna isotropic gain must be sufficient to achieve the required received carrier power-to-noise ratio. Second, the 3-db beamwidth for the uplink must be narrow enough that it does not cause interference with adjacent satellites, which are located every 2° in the geostationary orbit. The downlink earth station 3-dB beamwidth must be narrow enough not to receive signals from two or more adjacent satellites.

Whether the first consideration or the second consideration has greater influence over the antenna physical size is determined by the overall system design.

The angular distance between the first nulls on each side of the maximum antenna gain in an antenna radiation pattern is typically 2.5 times the 3-dB beamwidth for a parabolic reflector antenna. Therefore, an ideal condition would be to point the maximum on-axis gain of the earth station to the desired satellite position. The two adjacent satellites would fall in the first nulls of the antenna pattern on each side of the maximum on-axis gain. The null-to-null spacing would be 4°, which would result in a 3-dB beamwidth of 1.6°. Figure 12–4 illustrates the earth station antenna beamwidth requirement discussed above. The diameter of the earth station antenna that meets these criteria is

$$D = 70\frac{\lambda}{\theta_{3\text{dB}}} = \left(\frac{70}{1.6}\right)\lambda = 43.75\lambda \qquad \textbf{(12–11)}$$

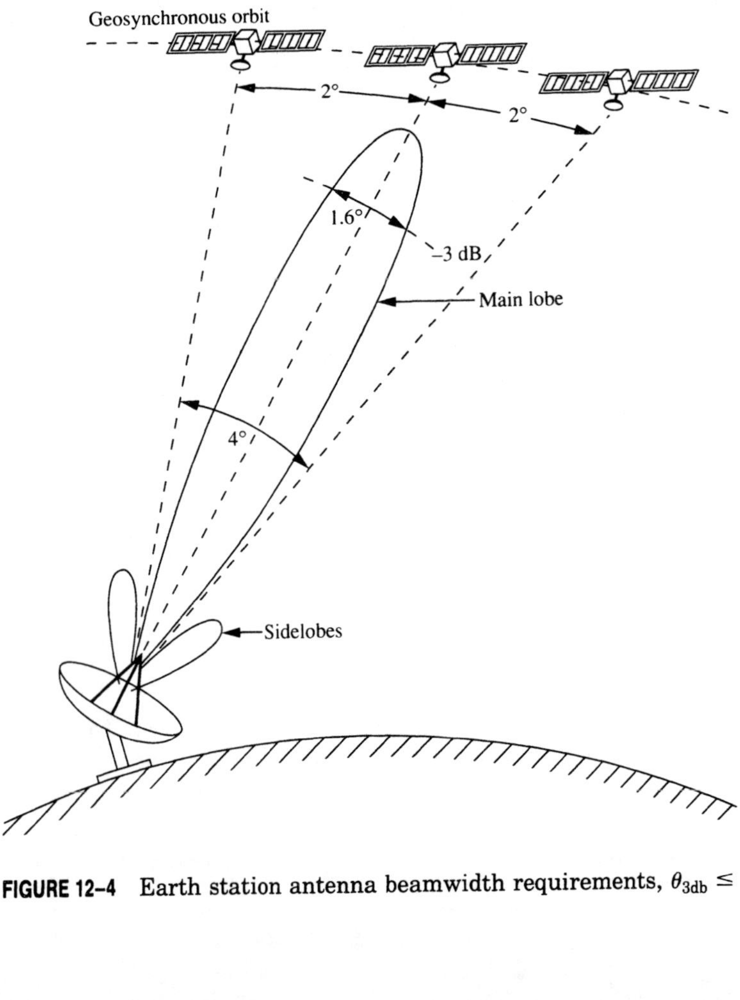

FIGURE 12-4 Earth station antenna beamwidth requirements, $\theta_{3\text{db}} \leq 1.6°$.

The isotropic gain of a lossless antenna is a function of the 3-dB beamwidth. Substitution of (12–1) into (12–2) yields

$$G_i = \eta_I\left(\frac{70\pi}{\theta}\right)^2 = (4.836 \times 10^4)\frac{\eta_I}{\theta^2} \qquad \textbf{(12–12)}$$

where the 3-dB beamwidth is expressed in degrees.

The isotropic gain for a 3-dB beamwidth of 1.6° is shown to be 41.21 dB in Example 12–2 which follows this discussion. If more gain is required, the diameter of the antenna can be increased. However, the 3-dB beamwidth decreases as the gain is increased. Due to problems of pointing (pointing losses and reflector surface accuracy), the upper limit to reducing antenna beamwidth is a beamwidth of approximately 0.3°. Substitution of 0.3° into (12–12) yields an upper limit to earth station antenna isotropic gains of 55.75 dB.

The development of the Ku-band satellite communication frequency bands provided a significant reduction in the size of the earth station antennas. Referring to Examples 12–2 and 12–3, note that the use of Ku-band instead of C-band reduces the antenna diameters by a factor of approximately 3, thereby reducing the area by a factor of 9. Therefore, not only are the antennas less expensive; they are also more portable and easier to handle. The future use of K-band reduces the antenna diameter by a factor of approximately 5 over that of C-band, and it reduces the area by a factor of 25. This is significant reduction in area and will enhance the use of ultrasmall mobile earth stations.

EXAMPLE 12–2

Determine the earth station antenna diameters and isotropic gains for C-band, Ku-band, and K-band uplinks and downlinks to achieve a 3-dB beamwidth of 1.6°. Assume the illumination efficiency to be 70% in all cases. For each frequency band, use the lowest frequency that results in the largest antenna diameter for that band.

Solution

The lowest downlink frequency for C-band, Ku-band, and K-band is 3.7 GHz, 10.7 GHz, and 17.7 GHz, respectively. The lowest uplink frequencies are 5.85 GHz, 14.0 GHz, and 27.5 GHz for C-band, Ku-band, and K-band, respectively. The free-space wavelength for 3.7 GHz is

$$\lambda = \frac{c}{f} = \frac{3 \times 10^8}{3.7 \times 10^9} = 8.11 \text{ cm} \qquad \textbf{(12–13)}$$

Using (12–13), the free-space wavelengths for 5.85 GHz, 10.7 GHz, 14.0 GHz, 17.7 GHz, and 27.5 GHz are 5.13 cm, 2.80 cm, 2.14 cm, 1.69 cm, and 1.09 cm. The diameter as a function of a 1.6° 3-dB beamwidth and free-space wavelength is

$$D = 70\frac{\lambda}{\theta_{3\text{dB}}} = \left(\frac{70}{1.6}\right)\lambda = 43.75\lambda \qquad \textbf{(12–14)}$$

Substituting the free-space wavelengths into (12–14) results in the following diameters:

	C-band	Ku-band	K-band
Downlink	3.55 m	1.23 m	0.739 m
Uplink	2.24 m	0.936 m	0.477 m

The isotropic gain is a function only of the illumination efficiency and the 3-dB beamwidth. Therefore, the isotropic gain can be determined from (12–12):

$$G_i = (4.836 \times 10^4)\frac{\eta_I}{\theta^2} = (4.836 \times 10^4)\frac{0.7}{1.6^2} \tag{12–15a}$$

$$= 1.322 \times 10^4 \tag{12–15b}$$

Expressing the isotropic gain in decibels, we have

$$G_{i_{\text{dB}}} = 10 \log G_i = 10 \log(1.322 \times 10^4) = 41.21 \text{ dB} \tag{12–16}$$

EXAMPLE 12–3

Determine the earth station downlink and uplink antenna diameters required to obtain an isotropic gain of 55 dB, which is the practical upper limit for communication satellite earth stations. Determine for the lowest downlink and uplink frequency in the C-band, Ku-band, and K-band. Refer to Example 12–2 for the frequencies and corresponding wavelengths. Assume an illumination efficiency of 70%.

Solution

Rearranging (12–2) to obtain the antenna diameter as a function of the isotropic antenna gain, we obtain

$$D = \sqrt{\frac{G_i}{\eta_I}\frac{\lambda}{\pi}} \tag{12–17}$$

The isotropic gain of 55 dB must be expressed as a factor:

$$G_i = 10^{G_{i_{\text{dB}}}/10} = 10^{55/10} = 3.16 \times 10^5 \tag{12–18}$$

Substitution of the illumination efficiency, isotropic antenna gain, and wavelength into (12–17) for the 3.7-GHz C-band downlink frequency yields

$$D = \sqrt{\frac{G_i}{\eta_I}\frac{\lambda}{\pi}} = \sqrt{\frac{(3.16 \times 10^5)}{0.70}\left(\frac{8.11 \times 10^{-2}}{\pi}\right)} \tag{12–19a}$$

$$= 17.34 \text{ m} \tag{12–19b}$$

The free-space wavelengths for 5.85 GHz, 10.7 GHz, 14.0 GHz, 17.7 GHz, and 27.5 GHz are 5.13 cm, 2.80 cm, 2.14 cm, 1.69 cm, and 1.09 cm. Substituting the free-space wavelengths into (12–19) results in the following diameters:

	C-band	Ku-band	K-band
Downlink	17.34 m	5.99 m	3.61 m
Uplink	10.97 m	4.58 m	2.33 m

12–6 ■ SATELLITE LINK ANALYSIS

Two principal communication links are involved in most satellite communication systems. First, there is the **uplink** which is the transmission from the transmitting station located on the earth's surface to the communication satellite in orbit about the earth. The second is the **downlink** which is the transmission from the communication satellite to the receiving station on the earth's surface. Both of these communication links are point-to-point links, and therefore the Friis transmission formula described in Section 10–3 is applicable:

$$P_r = \frac{\lambda^2 G_t G_r P_t}{(4\pi)^2 r^2} \tag{12–20}$$

The level of the received carrier power C must exceed the received noise power level N by an amount determined by system performance criteria. These criteria might be the quality of a television picture or the bit error rate in a digital communication system. These requirements usually translate into a required carrier power-to-noise power ratio or C/N. The required carrier-to-noise power ratio typically varies from 7 to 25 dB. The received carrier power is determined from the Friis transmission equation (12–20).

The noise power is a function of the total system noise and the bandwidth of the receiver. The system noise power level was determined in Chapter 10 by (10–40):

$$N_{sys} = kT_{sys}B \tag{12–21}$$

In (12–21), T_{sys} is the system noise temperature determined by (10–39) as follows:

$$T_{sys} = T_e + T_{ext} + T_{ohm} \tag{12–22}$$

where

T_e = effective input noise temperature of the receiver

T_{ext} = external atmospheric noise (as shown in Fig. 12–2)

T_{ohm} = noise from ohmic losses in the antenna and transmission line between the antenna and the first amplifier in the receiver

The noise specifications for receivers are usually given in noise figure decibels rather than effective input noise temperature in kelvins. The relationship between the noise figure in dB and the noise figure as a factor is

$$F = 10^{F_{dB}/10} \tag{12–23}$$

The effective input noise temperature is

$$T_e = (F - 1)290 \tag{12–24}$$

The carrier power-to-system noise power ratio can be determined from (12–20), (12–21), and (12–22):

$$\frac{C}{N} = \frac{P_r}{kT_{\text{sys}}B} = \frac{\lambda^2 G_t G_r P_t}{(4\pi r)^2 kT_{\text{sys}}B} \tag{12–25}$$

This equation is usually rearranged as follows:

$$\frac{C}{N} = \left(\frac{P_t G_t}{kB}\right)\left(\frac{\lambda}{4\pi r}\right)^2\left(\frac{G_r}{T_{\text{sys}}}\right) \tag{12–26}$$

where the first term is the gain function of the transmitting station. The product of the transmitter power and transmitting antenna gain is the effective isotropic radiated power (EIRP) as defined by (10–17).

Figure 12–5 provides typical satellite downlink effective isotropic radiated power contours over the continental United States. The figure is for the vertically polarized Ku-band downlink on the *SBS 4* satellite located at 91° west longitude. The bandwidth B is determined by the satellite communication channels being used. The constant k is Boltzmann's constant ($k = 1.38 \times 10^{-23}$ J/K). The second term in (12–26) is the free-space path loss defined by (10–22); it is a function of the range r and the free-space wavelength λ.

The third term in (12–26) is the gain function of the receiving station. It is equal to the ratio of the receiving antenna isotropic gain to the system noise temperature. The G_r/T_{sys} ratio is a specification used in satellite communication systems, especially for the downlink earth receiving station. Typically the first amplifier in the receiver, referred to as the **low noise amplifier (LNA),** is physically integrated into the receiving antenna feed system. This eliminates the transmission line ohmic loss noise term and reduces the system noise power.

The C/N ratio is typically expressed in decibels. Therefore, link analysis is normally accomplished by adding or subtracting decibels rather than by multiplying or dividing factors. The standard procedure is to add system gains and subtract system losses. The first and third factors of (12–26) must be expressed as decibels and are gains. The second term is the reciprocal of the free-space path loss as defined in (10–22); it is subtracted when the free-space path loss is expressed in decibels. Other losses, such as atmospheric loss, rain loss, and antenna pointing loss, must also be subtracted when determining the C/N ratio.

The link analysis equation for the C/N ratio in decibels is

$$\left(\frac{C}{N}\right)_{\text{dB}} = [10\log(P_t G_t) - 10\log(kB)] + [10\log(G_r) - 10\log(T_{\text{sys}})] - \left[20\log\left(\frac{4\pi r}{\lambda}\right)\right] - \text{other losses in dB} \tag{12–27}$$

SBS 4

SBS 4 (91° west longitude)

Typical Effective Isotropic Radiated Power (EIRP)

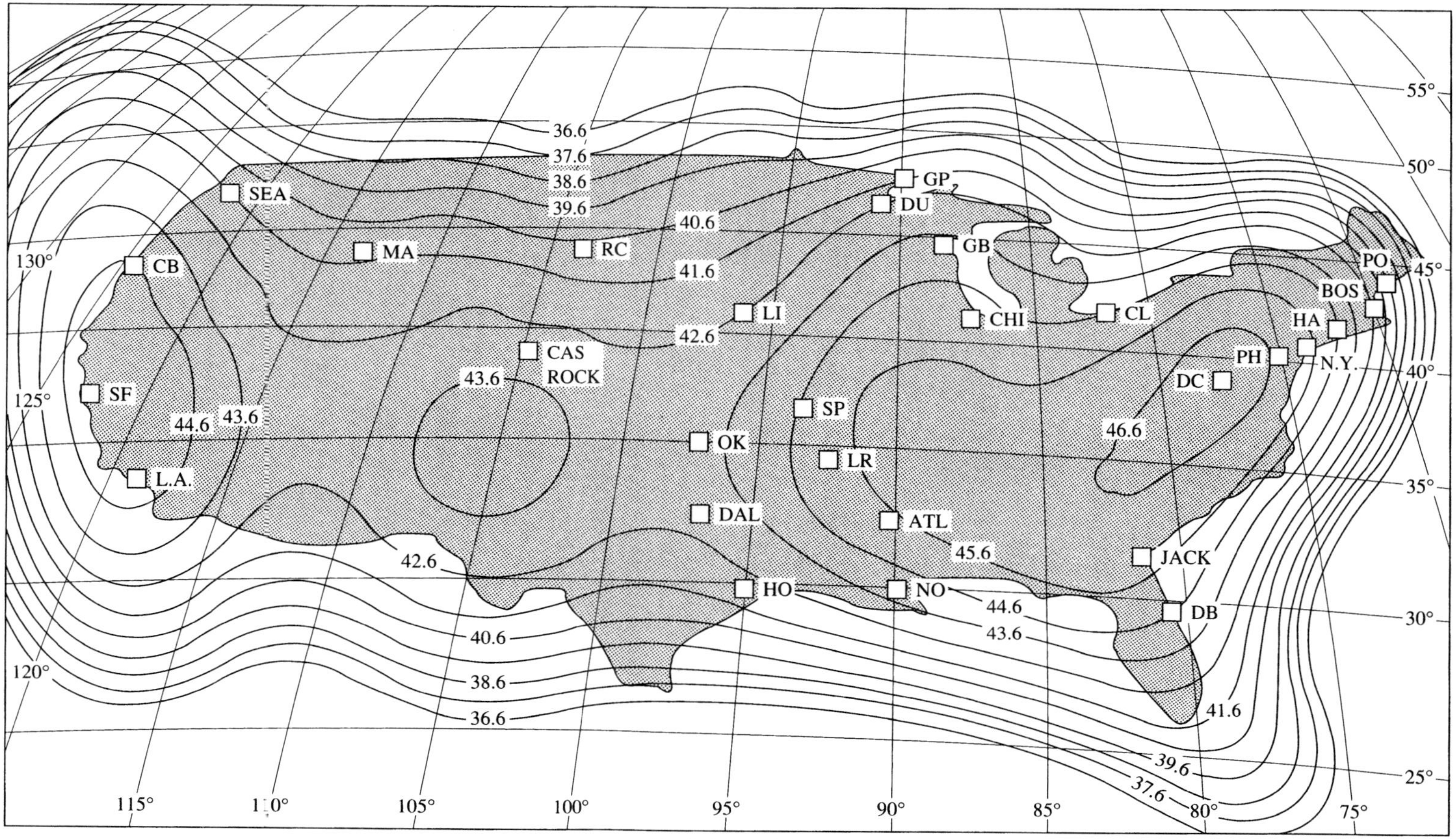

FIGURE 12–5 Satellite downlink effective isotropic radiated power footprint. (Hughes Aircraft Company, Hughes Telecommunications/Space Sector. Used with permission.)

The link analysis equation for the C/N ratio with the factor expressed in decibels units is

$$\left(\frac{C}{N}\right)_{\text{dB}} = 228.6 + \text{EIRP} - 10\log(B) + G_r - 10\log(T_{\text{sys}}) - L_p - L_a \quad \textbf{(12–28)}$$

where

B = bandwidth in Hz
T_{sys} = system noise temperature in K
L_p = free-space path loss in dB
L_a = atmospheric loss in dB
EIRP = $P_t G_t$ = effective isotropic radiated power in dBW
G_r = receiving antenna gain in dB

The distance from the earth station to the satellite in the geosynchronous orbit is a function of the latitude and longitude of the earth station and the longitude of the communication satellite. The range can be determined from the following equation:

$$r = \sqrt{R_o^2 + 2R_e(R_o + R_e)(1 - \cos l \cos L)} \quad \textbf{(12–29)}$$

where

R_o = satellite altitude (35,786 km)
R_e = radius of the earth (6378 km)
l = latitude of the earth station
L = difference between longitudes of the earth station and the satellite

Factoring the satellite altitude out of (12–29) and substituting the values of R_e and R_o into (12–29) results in

$$r = (3.5786 \times 10^4)\sqrt{1 + 0.42(1 - \cos l \cos L)} \quad \textbf{(12–30)}$$

The maximum value of the multiplication factor, due to the term under the square root, is 1.1645. This value will increase the free-space loss from that of the shortest distance by only 1.32 dB. The range varies from 35,786 km to 41,673 km.

The free-space path loss L_p is a function of the range and the free-space wavelength. Both the range and the free-space wavelength must be in the same units of length. The free-space path loss expressed in decibels can be determined from

$$L_p = 20\log(4\pi) + 20\log r - 20\log\lambda \quad \textbf{(12–31a)}$$
$$= 21.98 + 20\log r - 20\log\lambda \text{ dB} \quad \textbf{(12–31b)}$$

EXAMPLE 12-4

Determine the range from an earth station located in New York City to the *SBS 5* communication satellite located at 123° west longitude. The latitude and longitude of New York are 40.5° north and 70.2° west.

Solution

The difference in longitudes is 52.8°. Substitution into (12–30) yields

$$r = (3.5786 \times 10^4)\sqrt{1 + 0.42(1 - \cos l \cos L)} \quad \textbf{(12–32a)}$$

$$= (3.5786 \times 10^4)\sqrt{1 + 0.42(1 - \cos(40.5°)\cos(52.8°))} \quad \textbf{(12–32b)}$$

$$= 3.9639 \times 10^4 \text{ km} \quad \textbf{(12–32c)}$$

EXAMPLE 12-5

Determine the carrier power-to-system noise power C/N ratio for both the uplink and the downlink of a C-band satellite communication system with an uplink frequency of 6.125 GHz and a downlink frequency of 3.9 GHz. The bandwidth B of the satellite transponder is 36 MHz. The communication satellite *Galaxy VII,* located at 91° west longitude, radiates an EIRP of 47.8 dBW toward Atlanta. Atlanta is located at 32° north latitude and 90° west longitude. The uplink earth station is located at New York City, whose location is 40.5° north latitude and 70.2° west longitude. The noise figure of the satellite's receiver is 3 dB, and the noise figure of the earth station is 2.5 dB. The external noise temperature for the earth station receiver is 100 K, while that for the satellite receiver is 300 K. The ohmic loss noise temperature for the earth station is 50 K and 150 K for the satellite. The atmospheric loss at 3.9 GHz is 1.0 dB and at 6.125 GHz is 1.6 dB. The uplink earth station has an isotropic gain of 55 dB and a power output of 100 W. The gain of the uplink receiving antenna on the satellite is 27 dB. The gain of the downlink receiving antenna is 42 dB.

Solution

The range for the uplink from New York City to the *SBS 4* satellite can be determined from (12–30), where l is 40.5° and the difference in longitudes is 20.8°:

$$r = (3.5786 \times 10^4)\sqrt{1 + 0.42(1 - \cos l \cos L)} \quad \textbf{(12–33a)}$$

$$= (3.5786 \times 10^4)\sqrt{1 + 0.42(1 - \cos 40.2° \cos 20.8°)} \quad \textbf{(12–33b)}$$

$$= 3.7874 \times 10^4 \text{ km} \quad \textbf{(12–33c)}$$

The range for the downlink from the *SBS 4* satellite to Atlanta also can be determined from (12–30), where l is 32° and the difference in longitudes is 1°.

$$r = (3.5786 \times 10^4)\sqrt{1 + 0.42(1 - \cos l \cos L)} \quad \textbf{(12–34a)}$$

$$= (3.5786 \times 10^4)\sqrt{1 + 0.42(1 - \cos 32° \cos 1°)} \quad \textbf{(12–34b)}$$

$$= 3.8072 \times 10^4 \text{ km} \quad \textbf{(12–34c)}$$

The free-space wavelength must be determined for the downlinks and uplinks whose frequencies are 3.9 GHz and 6.125 GHz, respectively. The wavelengths are

$$\lambda = \frac{c}{f} = \frac{3 \times 10^8}{3.9 \times 10^9} = 7.69 \times 10^{-2}\,\text{m} \tag{12–35}$$

$$\lambda = \frac{c}{f} = \frac{3 \times 10^8}{6.125 \times 10^9} = 4.90 \times 10^{-2}\,\text{m} \tag{12–36}$$

The free-space path loss for the uplink can be determined from (12–31):

$$L_p = 21.98 + 20 \log r - 20 \log \lambda \tag{12–37a}$$
$$= 21.98 + 20 \log(3.7874 \times 10^7) - 20 \log(4.90 \times 10^{-2}) \tag{12–37b}$$
$$= 195.62\ \text{dB} \tag{12–37c}$$

The free-space path loss for the downlink is

$$L_p = 21.98 + 20 \log r - 20 \log \lambda \tag{12–38a}$$
$$= 21.98 + 20 \log(3.8072 \times 10^7) - 20 \log(7.69 \times 10^{-2}) \tag{12–38b}$$
$$= 195.88\ \text{dB} \tag{12–38c}$$

The system noise temperature for the uplink is determined by the low noise amplifier (LNA) within the receiving system on the satellite. Since the satellite receiving antenna is looking at the surface of the earth, the external noise temperature of the earth and intervening atmosphere is 300 K. The effective input noise temperature of the uplink LNA is determined from (12–23) and (12–24):

$$F = 10^{F_{\text{dB}}/10} = 10^{3.0/10} = 1.995 \tag{12–39}$$

$$T_e = (F - 1)290 = (1.995 - 1)(290) = 288.6\ \text{K} \tag{12–40}$$

The effective input noise temperature of the downlink LNA is

$$F = 10^{F_{\text{dB}}/10} = 10^{2.5/10} = 1.778 \tag{12–41}$$

$$T_e = (F - 1)290 = (1.778 - 1)(290) = 225.7\ \text{K} \tag{12–42}$$

The uplink system noise temperature can be determined from (12–22):

$$T_{\text{sys}} = T_e + T_{\text{ext}} + T_{\text{ohm}} = 288.6 + 300 + 150 = 738.6\ \text{K} \tag{12–43}$$

The downlink system noise temperature is

$$T_{\text{sys}} = T_e + T_{\text{ext}} + T_{\text{ohm}} = 225.7 + 100 + 50 = 375.7\ \text{K} \tag{12–44}$$

The C/N ratio for the uplink can be determined from (12–28):

$$\left(\frac{C}{N}\right)_{\text{dB}} = 228.6 + \text{EIRP} - 10 \log(B) + G_r - 10 \log(T_{\text{sys}}) - L_p - L_a \tag{12–45a}$$

$$= 228.6 + 75.0 - 10 \log(3.6 \times 10^7) + 27 - 10 \log(738.6) - 195.62 - 1.6 \tag{12–45b}$$
$$= 228.6 + 75.0 - 75.6 + 27 - 28.7 - 195.6 - 1.6 \tag{12–45c}$$
$$= 5.0\ \text{dB} \tag{12–45d}$$

The C/N ratio for the downlink is

$$\left(\frac{C}{N}\right)_{\text{dB}} = 228.6 + \text{EIRP} - 10\log(B) + G_r - 10\log(T_{\text{sys}}) - L_p - L_a \qquad \textbf{(12–46a)}$$

$$= 228.6 + 47.8 - 10\log(3.6 \times 10^7) + 42$$
$$- 10\log(375.7) - 195.9 - 1.2 \qquad \textbf{(12–46b)}$$
$$= 228.6 + 47.8 - 75.6 + 42 - 25.8 - 195.9 - 1.2 = 19.9 \text{ dB} \qquad \textbf{(12–46c)}$$

■ PROBLEMS

12–1. Determine the diameter and the isotropic gain of uplink and downlink antennas on a communication satellite that provides global coverage of the earth with a 3-dB beamwidth of 18°. The downlink frequency is 3.720 GHz, and the uplink frequency is 5.945 GHz. The illumination efficiency is 70%.

12–2. Determine the diameter and the isotropic gain of uplink and downlink antennas on a communication satellite that provides hemispheric coverage of the continental United States with a 3-dB beamwidth of 8°. The downlink frequency is 10.95 GHz, and the uplink frequency is 14.00 GHz. The illumination efficiency is 70%.

12–3. Determine the earth station antenna diameters and the isotropic gains for C-band, Ku-band, and K-band uplinks and downlinks to achieve a 3-dB beamwidth of 1.0°. Assume the illumination efficiency to be 70% in all cases. For each frequency band, use the lowest frequency that results in the largest antenna diameter for that band.

12–4. Determine the earth station antenna diameters and the isotropic gains for C-band, Ku-band, and K-band uplinks and downlinks to achieve a 3-dB beamwidth of 0.6°. Assume the illumination efficiency to be 70% in all cases. For each frequency band, use the lowest frequency that results in the largest antenna diameter for that band.

12–5. Repeat Problem 12–3 for the highest frequency in each band.

12–6. Repeat Problem 12–4 for the highest frequency in each band.

12–7. An antenna commonly used in C-band satellite ground stations has a diameter of 12 ft. Determine the 3-dB beamwidth and the isotropic antenna gain for both the highest and the lowest frequency in both the downlink and the uplink. Assume an illumination efficiency of 70%.

12–8. An antenna commonly used in Ku-band satellite ground stations has a diameter of 4 ft. Determine the 3-dB beamwidth and the isotropic antenna gain for both the highest and the lowest frequency in both the downlink and the uplink. Assume an illumination efficiency of 70%.

12–9. Determine the range from an earth station located in New York City to the *GSTAR IV* communication satellite located at 105° west longitude. The latitude and longitude of New York are 40.5° north and 70.2° west.

12–10. Determine the range from an earth station located in New York City to the *SATCOM K1* communication satellite located at 85° west longitude. The latitude and longitude of New York are 40.5° north and 70.2° west.

12–11. Determine the carrier power-to-system noise power ratio C/N for both the uplink and the downlink of the Ku-band satellite communication system with an uplink frequency of 14.02 GHz and a downlink frequency of 11.72 GHz. The bandwidth of the satellite transponder is 43 MHz. The communication satellite *SBS 6* located at 99° west longitude radiates an EIRP of 50.4 dBW toward Atlanta. Atlanta is located at 32° north latitude and 90° west longitude. The uplink earth station is located at New York City, whose location is 40.5° north latitude and 70.2° west longitude. The noise figure of the satellite's receiver is 4 dB, and the noise figure of the earth station is 3.5 dB. The external noise temperature for the earth station receiver is 150 K, while the external noise temperature for the satellite receiver is 300 K. The ohmic loss noise temperature for the earth station is 50 K and 150 K for the satellite. The atmospheric loss at 11.72 GHz is 1.5 dB and at 14.02 GHz is 2.2 dB. The uplink earth station has an isotropic gain of 55 dB and a power output of 500 W. The gain of the uplink receiving antenna on the satellite is 30 dB. The gain of the downlink receiving antenna is 46 dB.

12–12. Determine the carrier power-to-system noise power C/N ratio for both the uplink and the downlink of a C-band satellite communication system with an uplink frequency of 6.405 GHz and a downlink frequency of 4.18 GHz. The bandwidth of the satellite transponder is 36 MHz. The communication satellite *Galaxy II,* located at 74° west longitude, radiates an EIRP of 37.7 dBW toward Atlanta. Atlanta is located at 32° north latitude and 90° west longitude. The uplink earth station is located at New York City, whose location is 40.5° north latitude and 70.2° west longitude. The noise figure of the satellite's receiver is 2.5 dB, and the noise figure of the earth station is 1.5 dB. The external noise temperature for the earth station receiver is 100 K, while the external noise temperature for the satellite receiver is 300 K. The ohmic loss noise temperature for the earth station is 25 K and 100 K for the satellite. The atmospheric loss at 4.18 GHz is 1.2 dB and at 6.405 GHz is 1.9 dB. The uplink earth station has an isotropic gain of 55 dB and a power output of 1000 W. The gain of the uplink receiving antenna on the satellite is 27 dB. The gain of the downlink receiving antenna is 48 dB.

Answers to Odd-Numbered Problems

CHAPTER 1

1–1. **(a)** 150 km
(b) 1.5 km
(c) 15 m
(d) 15 cm

1–3. 465 mi

1–5. 3.75 MHz

1–7. 9 cm

1–9. 3.75 cm

1–11. 500 μH

1–13. 250 nF

1–15. 74.93 Ω

1–17. 53.05 pF/m

1–19. 1.005 mH/m

1–21. 2.341×10^8 m/s

1–23. 1.384×10^8 m/s

1–25. 347.6 nH/m, 65.23 pF/m

1–27. 136.4 m

1–29. 1.619 m

CHAPTER 2

2–1. **(a)** 40 V starting at $t = 0$, 40 V starting at $t = 0.4\ \mu s$
(b) 0.8 A starting at $t = 0$, 0.8 A starting at $t = 0.4\ \mu s$

2–3. **(a)** $v(0,t) = 40$ V for $0 < t < 0.8\ \mu s$
$= 30$ V for $t > 0.8\ \mu s$
$v(80,t) = 30$ V for $t > 0.4\ \mu s$
(b) $i(0,t) = 0.8$ A for $0 < t < 0.8\ \mu s$
$= 1$ A for $t > 0.8\ \mu s$
$i(80,t) = 1$ A for $t > 0.4\ \mu s$

2–5. **(a)** $v(0,t) = 50$ V for $0 < t < 0.4\ \mu s$
$= 5$ V for $0.4\ \mu s < t < 0.8\ \mu s$

$= 18.5$ V for $0.8\ \mu s < t < 1.2\ \mu s$
$= 14.45$ V for $1.2\ \mu s < t < 1.6\ \mu s$, etc.
$v(0,\infty) = 15.38$ V
$v(40,t) = 0$ V for $0 < t < 0.2\ \mu s$
$= 20$ V for $0.2\ \mu s < t < 0.6\ \mu s$
$= 14$ V for $0.6\ \mu s < t < 1\ \mu s$
$= 15.8$ V for $1\ \mu s < t < 1.4\ \mu s$, etc.
$v(40,\infty) = 15.38$ V

(b) $i(0,t) = 1$ A for $0 < t < 0.4\ \mu s$
$= 1.3$ A for $0.4\ \mu s < t < 0.8\ \mu s$
$= 1.21$ A for $0.8\ \mu s < t < 1.2\ \mu s$
$= 1.237$ A for $1.2\ \mu s < t < 1.6\ \mu s$, etc.
$i(0,\infty) = 1.231$ A
$i(40,t) = 0$ A for $0 < t < 0.2\ \mu s$
$= 1.6$ A for $0.2\ \mu s < t < 0.6\ \mu s$
$= 1.12$ A for $0.6\ \mu s < t < 1\ \mu s$
$= 1.264$ A for $1\ \mu s < t < 1.4\ \mu s$, etc.
$i(40,\infty) = 1.231$ A

2–7. (a) $v(0,t) = 40$ V for $0 < t < 0.1\ \mu s$
$= -10$ V for $0.8\ \mu s < t < 0.9\ \mu s$
$= 0$ otherwise
$v(80,t) = 30$ V for $0.4\ \mu s < t < 0.5\ \mu s$
$= 0$ otherwise

(b) $i(0,t) = 0.8$ A for $0 < t < 0.1\ \mu s$
$= 0.2$ A for $0.8\ \mu s < t < 0.9\ \mu s$
$= 0$ otherwise
$i(80,t) = 1$ A for $0.4\ \mu s < t < 0.5\ \mu s$
$= 0$ otherwise

2–9. All results that follow are a series of pulses starting at the times indicated, each of which has a width of 0.1 μs.

(a) $v(0,t)$: $t = 0$, 50 V; $t = 0.4\ \mu s$, -45 V; $t = 0.8\ \mu s$, 13.5 V; $t = 1.2\ \mu s$, -4.05 V; etc.
$v(40,t)$: $t = 0.2\ \mu s$, 20 V; $t = 0.6\ \mu s$, -6 V; $t = 1\ \mu s$, 1.8 V; etc.

(b) $i(0,t)$: $t = 0$, 1 A; $t = 0.4\ \mu s$, 0.3 A; $t = 0.8\ \mu s$, -0.09 A; $t = 1.2\ \mu s$, 0.027 A; etc.
$i(40,t)$: $t = 0.2\ \mu s$, 1.6 A; $t = 0.6\ \mu s$, -0.48 A; $t = 1\ \mu s$, 0.144 A; etc.

2–11. 10-V pulse of width 1 μs starting at $t = 0$

2–13. $v(0,t) = 10$ V for $0 < t < 2\ \mu s$
$v(0,t) = 20$ V for $t > 2\ \mu s$

2–15. 10-V pulse of width 0.1 μs starting at $t = 0$ followed by a -10-V pulse of width 0.1 μs starting at $t = 1\ \mu s$

2–17. 10-V pulse of width 0.4 μs starting at $t = 0$ followed by a 10-V pulse of width 0.4 μs starting at $t = 2\ \mu s$

2–19. $v(0,t) = 10$ V for $0 < t < 1\ \mu s$
$v(0,t) = -5$ V for $1\ \mu s < t < 2\ \mu s$
$v(0,t) = 2.5$ V for $2\ \mu s < t < 3\ \mu s$
$v(0,t) = -1.25$ V for $3\ \mu s < t < 4\ \mu s$, etc.

2–21. All results that follow are a series of pulses starting at the times indicated, each of which has a width of 0.2 μs.
0, 10 V; 1 μs, −15 V; 2 μs, 7.5 V; 3 μs, −3.75 V; etc.

2–23. **(a)** 50 m
(b) 200 Ω

2–25. **(a)** 33.33 Ω
(b) −3.2 V

2–27. **(a)** 26.92 Ω
(b) 0.9 V

CHAPTER 3

3–1. **(a)** $+x$
(b) 8 mA
(c) 2π Mrad/s
(d) 0.025 rad/m
(e) 1 MHz
(f) 1 μs
(g) 251.3 m
(h) 2.513×10^8 m/s

3–3. **(a)** 20 ns
(b) 3.142×10^8 rad/s
(c) 1.571 rad/m
(d) 4 m
(e) $2\cos(3.142 \times 10^8 t - 1.571x)$

3–5. **(a)** $\bar{I}^+ = 8\underline{/0}$ A
(b) $8\underline{/-0.025x}$ A
(c) $8\underline{/-2.5}$ A

3–7. **(a)** $\bar{I}^+ = 8\underline{/1.5}$ A
(b) $8\underline{/-0.025x + 1.5}$ A
(c) $8\underline{/-1}$ A

3–9. $5.657\underline{/0}$ A, 1.6 kW

3–11. $5.657\underline{/1.5}$ A, yes

3–13. $100\underline{/0}$ V

3–15. $-4\underline{/0}$ A or $4\underline{/\pi}$ A

3–17. **(a)** 0.012 dB/ft
(b) 1.382×10^{-3} Np/ft
(c) 3.6 dB
(d) 0.4145 Np
(e) 0.6607

3–19. **(a)** 7.959 dB
(b) 0.9163 Np
(c) 0.03980 dB/m
(d) 4.581×10^{-3} Np/m

3–21. **(a)** $377.29\underline{/1.5310}$ Ω/m
(b) $3.1416 \times 10^{-3}\underline{/1.5695}$ S/m
(c) $22.36 \times 10^{-3} + j1.088$
(d) 0.1942 dB/m
(e) 2.886×10^{8} m/s
(f) $346.5\underline{/-0.01926}$ Ω

3–23. **(a)** $2.605 \times 10^{-3} + j0.03474$
(b) 0.02263 dB/m
(c) 1.809×10^{8} m/s
(d) $58.06\underline{/-0.07403}$ Ω

3–25. **(a)** $0.08431\underline{/0.1013}$ A
(b) $24.97\underline{/-0.1027}$ V
(c) 2.061 W
(d) $0.01881\underline{/-0.4987}$ A
(e) $5.571\underline{/-0.7027}$ V
(f) 0.1026 W
(g) 13.03 dB

CHAPTER 4

4–1. **(a)** $0.25\underline{/\pi}$
(b) $0.25\underline{/\pi/2}$, $0.25\underline{/0}$, $0.25\underline{/-\pi}$

4–3. **(a)** $0.3333\underline{/-1.5708}$
(b) $0.3333\underline{/-\pi}$
(c) $0.3333\underline{/-4.712}$

4–5. $94.94\underline{/0.7560}$ Ω

4–7. 1.667

4–9. 2

4–11. 3

4–13. 0.5

4–15. **(a)** 30 Ω
(b) $30.30 + j4.019$ Ω
(c) $44.12 + j23.53$ Ω
(d) 83.33 Ω
(e) 30 Ω

4–17. $27.36 + j13.08$ Ω

4–19. **(a)** 108.84 W
(b) 109.15 W
(c) 119.71 W
(d) 132.96 W
(e) 108.84 W

4–21. **(a)** $j230.8$ Ω
(b) $-j54.49$ Ω

4–23. 183.7 nH

4–25. 6.056 cm

4–27. 35.24 cm

4–29. **(a)** $-j24.37\ \Omega$
(b) $j103.2\ \Omega$

■ CHAPTER 5

5–1. **(b)** $1.18 + j1.6$
(c) $23.6 \times 10^{-3} + j32 \times 10^{-3}$ S

5–3. 3.9, $0.592\underline{/-133^\circ}$

5–5. **(a)** $0.48 - j0.61$, $24 - j30.5\ \Omega$
(b) 0.33, 16.5 Ω
(c) $0.47 + j0.62$, $23.5 + j31\ \Omega$
(d) $1.7 + j1.35$, $85 + j67.5\ \Omega$

5–7. **(a)** $0.33 - j0.68$, $16.6 - j34\ \Omega$
(b) 4.7, $0.65\underline{/\ 37.5^\circ}$

■ CHAPTER 6

6–1. 40,000 V

6–3. 80 V

6–5. 0.2653 A/m

6–7. 39.789×10^{-9} C/m^2

6–9. 3.33×10^{-7} Wb/m^2

6–11. 2.4×10^{-5} C

6–13. 3.1416×10^{-7} Wb

6–15. 4.00×10^{5} A/m^2

6–17. 0.08 V/m

6–19. **(a)** 7.54×10^{-2} V/m
(b) 1.508×10^{-5} W/m^2
(c) 2.960×10^{-2} W

6–21. **(a)** 120 Ω
(b) 3.0×10^{-6} W/m^2
(c) 9.870

6–23. **(a)** $3.159 + j3.124\ \Omega$
(b) $12.50 + j12.64$
(c) 12.495 Np/m or 108.53 dB/m
(d) 4.972×10^{6} m/s

6–25. **(a)** $42.149 + j0.005\ \Omega$
(b) $0.0149 + j18.733$
(c) 0.0149 Np/m or 0.1295 dB/m
(d) 3.354×10^7 m/s

6–27. 1, quasi-conductor

6–29. 2.250, quasi-conductor

6–31. 22.50, quasi-conductor

6–33. 9 MHz

6–35. 90 GHz

CHAPTER 7

7–1. $454\angle 342.5^\circ$ V/m

7–3. $9.76\angle 81.68^\circ$ A/m

7–5. **(a)** 16 A/m, 5.305×10^{-9} C/m^2
(b) 5.305×10^{-9} C/m^2
(c) 16 A/m

7–7. **(a)** 1.48×10^{-6} H/m
(b) 17.32×10^{-12} F/m
(c) $291.97\ \Omega$
(d) 197.8×10^6 m/s

7–9. **(a)** 0.277×10^{-6} H/m
(b) 0.108×10^{-9} F/m
(c) $50.62\ \Omega$
(d) 182.6×10^6 m/s

7–11. **(a)** 0.2525×10^{-6} H/m
(b) 0.101×10^{-9} F/m
(c) 2.3

7–13. **(a)** $104.47\ \Omega$
(b) 1.970×10^8 m/s

7–15. **(a)** $11.79\ \Omega$
(b) 1.970×10^8 m/s

7–17. **(a)** 7.295
(b) $25.92\ \Omega$

7–19. **(a)** 6.02
(b) $90.24\ \Omega$

7–21. 6.241×10^{-5}

CHAPTER 8

8–1. 9.49 GHz

8–3. 18.99 GHz, 18.99 GHz, 21.23 GHz

8–5. 4.18 cm, 1.96 cm

8–7. Group: 1.809×10^8 m/s, 2.549×10^8 m/s
Phase: 4.976×10^8 m/s, 3.531×10^8 m/s

8–9. 625.34 Ω, 443.75 Ω

8–11. 280.99 Ω

8–13. 0.219 dB/m, 0.149 dB/m

CHAPTER 9

9–1. 87.5 W/sr

9–3. 0.8 $\mu W/m^2$

9–5. 1.705 m

9–7. 6.144 km

9–9. 20,106; 43.03 dB

9–11. 2.39×10^{-4} W/m^2

9–13. 36,626; 45.64 dB

9–15. 13.04 cm; 19,952.6; 27.00 m^2

9–17. 100 Ω

9–19. **(a)** 114.86 m^2
(b) 58.17 dBi
(c) 0.219°

9–21. 769.23 m^2

CHAPTER 10

10–1. 1.517

10–3. 1.667×10^8 m/s

10–5. 20.17°

10–7. 46.43°

10–9. 43.63°

10–11. 9.976 MW

10–13. 1.416×10^{-14} W or 14.16 fW

10–15. 206.49 dB

10–17. **(a)** 5.376×10^{-14} W, 53.76 fW
(b) 5.376×10^{-14} W, 53.76 fW or −102.7 dBm
(c) 359.2 K

10–19. 7.667×10^{-13} W, 766.7 fW

CHAPTER 11

11–1. 2.3×10^{14} Hz

11–3. **(a)** 68.89°
(b) 32.46°
(c) 0.537
(d) 0.067 or 0.097

11–5. 3.767 μm

11–7. 0.956 eV

11–9. 32.04 dB

11–11. 23.49 dB

11–13. 22.00 km

CHAPTER 12

12–1. 3.720 GHz: 0.3136 m, 104.48, 20.19 dBi
5.945 GHz: 0.1962 m, 104.48, 20.19 dBi

12–3. Uplink: C-band; 5.85 GHz, 3.591 m
Ku-band; 14.0 GHz, 1.498 m
K-band; 27.5 GHz, 0.763 m
Downlink: C-band; 3.7 GHz, 5.677 m
Ku-band; 10.7 GHz, 1.960 m
K-band; 17.7 GHz, 1.183 m
$G_i = 3.385 \times 10^4$ or 45.30 dBi at all frequencies

12–5. Uplink: C-band; 7.075 GHz, 2.968 m
Ku-band; 14.5 GHz, 1.448 m
K-band; 30.0 GHz, 0.700 m
Downlink: C-band; 4.2 GHz, 5.000 m
Ku-band; 12.2 GHz, 1.721 m
K-band; 20.2 GHz, 1.040 m
$G_i = 3.385 \times 10^4$ or 45.30 dBi at all frequencies

12–7. Uplink: 3.7 GHz, 1.55°, 41.48 dBi
4.2 GHz, 1.37°, 42.56 dBi
5.85 GHz, 0.98°, 45.47 dBi
7.075 GHz, 0.81°, 47.13 dBi

12–9. 38,505 km

12–11. Uplink: 25.53 dB; Downlink: 14.48 dB

Index